Ethics, Technology, and Engineering

Ethics, Technology, and Engineering

An Introduction

Ibo van de Poel
and Lambèr Royakkers

A John Wiley & Sons, Ltd., Publication

This edition first published 2011

Blackwell Publishing was acquired by John Wiley & Sons in February 2007. Blackwell's publishing program has been merged with Wiley's global Scientific, Technical, and Medical business to form Wiley-Blackwell.

Registered Office
John Wiley & Sons Ltd, The Atrium, Southern Gate, Chichester, West Sussex, PO19 8SQ, United Kingdom

Editorial Offices
350 Main Street, Malden, MA 02148-5020, USA
9600 Garsington Road, Oxford, OX4 2DQ, UK
The Atrium, Southern Gate, Chichester, West Sussex, PO19 8SQ, UK

For details of our global editorial offices, for customer services, and for information about how to apply for permission to reuse the copyright material in this book please see our website at www.wiley.com/wiley-blackwell.

Library of Congress Cataloging-in-Publication Data

Poel, Ibo van de, 1966–
Ethics, Technology, and Engineering : An Introduction / by Ibo van de Poel and Lambèr Royakkers.
p. cm.
Includes bibliographical references and index.
ISBN 978-1-4443-3094-6 (hardcover : alk. paper) – ISBN 978-1-4443-3095-3 (pbk. : alk. paper)
1. Technology–Moral and ethical aspects. I. Royakkers, Lambèr M. M. II. Title.
BJ59.P63 2011
174′.96–dc22

2010042204

A catalogue record for this book is available from the British Library.

This book is published in the following electronic formats: eBook 978-1-4443-9570-9; ePub 978-1-4443-9571-6

Set in 10/12.5pt Galliard by SPi Publisher Services, Pondicherry, India
Printed and bound in the United Kingdom by TJ International Ltd, Padstow, Cornwall

2 2013

Contents

Acknowledgments

This book is based on our Dutch text book Royakkers, L., van de Poel, I. and Pieters, A. (eds) (2004). *Ethiek & techniek. Morele overwegingen in de ingenieurspraktijk*, HBuitgevers, Baarn. Most of the chapters have been thoroughly revised. Some chapters from the Dutch text book are not included and this book contains some new chapters.

Section 1.4 contains excerpts from Van de Poel, Ibo. 2007. De vermeende neutraliteit van techniek. De professionele idealen van ingenieurs, in *Werkzame idealen. Ethische reflecties op professionaliteit* (eds J. Kole and D. de Ruyter), Van Gorcum, Assen, pp. 11–23. [translated from Dutch].

Section 3.11 and large parts of Chapter 5 are drawn from Van de Poel, I., and Royakkers, L. (2007). The ethical cycle. *Journal of Business Ethics*, 71 (1), 1–13.

Section 6.2.4. contains excerpts from Devon, R. and Van de Poel, I. (2004). Design ethics: The social ethics paradigm. *International Journal of Engineering Education*, 20 (3), 461–469.

Section 6.3 contains excerpts from Van de Poel, I. (2009). Values in engineering design, in *Handbook of the Philosophy of Science. Vol. 9: Philosophy of Technology and Engineering Sciences* (ed. A. Meijers), Elsevier, Amsterdam, pp. 973–1006.

Chapter 7, which is written by Peter-Paul Verbeek is based on Verbeek, P.P. (2006a). Materializing morality – Design ethics and technological mediation. *Science, Technology and Human Values*, 31 (3), 361–380; Verbeek, P.P. (2006b), The morality of things – A postphenomenological inquiry, in *Postphenomenology: A Critical Companion to Ihde* (ed. E. Selinger), State University of New York Press, New York, pp. 117–130; and Verbeek, P.P. (2008), Morality in design: Design ethics and the morality of technological artifacts, in *Philosophy And Design: From Engineering to Architecture* (eds P.E. Vermaas, P. Kroes, A. Light, and S.A. Moore), Springer, Dordrecht, pp. 91–103.

Section 8.7 contains excerpts from Van de Poel, I. (2009). The introduction of nanotechnology as a societal experiment, in *Technoscience in Progress. Managing the Uncertainty of Nanotechnology* (eds S. Arnaldi, A. Lorenzet and F. Russo), IOS Press, Amsterdam, pp. 129–142.

Section 9.2 contains excerpts from van de Poel, I., Fahlquist, J.N., de Lima, T., Doorn, N., Royakkers, L. and Zwart, S. Fairness and completeness in distributing responsibility: The case of engineering. Manuscript.

Introduction

One of the main differences between science and engineering is that engineering is not just about better understanding the world but also about changing it. Many engineers believe that such change improves, or at least should improve, the world. In this sense engineering is an inherently morally motivated activity. Changing the world for the better is, however, no easy task and also not one that can be achieved on the basis of engineering knowledge alone. It also requires, among other things, ethical reflection and knowledge. This book aims at contributing to such reflection and knowledge, not just in a theoretical sense but also more practically.

This book takes an innovative approach to engineering ethics in several respects. It provides a rather unique approach to ethical decision-making: the ethical cycle. This approach is illustrated by an abundance of cases studies and examples, not only from the US but also from Europe and the rest of the world. The book is also innovative in paying more attention than most traditional introductions in engineering ethics to such topics as ethics in engineering design, the organizational context of engineering, the distribution of responsibility, sustainability, and new technologies such as nanotechnology.

There is an increasing attention to ethics in the engineering curricula. Engineers are supposed not only to carry out their work competently and skillfully but also to be aware of the broader ethical and social implications of engineering and to be able to reflect on these. According to the Engineering Criteria 2000 of the Accreditation Board for Engineering and Technology (ABET) in the US, engineering graduates must have "an understanding of professional and ethical responsibility" and "the broad education necessary to understand the impact of engineering solutions in a global and societal context" (Herkert 1999).

This book provides an undergraduate introduction to ethics in engineering and technology. It helps students to acquire the competences mentioned in the ABET

Ethics, Technology, and Engineering: An Introduction, First Edition.
Ibo van de Poel and Lambèr Royakkers.

criteria or comparable criteria formulated in other countries. More specifically, this book helps students to acquire the following moral competencies:

- *Moral sensibility*: the ability to recognize social and ethical issues in engineering;
- *Moral analysis skills*: the ability to analyze moral problems in terms of facts, values, stakeholders and their interests;
- *Moral creativity*: the ability to think out different options for action in the light of (conflicting) moral values and the relevant facts;
- *Moral judgment skills*: the ability to give a moral judgment on the basis of different ethical theories or frameworks including professional ethics and common sense morality;
- *Moral decision-making skills*: the ability to reflect on different ethical theories and frameworks and to make a decision based on that reflection; and
- *Moral argumentation skills*: the ability to morally justify one's actions and to discuss and evaluate them together with other engineers and non-engineers.

With respect to these competencies, our focus is on the concrete moral problems that students will encounter in their future professional practice. With the help of concrete cases we show how the decision to develop a technology, as well as the process of design and production, is inherently moral. The attention of students is drawn towards the specific moral choices that engineers face. In relation to these concrete choices students will encounter different reasons for and against certain actions, and they will discover that these reasons can be discussed. In this way, students become aware of the moral dimensions of technology and acquire the argumentative capacities that are needed in moral debates.

In addition to an emphasis on cases – which is common to most other introductory text books in engineering ethics as well – we would like to mention three further characteristics of the approach to engineering ethics we have chosen in this text book.

First, we take a broad approach to ethical issues in engineering and technology and the engineer's responsibility for these. Some of the issues we discuss in this book extend beyond the issues traditionally dealt with in engineering ethics like safety, honesty, and conflicts of interest. We also include, for example, ethical issues in engineering design (Chapters 6 and 7) and sustainability (Chapter 10). We also pay attention to such technologies as the atomic bomb and nanotechnology. While we address such "macro-ethical" issues (Herkert 2001) in engineering and technology, our approach to these issues may be characterized as inside-out, that is to say: we start with ethical issues that emerge in the practice of engineers and we show how they arise or are entangled with broader issues.

A second characteristic of our approach is that we pay attention to the broader contexts in which individual engineers do their work, such as the project team, the company, the engineering profession and, ultimately, society. We have devoted a chapter to the issues this raises with respect to organizing responsibility in engineering (Chapter 9). Where appropriate we also pay attention to other actors and stakeholders in these broader contexts. Again our approach is mainly inside-out, starting from concrete examples and the day-to-day work of engineers. It is sometimes thought that paying attention to such broader contexts diminishes the responsibility of engineers, because it shows that engineers lack the control needed to be responsible.[1] Although there is some truth in this, we argue that the broader contexts also change the content of the responsibility of engineers and in some respects increase their responsibility. Engineers, for example, need to take into account the view points, values and interests of relevant stakeholders (Chapter 1). This also implies including such stakeholders, and their viewpoints, in relevant discussion and decision making, for example in design (Chapters 5 and 6). Engineers also need to inform managers, politicians, and the public not only of technological risks but also of uncertainties and potential ignorance (Chapter 8).

A third characteristic of our approach is our attention to ethical theories. We consider these theories important because they introduce a richness of moral perspectives, which forces students to look beyond what seems obvious or beyond debate. Although we consider it important that students get some feeling for the diversity and backgrounds of ethical views and theories, our approach is very much practice-oriented. The main didactical tool here is what we call the "ethical cycle" (Van de Poel and Royakkers 2007). This is an approach for dealing with ethical problems that systematically encourages students to consider a diversity of ethical points of view and helps them to come to a reasoned and justified judgment on ethical issues that they can discuss with others. The ethical cycle is explained in Chapter 5, but Chapters 2, 3, and 4 introduce important elements of it.

The development of the ethical cycle was largely inspired by the ten years of experiences we both have in teaching engineering ethics to large groups of students in the Netherlands, and the didactical problems we and our colleagues encountered in doing so (Van der Burg and Van de Poel 2005; Van de Poel, Zandvoort, and Brumsen 2001). We noticed that students often work in an unstructured way when they analyze moral cases, and they tend to jump to conclusions. Relevant facts or moral considerations were overlooked, or the argumentation was lacking. Ethical theories were often used in an instrumental way by applying them to cases in an unreflective way. Some students considered a judgment about a moral case as an opinion about which no (rational) discussion is possible.

The ethical cycle is intended as a didactical tool to deal with these problems. It provides students a guide for dealing with ethical issues that is systematic without assuming an instrumental notion of ethics. After all, what is sometimes called applied ethics is not a straightforward application of general ethical theories or principles to practical problem in an area. Rather, it is a working back and forth between a concrete moral problem, intuitions about this problem, more general moral principles, and a diversity of ethical theories and view points. This is perhaps best captured in John Rawls' notion of wide reflective equilibrium (Rawls 1971). (For a more detailed discussion, the reader is referred to Chapter 5.)

The ethical cycle provides a tool that does justice to this complexity of ethical judgment but at the same time is practical so that students do not get overwhelmed by the complexity and diversity of ethical theories. By applying the ethical cycle students will acquire the moral competencies that are needed for dealing with ethical issues in engineering and technology (see Figure I.1).

In conjunction with the ethical cycle, we, together with some colleagues have developed a software tool for analyzing ethical issues in engineering and

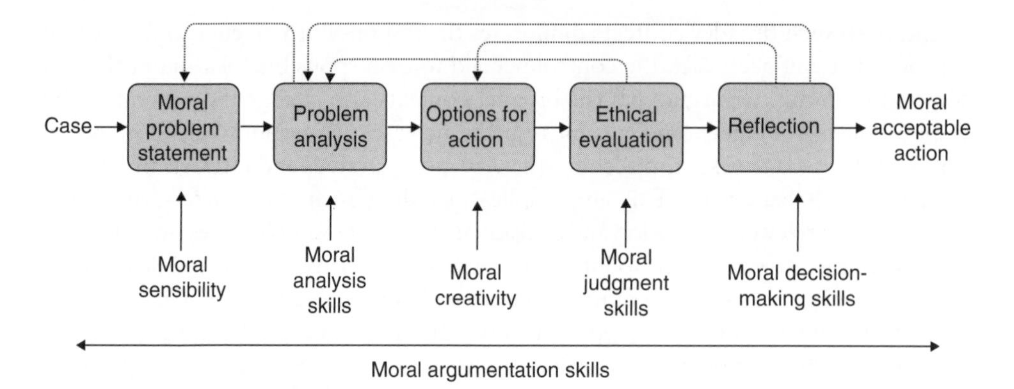

Figure I.1 Ethical issues in engineering and technology.

technology: AGORA (Van der Burg and Van de Poel 2005). The approach on which AGORA is based is basically the same as the ethical cycle. AGORA would therefore be a useful software platform to use in combination with this text book. The program contains a number of standard exercises that correspond to chapters in this book. In addition, teachers can develop their own exercises. For more information about AGORA, the reader is referred to the website www.ethicsandtechnology.com

This book consists of two parts. Part I introduces the ethical cycle. After an introductory chapter on the responsibility of engineers, it introduces the main elements of the ethical cycle: professional and corporate codes of conduct (Chapter 2), ethical theories (Chapter 3) and argumentation schemes that are used in ethical reasoning (Chapter 4). Chapter 5 then introduces the ethical cycle and offers an extensive illustration of the application of the cycle to an ethical issue in engineering.

Part II focuses on more specific ethical issues in engineering and technology. Chapters 6 and 7 deal with ethical issues in engineering design. Chapter 6 focuses on ethical issues that may arise during the various phases of the design process and pays special attention to how engineers are confronted with and can deal with conflicting values in design. Chapter 7 takes a broader look at how technologies influence the perceptions and actions of users and considers how such considerations can be taken into account in design. Chapter 8 deals with technological risks, and questions about how to assess such risks, the moral acceptability of risks, risk communication, and dealing with uncertainty and ignorance. Chapter 9 discusses issues of responsibility that arise due to the social organization of engineering. It discusses in particular the problem of many hands, the difficulty of pinpointing who is responsible if a large number of people are involved in an activity, and it discusses ways of dealing with this problem in engineering. Chapter 10 discusses sustainability, both in more general terms and how it affects the work of engineers and can be taken into account in, for example, the design process.

To a large extent, Parts I and II can be used independently from each other. Teachers who have only limited course hours available can, for example, choose to teach a basic introduction and only use the first five chapters. Conversely, students who have earlier followed some basic introduction to engineering ethics can be offered a course that uses some or all of the chapters from Part II. Although the chapters in Part II are consistent with the ethical cycle introduced in Part I, they contain hardly any explicit references to it and most of the necessary background would also be covered by any other basic course in engineering ethics. In fact the chapters in Part II can also largely be used independent of each other, so that they could be used for smaller teaching modules.

Teachers, who want to offer their students an introduction to engineering ethics without discussing the various ethical theories and the ethical cycle, could choose to use the first two chapters and a selection of the chapters from Part II that deal with more specific issues. Any set-up that aims at introducing the ethical cycle should, we feel, at least include Chapters 2, 3 and 5. Chapter 4 is more optional because it provides moral argumentation schemes which will improve the student's ability to use the ethical cycle but are not strictly necessary.

Each of the chapters starts with an illustrative case study that introduces some of the main issues that are covered in the chapter. Each chapter introduction also indicates the learning objectives so that students know what they should know and be able to do after reading the chapter. Each chapter also contains key terms and a summary that provide a further guide for getting to the core of the subject matter. Study questions provide further help in rehearsing the main points and in applying the main notions to concrete examples. AGORA exercises (see above) may be a further helpful tool to teach students how to apply what they have learned to more complex cases.

A book like this is impossible without the help of a lot of people. First of all we like to thank everybody who contributed to the composition of the Dutch textbook *Ethiek en Techniek. Morele overwegingen in de Ingenieurspraktijk* that formed the basis for this book. In particular we would like to thank Angèle Pieters, our co-editor of the Dutch textbook and Stella de Jager of HB Uitgevers. We also like to thank Peter-Paul Verbeek and Michiel Brumsen for contributing a chapter to this book. We thank Steven Ralston and Diane Butterman for translating parts of our Dutch texts. Jessica Nihlén Fahlquist, Tiago de Lima, Sjoerd Zwart, and Neelke Doorn were so kind to allow us to use a part of a common manuscript in chapter 9 of this book. We would also like to thank the people of Wiley-Blackwell for their comments and support, in particular Nick Bellorini, Ian Lague, Louise Butler, Tiffany Mok, Dave Nash, and Mervyn Thomas. Finally we would like to thank the anonymous reviewers and the people who anonymously filled in a questionnaire about the scope of the book for their comments and suggestions.

Ibo van de Poel is grateful to NIAS, the Netherlands Institute for Advanced Study, for providing him with the opportunity, as a Fellow-in-Residence, to finish this book.

Ibo van de Poel and Lambèr Royakkers

Note

1 Michael Davis, for example has expressed the concern that what he calls a sociological approach to the wider contexts that engineers face may in effect free engineers from any responsibility (see Davis 2006).

1

The Responsibilities of Engineers

Having read this chapter and completed its associated questions, readers should be able to:

- Describe passive responsibility, and distinguish it from active responsibility;
- Describe the four conditions of blameworthiness and apply these to concrete cases;
- Describe the professional ideals: technological enthusiasm, effectiveness and efficiency, and human welfare;
- Debate the role of the professional ideals of engineering for professional responsibility;
- Show an awareness that professional responsibility can sometimes conflict with the responsibility as employee and how to deal with this;
- Discuss the impact of social context of technological development for the responsibility of engineers.

Contents

Ethics, Technology, and Engineering: An Introduction, First Edition.
Ibo van de Poel and Lambèr Royakkers.

1.1 Introduction

Case Challenger

The 25th launching of the space shuttle was to be something special. It was the first time that a civilian, the teacher Christa McAuliffe, or as President Ronald Reagan put it: "one of America's finest" would go into space. There was, therefore, more media attention than usual at cold Cape Canaveral (Florida, United States). When, on the morning of January 28, 1986, the mission controllers' countdown began it was almost four degrees Celsius below freezing point (or about 25 degrees Fahrenheit). After 73 seconds the Challenger space shuttle exploded 11 kilometers above the Atlantic Ocean. All seven astronauts were killed. At the time it was the biggest disaster ever in the history of American space travel.

Figure 1.1 Challenger Space Shuttle. Photo: © Bob Pearson / AFP / Getty Images.

After the accident an investigation committee was set up to establish the exact cause of the explosion. The committee concluded that the explosion leading to the loss of the 1.2 billion dollar spaceship was attributable to the failure of the rubber sealing ring (the O-ring). As the component was unable to function properly at low temperatures fuel had started to leak from the booster rocket. The fuel then caught fire, causing the Challenger to explode.

Morton Thiokol, a NASA supplier, was the company responsible for the construction of the rocket boosters designed to propel the Shuttle into space. In January 1985 Roger Boisjoly, an engineer at the Morton Thiokol company, had aired his doubts about the reliability of the O-rings. In July 1985 he had sent a confidential memo to the Morton Thiokol management board. In that memo he had expressed his concerns about the effectiveness of the O-rings at low temperatures: "I am really afraid that if we do not take immediate steps we will place both the flight and the launching pad in serious danger. The consequences would be catastrophic and human lives would be put at risk." The memo instantly led to a project group being set up in order to investigate the problem. However, the project group received from the management insufficient material and funding to carry out its work properly. Even after one of the project group managers had sent a memo headed "Help!" and ending with the words: "This is a red flag!" to Morton Thiokol's vice-chairman nothing concrete was actually undertaken.

On the day of the fatal flight the launching was delayed five times, partly for weather-related reasons. The night preceding the launching was very cold; it froze 10 degrees Celsius (or 14 degrees Fahrenheit). NASA engineers confessed to remembering having heard that it would not be safe to launch at very low temperatures. They therefore decided to have a telephone conference on the eve of the launching between NASA and Morton Thiokol representatives, Boisjoly also participated. The Morton Thiokol Company underlined the risk of the O-rings eroding at low temperatures. They had never been tested in subzero conditions. The engineers recommended that if the temperature fell below 11 degrees Celsius (or 52 degrees Fahrenheit) then the launch should not go ahead. The weather forecast indicated that the temperature would not rise above freezing point on the morning of the launch. That was the main reason why Morton Thiokol initially recommended that the launch should not be allowed to go ahead.

The people at NASA claimed that the data did not provide sufficient grounds for them to declare the launching, which was extremely important to NASA, unsafe. What was rather curious was the fact that the burden of proof was placed with those who were opposed to the launching; they were requested to prove that the flight would be unsafe. The official NASA policy, though, was that it had to be proved that it would be safe to make the flight.

A brief consultation session was convened so that the data could once again be examined. While the connection was broken for five minutes the General Manager of Thiokol commented that a "management decision" had to be made. Later on several employees actually stated that shortly after the launching NASA would make a decision regarding a possible contract extension with the company. It was at least the case that Boisjoly felt that people were no longer listening to his arguments. For Morton Thiokol it was too much of a political and financial risk to postpone the launch. After discussing matters amongst themselves the four managers present, the engineers excluded, put it to the vote. They were reconnected and Thiokol, ignoring the advice of Boisjoly, announced to NASA its positive recommendations concerning the launching of the Challenger. It was a decision that was immediately followed by NASA without any further questioning. As agreement had been reached, the whole problem surrounding the inadequate operating of the O-ring at low temperatures was not passed on to NASA's higher management level. Several minutes after the launch someone of the mission control team concluded that there had: "obviously been … a major malfunction."

A Presidential Commission determined that the whole disaster was due to inadequate communication at NASA. At the same time, they argued for a change in system and ethos that would ensure transparency and encourage whistle blowing. As a consequence, the entire space program was stopped for two years so that the safety of the Shuttle could be improved. Morton Thiokol did not lose its contract with NASA but helped, instead, to work on finding a solution to the O-ring problem. Engineers were given more of a say in matters. In the future, they will have the power to halt a flight if they had their doubts.

Source: Based on Wirtz (2007, p. 32), Vaughan (1996), and the BBC documentary *Challenger: Go for Launch* of Blast!Films.

In this case we see how the Challenger disaster was caused by technical error and inadequate communication. For the designers of the O-rings, the engineers at Morton Thiokol, the disaster did not have legal implications. Does that mean that the case is thus closed or do they bear some kind of responsibility? If so, what then is their responsibility? This chapter first investigates what exactly responsibility is (Section 1.2), distinguishing between passive responsibility for things that happened in the past (Section 1.3) and active responsibility for things not yet attained (Section 1.4). The final two sections discuss the position of engineers vis-à-vis managers, which was obviously important in the Challenger case, the wider context of technological development, and examine the consequences for the responsibility of engineers of this wider context.

1.2 Responsibility

Whenever something goes wrong or there is a disaster like that of the Challenger then the question who is responsible for it often quickly arises. Here responsibility means in the first place being held accountable for your actions and for the effects of your actions. The making of choices, the taking of decisions but also failing to act are all things that we regard as types of actions. Failing to save a child who is drowning is therefore also a type of action. There are different kinds of responsibility that can be distinguished. A common distinction is between active responsibility and passive responsibility. Active responsibility is responsibility before something has happened. It refers to a duty or task to care for certain state-of-affairs or persons. Passive responsibility is applicable after something (undesirable) has happened.

Responsibility (both active and passive) is often linked to the role that you have in a particular situation. In the case described here Boisjoly fulfilled the role of engineer and not that of, for example, family member. You often have to fulfill a number of roles simultaneously such as those of friend, parent, citizen, employee, engineer, expert, and colleague. In a role you have a relationship with others, for instance, as an employee you have a relationship with your employer, as an expert you have a relationship with your customers and as a colleague you have relationships with other colleagues. Each role brings with it certain responsibilities. A parent, for example, is expected to care for his child. In the role of employee it is expected that you will execute your job properly, as laid down in collaboration with your employer; in the role of expert it will be presumed that you furnish your customer with information that is true and relevant and in the role of colleague you will be expected to behave in a collegial fashion with others in the same work situation. An engineer is expected to carry out his work in a competent way. Roles and their accompanying responsibilities can be formally laid down, for instance legally, in a contract or in professional or corporate codes of conduct (see Chapter 2). In addition, there are more informal roles and responsibilities, like the obligations one has within a family or towards friends. Here, too, agreements are often made and rules are assumed but they are not usually put down in writing. We will call the responsibility that is based on a role you play in a certain context **role responsibility**.

Role responsibility The responsibility that is based on the role one has or plays in a certain situation.

Since a person often has different roles in life he/she has various role responsibilities. One role may have responsibilities that conflict with the responsibilities that accompany another role. Boisjoly for example in the Challenger case both had a role as an employee and as an engineer. As an employee he was expected to be loyal to his company and to listen to his superiors, who eventually decided to give positive advice about the launch. As an engineer he was expected to give technically sound advice taking into account the possible risks to the astronauts and, in his view, this implied a negative advice with respect to the launch.

Although roles define responsibilities, **moral responsibility** is not confined to the roles one plays in a situation. Rather it is based on the obligations, norms, and duties that arise from *moral* considerations. In Chapter 3, we will discuss in more detail what we mean with terms like morality and ethics, and what different kinds of ethical theories can be distinguished. Moral responsibility can extend beyond roles. In the Challenger case, it was part of Boisjoly's moral responsibility to care for the consequences of his advice for the astronauts and for others. Moral responsibility can, however, also limit role responsibilities because with some roles immoral responsibilities may be associated. (Think of the role of Mafioso.) In this and the next chapter we are mainly interested in the **professional responsibility** of engineers. Professional responsibility is the responsibility that is based on your role as a professional engineer in as far it stays within the limits of what is morally allowed. Professional responsibilities are not just passive but they also contain an active component. We will examine the content of the professional responsibility of engineers in more detail in Section 1.4, but first we turn to a more detailed description of passive responsibility.

Moral responsibility Responsibility that is based on moral obligations, moral norms or moral duties.

Professional responsibility The responsibility that is based on one's role as professional in as far it stays within the limits of what is morally allowed.

1.3 Passive Responsibility

Typical for **passive responsibility** is that the person who is held responsible must be able to provide an account why he/she followed a particular course of action and why he/she made certain decisions. In particular, the person is held to justify his/her actions towards those who are in a position to

Passive responsibility Backward-looking responsibility, relevant after something undesirable occurred; specific forms are accountability, blameworthiness, and liability.

demand that the individual in question accounts for his/her actions. In the case of the Challenger, NASA had to be able to render account for its actions to the families of the victims, to society, and to the sitting judge. We will call this type of passive responsibility **accountability**.

Passive responsibility often involves not just accountability but also **blameworthiness**. Blameworthiness means that it is proper to blame someone for his/her actions or the consequences of those actions. You are not always blameworthy for the consequences of your actions or for your actions themselves. Usually, four conditions need to apply: wrong-doing, causal contribution, foreseeability, and freedom. The extent to which you can be blamed is determined by the degree to which these conditions are fulfilled. The four conditions will be illustrated on the basis of the Challenger disaster.

Accountability Backward-looking responsibility in the sense of being held to account for, or justify one's actions towards others.

Blameworthiness Backward-looking responsibility in the sense of being a proper target of blame for one's actions or the consequences of one's actions. In order for someone to be blameworthy, usually the following conditions need to apply: wrong-doing, causal contribution, foreseeability, and freedom.

Wrong-doing

Whenever one blames a person or institution one usually maintains that in carrying out a certain action the individual or the institution in question has violated a norm or did something wrong. This can be a norm that is laid down in the law or that is common in the organization. In the Challenger case, for example, NASA violated the norm that a flight had to be proven to be safe. Instead the burden of proof was reversed in this case. In this book, we are not just interested in legal and organizational norms, but in moral ones. We will therefore investigate different kind of ethical frameworks that can be applied in judging the moral rightness or wrongness of actions and their consequences. This includes ethical frameworks such as your own conscience and moral beliefs but also codes of conduct (Chapter 2) and ethical theories (Chapter 3). Together these frameworks form a means of thinking about how one can arrive at what is good, and how one can act in the right way.

Causal contribution

A second criterion is that the person who is held responsible must have made a causal contribution to the consequences for which he or she is held responsible. Two things are to be kept in mind when judging whether someone made a causal contribution to a certain consequence. First, not only an action, but also a failure to act may often be considered a causal contribution, like in the case of the Challenger the failure to stop the launch. Second, a causal contribution is usually not a sufficient condition for the occurrence of the consequence under consideration. Often, a range of causal contributions will have to be present for the consequence to occur. A causal contribution will often be a necessary ingredient in the actual chain of events that led to the consequence, that is, without the causal contribution the consequence would not have occurred.

Both the NASA project team and the Morton Thiokol management team made a causal contribution to the disaster because both could have averted the disaster by postponing the launch. In fact, before the Challenger could be launched, both teams needed to make a positive decision. The engineer, Boisjoly, maintained that he no longer had the chance to take action. Internally he had done everything in his power to prevent the consequences but he did not have enough influence. In retrospect he could possibly have gone public by informing the press. He should also possibly have intervened earlier on in the process – before the telephone conference – to ensure that the O-ring problem had been tackled more successfully.

Foreseeability

A person who is held responsible for something must have been able to know the consequences of his or her actions. The consequences are the harm actually arising from transgressing a norm. People cannot be held responsible if it is totally unreasonable to expect that they could possibly have been aware of the consequences. What we do expect is that people do everything that is reasonably possible to become acquainted with the possible consequences.

In the Challenger case engineer Boisjoly, the Morton Thiokol management team and the NASA representatives (the project team) could all have expected the Challenger disaster because all three were aware of the risks of erosion when the O-rings are exposed to low temperatures, a factor which thus meant that safe launching could not be guaranteed under such conditions. Though there was no conclusive scientific evidence that the launching was unsafe, all parties were certainly aware of the danger of a possible disaster, which means that the condition of foreseeability was fulfilled.

Freedom of action

Finally, the one who is held responsible must have had freedom of action, that is, he or she must not have acted under compulsion. Individuals are either not responsible or are responsible to a lesser degree if they are, for instance, coerced to take certain decisions. The question is, however, what exactly counts as coercion. A person can, for example, be "forced" or manipulated to work on the development of a particular technology under the threat that if he does not cooperate he will sacrifice his chances of promotion. In this case, this person is strictly speaking not coerced to work on the development of the particular technology, he can still act differently. Therefore the person remains responsible for his actions. However, since he is also not entirely free we could say that his responsibility is somewhat smaller than in the case where he had freely chosen to be involved in the development of this technology.

The NASA project team was under pressure. The launch had already been postponed several times, which meant that the time available for other space missions was becoming very limited. There was also the pressure of the eager public, largely because of the presence of McAuliffe. Morton Thiokol might also have felt the pressure of NASA because negative recommendations could well have prevented further cooperation with NASA and that would have had its financial consequences. The possibilities open to the engineer Boisjoly were limited. The only thing he could have possibly done to prevent the disaster was inform the press but that would have had negative consequences (e.g., dismissal) for him and his family. In all three cases, the pressure was probably not strong enough to say that NASA, Morton Thiokol, or Boisjoly lacked freedom of action; they could have done other things than they actually did, they were not compelled to act as they did. Nevertheless, especially in the case of Boisjoly you could argue that the negative personal consequences he could expect diminished his responsibility.

1.4 Active Responsibility and the Ideals of Engineers

We considered above questions of responsibility when something has gone wrong. Responsibility is also something that comes into play beforehand, if nothing has yet gone wrong or if there is the chance to realize something good. We will refer to this as **active responsibility**. If someone is actively responsible for something he/she is expected to act in such a way that undesired consequences are avoided as much as possible and so that positive consequences are realized. Active responsibility is not primarily about blame but requires a certain positive attitude or character trait of dealing with matters. Philosophers call such positive attitudes or character traits virtues (see Chapter 3). Active responsibility, moreover, is not only about preventing the negative effects of technology as about realizing certain positive effects.

Active responsibility Responsibility before something has happened referring to a duty or task to care for certain state-of-affairs or persons.

Active Responsibility

Mark Bovens mentions the following features of active responsibility:

- Adequate perception of threatened violations of norms;
- Consideration of the consequences;
- Autonomy, i.e. the ability to make one's own independent moral decisions;
- Displaying conduct that is based on a verifiable and consistent code; and
- Taking role obligations seriously. (Bovens, 1998)

One way in which the active responsibility of engineers can be understood is by looking at the **ideals** of engineers. Ideals, as we will understand the notion here, have two specific characteristics. First ideals are ideas or strivings which are particularly motivating and inspiring for the person having them. Second, it is typical for ideals that they aim at achieving an optimum or maximum. Often, therefore, ideals cannot be entirely achieved but are strived for. In the course of practicing their profession engineers can be driven by several ideals. Those can be personal ideals such as the desire to earn a lot of money or to satisfy a certain degree of curiosity but they can also be social or moral ideals, such as wanting to implement technological ends to improve the world. Those are also the types of ideals

Ideals Ideas or strivings which are particularly motivating and inspiring for the person having them, and which aim at achieving an optimum or maximum.

Professional ideals Ideals that are closely allied to a profession or can only be aspired to by carrying out the profession.

that can spur people on to opt for an engineering field of study and career. Some of these ideals are directly linked to professional practice because they are closely allied to the engineering profession or can only be aspired to by carrying out the profession of engineer. We call such ideals **professional ideals**. As *professional* ideals, these ideals are part of professional responsibility in as far they stay within the limits of what is morally allowed. Below, we shall therefore discuss three different professional ideals of engineers and we shall establish whether these ideals are also morally commendable.

1.4.1 Technological enthusiasm

Technological enthusiasm The ideal of wanting to develop new technological possibilities and taking up technological challenges.

Technological enthusiasm pertains to the ideal of wanting to develop new technological possibilities and take up technological challenges. This is an ideal that motivates many engineers. It is fitting that Samuel Florman refers to this as "the existential pleasures of engineering" (Florman, 1976). One good example of technological enthusiasm is the development of Google Earth, a program with which, via the Internet, it is possible to zoom in on the earth's surface. It is a beautiful concept but it gives rise to all kinds of moral questions, for instance in the area of privacy (you can study the opposite neighbor's garden in great detail) and in the field of security (terrorists could use it to plan attacks). In a recent documentary on the subject of Google Earth one of the program developers admitted that these are important questions.[1] Nevertheless, when developing the program these were matters that the developers had failed to consider because they were so driven by the challenge of making it technologically possible for everyone to be able to study the earth from behind his or her PC.

Technological enthusiasm in itself is not morally improper; it is in fact positive for engineers to be intrinsically motivated as far as their work is concerned. The inherent danger of technological enthusiasm lies in the possible negative effects of technology and the relevant social constraints being easily overlooked. This has been exemplified by the Google Earth example. It is exemplified to an extreme extent by the example of Wernher von Braun (see box).

Wernher von Braun (1912–77)

Wernher von Braun is famous for being the creator of the space program that made it possible to put the first person on the moon on July 20, 1969. A couple of days before, on July 16, the Apollo 11 spaceship used by the astronauts to travel from the earth had been launched with the help of a Saturn V rocket and Von Braun had been the main designer of that rocket. Sam Phillips, the director of the American Apollo program, was reported to have said that without

Von Braun the Americans would never have been able to reach the moon as soon as they did. Later, after having spoken to colleagues, he reviewed his comment by claiming that without Von Braun the Americans would never have landed on the moon full stop.

Figure 1.2 Wernher von Braun. Photo: NASA Archives.

Von Braun grew up in Germany. From an early age he was fascinated by rocket technology. According to one anecdote Von Braun was not particularly brilliant in physics and mathematics until he read a book entitled *Die Rakete zu den Planetenraümen* by Hermannn Oberth and realized that those were the subjects he would have to get to grips with if he was later going to be able to construct rockets. In the 1930s Von Braun was involved in developing rockets for the German army. In 1937 he joined Hitler's National Socialist Party and in 1940 he became a member of the SS. Later he explained that he had been forced to join that party and that he had never participated in any political activities, a matter that is historically disputed. What is in any case striking is the argument that he in retrospect gave for joining the National Socialist Party which was this: "My refusal to join the party would have meant that I would have had to abandon the work of my life. Therefore, I decided to join" (Piszkiewicz (1995, p. 43). His life's work was, of course, rocket technology and a devotion to that cause was a constant feature of Von Braun's life.

During World War II Von Braun played a major role in the development of the V2 rocket, which was deployed from 1944 onwards to bomb, amongst other targets, the city of London. Incidentally more were killed during the V2-rocket's development and production – an estimated 10 000 people – than during the actual bombings (Neufeld, 1995, p. 264). The Germans had deployed prisoners from the Mittelbau-Dora concentration camp to help in the production of the V2 rockets. Von Braun was probably aware of those people's abominable working conditions.

There is, therefore, much to indicate that Von Braun's main reason for wanting to join the SS was carefully calculated: in that way he would be able to continue his important work in the field of rocket technology. In 1943 he was arrested by the Nazis and later released. It was claimed that he had allegedly sabotaged the V2 program. One of the pieces of evidence used against him was that he had apparently said that after the war the V2 technology should be further developed in the interests of space travel – and that is indeed what ultimately happened when he later started to work for the Americans. When, in 1945, Von Braun realized that the Germans were going to lose the war he arranged for his team to be handed over to the Americans.

In the United States Von Braun originally worked on the development of rockets for military purposes but later he fulfilled a key role in the space travel program, a program that was ultimately to culminate in man's first steps on the moon. Von Braun's big dream did therefore ultimately come true.

Source: Based on Stuhlinger and Ordway (1994), Neufeld (1995), and Piszkiewicz (1995).

1.4.2 Effectiveness and efficiency

Effectiveness The extent to which an established goal is achieved.

Efficiency The ratio between the goal achieved and the effort required.

Engineers tend to strive for effectiveness and efficiency. **Effectiveness** can be defined as the extent to which an established goal is achieved; **efficiency** as the ratio between the goal achieved and the effort required. The drive to strive towards effectiveness and efficiency is an attractive ideal for engineers because it is – apparently – so neutral and objective. It does not seem to involve any political or moral choices, which is something that many engineers experience as subjective and therefore wish to avoid. Efficiency is also something that in contrast, for example, to human welfare can be defined by engineers and is also often quantifiable. Engineers are, for example, able to define the efficiency of the energy production in an electrical power station and they can also measure and compare that efficiency. An example of an engineer who saw efficiency as an ideal was Frederick W. Taylor (see box).

Frederick W. Taylor (1856–1915)

Frederick Taylor was an American mechanical engineer. He became known as the founder of the efficiency movement and was specifically renowned for developing scientific management also known as Taylorism.

Out of all his research Taylor became best known for his time-and-motion studies. There he endeavored to scientifically establish which actions – movements – workers were required to carry out during the production process and how much time that took. He divided the relevant actions into separate movements, eliminated all that was superfluous and endeavored, with the aid of a stopwatch, to establish precisely how long the necessary movements took. His aim was to make the whole production process as efficient as possible on the basis of such insight. Taylorism is often seen as an attempt to squeeze as much as possible out of workers and in practice that was often what it amounted to but that had probably not been Taylor's primary goal. He believed that it was possible to determine, in a scientific fashion, just what would be the best way of carrying out production processes by organizing such processes in such a way that optimal use could be made of the opportunities provided by workers without having to demand too much of them. He maintained that his approach would put an end to the on-going conflict between the trade unions and the managerial echelons, thus making trade unions redundant. He was also critical of management which he found unscientific and inefficient. To his mind having the insight of engineers and their approach to things would culminate in a better and more efficient form of management.

Figure 1.3 Frederick Taylor. Photo: Bettmann Archive/Corbis.

In 1911 Taylor published his *The Principles of Scientific Management* in which he explained the four principles of scientific management:

- Replace the present rules of thumb for working methods with methods based on a scientific study of the work process.
- Select, train and develop every worker in a scientific fashion instead of allowing workers to do that themselves.
- Really work together with the workers so that the work can be completed according to the developed scientific principles.
- Work and responsibility are virtually equally divided between management and workers. The management does the work for which it is best equipped: applying scientific management principles to plan the work; and the workers actually perform the tasks.

Though Taylor was a prominent engineer – for a time he was, for instance, president of the influential American Society of Mechanical Engineers (ASME) – he only had a limited degree of success when it came to the matter of conveying his ideas to people. They were not embraced by all engineers but, thanks to a number of followers, they were ultimately very influential. They fitted in well with the mood of the age. In the United States the first two decades of the twentieth century were known as the "Progressive Era." It was a time when engineers clearly manifested themselves as a professional group capable of promoting the interests of industry and society. It was frequently implied that the engineering approach to social problems was somehow superior. Taylor's endeavors to achieve a form of management that was efficient and scientific fitted perfectly into that picture.

Source: Based on Taylor (1911), Layton (1971), and Nelson (1980).

Though many engineers would probably not have taken things as far as Taylor did, his attempt to efficiently design the whole production process – and ultimately society as a whole – constituted a typical engineering approach to matters. Efficiency is an ideal that endows engineers with authority because it is something that – at least at first sight – one can hardly oppose and that can seemingly be measured objectively. The aspiration among engineers to achieve authority played an important part in Taylor's time. In the United States the efficiency movement became an answer to the rise of large capitalistic companies where managers ruled and engineers were mere subordinate implementers. It constituted an effort to improve the position of the engineer in relation to the manager. What Taylor was really arguing was that engineers were the only really capable managers.

From a moral point of view, however, effectiveness and efficiency are not always worth pursuing. That is because effectiveness and efficiency suppose an external goal in relation to which they are measured. That external goal can be to consume a minimum amount of non-renewable natural resources to generate energy, but also war or even genocide. It was no coincidence that Nazi engineers like Eichmann were proud of the efficient way in which they were able to contribute to the so-called "resolving of the Jewish question" in Europe which was to lead to the murdering of six million Jews and other groups that were considered inferior by the Nazis like Gypsies and mental patients (Arendt, 1965). The matter of whether effectiveness or efficiency is morally worth pursuing therefore depends very much on the ends for which they are employed. So, although some engineers have maintained the opposite, the measurement of the effectiveness and efficiency of a technology is value-laden. It proposes a certain goal for which the technology is to be employed and that goal can be value-laden. Moreover, to measure efficiency one need to calculate the ratio between the output (the external goal) and the input, and also the choice of the input may be value-laden. A technology may for example be efficient in terms of costs but not in terms of energy consumption.

1.4.3 Human welfare

A third ideal of engineers is that of contributing to or augmenting human welfare. The professional code of the American Society of Mechanical Engineering (ASME) and of the American Society of Civil Engineers (ASCE) states that "engineers shall use their knowledge and skill for the enhancement of human welfare." This also includes values such as health, the environment, and sustainability. According to many professional codes that also means that: "Engineers shall hold paramount the safety, health and welfare of the public" (as, for example, stated by the code of the National Society of Professional Engineers, see Chapter 2). It is worth noting that the relevant values will differ somewhat depending on the particular engineering specialization. In the case of software engineers, for instance, values such as the environment and health will be less relevant whilst matters such as the privacy and reliability of systems will be more important. One of the most important values that falls under the pursuit of human welfare among engineers is safety. One of the engineers who was a great proponent of safety was the Dutch civil engineer Johan van Veen.

Johan van Veen (1893–1959)

Figure 1.4 Netherlands. Viewed from a US Army helicopter, a Zuid Beveland town gives a hint of the tremendous damage wrought by the 1953 flood to Dutch islands. Photo: Agency for International Development/National Archives, Washington (ARC Identifier 541705).

Johan van Veen is known as the father of the Delta Works, a massive plan devised to protect the coasts of the South-western part of the Netherlands which materialized after the flood disaster of 1953. During the disaster 1835 people died and more than 72 000 were forced to evacuate their homes.

Before the disaster occurred there were indications that the dykes were not up to standard. In 1934 it was discovered that a number of dykes were probably too low. In 1939 Wemelsfelder, a Public Works Agency employee working for the Research Service for the Estuaries, Lower River Reaches and Coasts sector, was able to support that assumption with a series of models. Even before the big disaster of 1953 Johan van Veen had emphasized the need to close off certain estuaries.

Van Veen studied civil engineering in Delft before then going on, in 1929, to work for the Research Service which he was later to head. On the basis of his interest in the history of hydraulic engineering and his activities with the Public Works Agency, he gradually became convinced that the danger posed by storm-driven flooding had been vastly underestimated and that the dykes were indeed too low. Van Veen was quite adamant about his beliefs which soon earned him the nickname, within the service, of "the new Cassandra" after the Trojan priestess who had perpetually predicted the fall of Troy. He even adopted the pseudonym Cassandra in the epilogue to the fourth edition of his book *Dredge, Drain, Reclaim* that was published in 1955. According to Van Veen, Cassandra had been warning people about the too low state of the dykes since 1937. In the fifth edition of his book, which appeared in 1962, Van Veen revealed that he was in fact Cassandra. Van Veen's reporting of the lowness of the dykes was not something that was welcomed. In fact it was deliberately kept secret to the public. It is even said that Van Veen was sworn to silence on the matter.

In 1939 Van Veen became secretary of the newly created Storm Flood Committee. In that capacity he was given the space to elaborate several of his plans for the further defense of the Netherlands. In public debates he consistently based his arguments for those plans on the need to combat silting up and the formation of salt-water basins. Undoubtedly that was because even then he was unable to publicly air his views about safety.

Even though pre-1953 there was growing doubt within the Public Works Agency as to the ability of the existing dykes to be able to withstand a storm-driven flood that was not a matter that became publicly known. It was not only the Public Works Agency and the relevant minister that kept quiet about the possibility of a flood disaster. At that time the press was not keen to publish such doom and gloom stories either. As there was little or no publicity about the inadequacy of the dykes the inhabitants of Zeeland were thus totally surprised by the disaster. There are no indications that in the period leading up to 1953 steps were taken to improve the storm warning systems and the aid networks. If that had happened then undoubtedly considerably fewer people would have lost their lives.

Source: Based on ten Horn-van Nispen (2002), Van der Ham (2003), and De Boer (1994).

From a moral point of view the professional ideal of human welfare is hardly contestable. One could maybe wonder whether serving human welfare is a moral obligation for engineers, but if they choose to do so this seems certainly laudable. Therefore from a moral angle, this ideal has another status than the other two ideals discussed above. As we have seen technological enthusiasm and effectiveness and efficiency are ideals that are not necessarily morally commendable, although they are also not always morally reprehensible; in both cases much depends on the goals for which technology is used and the side-effects so created. Both ideals, moreover, carry the danger of forgetting about the moral dimension of technology. On the other hand, the ideal of human welfare confirms that the professional practice of engineers is not something that is morally neutral and that engineers do more than merely develop neutral means for the goals of others.

1.5 Engineers versus Managers

Engineers are often salaried employees and they are usually hierarchically below managers. Just as with other professionals this can lead to situations of conflict because they have, on the one hand, a responsibility to the company in which they work and, on the other hand, a professional responsibility as engineers, including – as we have seen – a responsibility for human welfare. We will discuss below three models of dealing with this tension and the potential conflict between engineers and managers: separatism, technocracy, and whistle-blowing. These three models are positions that engineers can adopt versus managers in specific situations, but they also reflect more general social frameworks for dealing with the potential tension between engineers and managers.

1.5.1 Separatism

Several months after the Challenger disaster Boisjoly, the engineer, said the following: "I must emphasize, I had my say, and I never [would] take [away] any management right to take the input of an engineer and then make a decision based upon that input … I have worked at a lot of companies … and I truly believe that … there was no point in me doing anything further [other] than [what] I had already attempted to do" Goldberg (1987, p. 156). It is a view that fits into what might be termed **separatism**: "the notion that scientists and engineers should apply the technical inputs, but appropriate management and political organs should make the value decisions" (Goldberg, 1987, p. 156). Separatism is well illustrated by the **tripartite model**.

Separatism The notion that scientists and engineers should apply the technical inputs, but appropriate management and political organs should make the value decisions.

Tripartite model A model that maintains that engineers can only be held responsible for the design of products and not for wider social consequences or concerns. In the tripartite model three separate segments are distinguished: the segment of politicians; the segment of engineers; and the segment of users.

In the tripartite model three separate segments are distinguished (Figure 1.5). The first segment contains politicians, policy makers, and managers who establish the objectives for engineering projects and products and make available resources without intervening in engineering matters. They also stake out the ultimate boundaries of the engineering projects. The second segment relates to the engineers who take care of the designing, developing, creating, and executing of those projects or products. The final segment, the users, includes those who make use of the

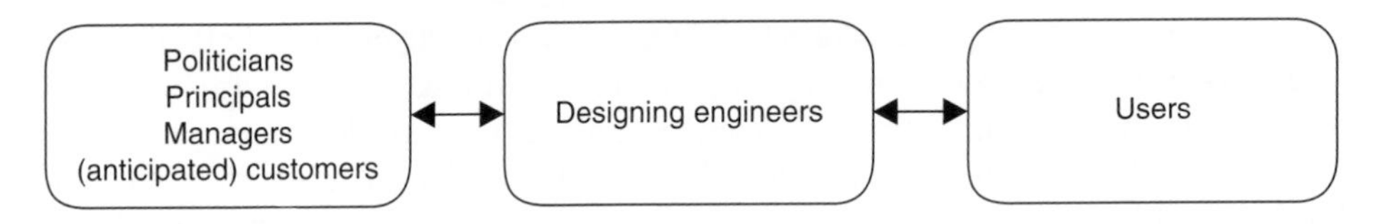

Figure 1.5 The tripartite model.

various technologies. According to this model engineers can only be held responsible for the technical creation of products.

The tripartite model (see, for example, Van de Poel (2001); originally based on Boers (1981)) is based on the assumption that the responsibility of engineers is confined to the engineering choices that they make. The formulation of the design assignment, the way in which the technology is used and the consequences of all of that are not thus considered to be part of the responsibility of engineers. According to this view the responsibility of engineers limits itself to the professional responsibility that they have to their employer, customer, and colleagues, excluding the general public. The case of Werner von Braun illustrates this well. Von Braun was reconciled to the subordinate role of engineers but perpetually sought ways of pursuing his technological ideals and, in doing so, displayed a degree of indifference to the social consequences of the application of his work and to the immoral intentions of those who had commissioned the task. His creed must have been: "In times of war, a man has to stand up for his country, as a combat soldier as a scientist or as an engineer, regardless of whether or not he agrees with the policy his government is pursuing" (Stuhlinger and Ordway, 1994, p. xiii). It is a role that might alternatively be described as being that of a **"hired gun."** The dangerous side of this role can perhaps best be summed up in the words of the song text of the American satirist Tom Lehrer[2]:

"Hired gun" Someone who is willing to carry out any task or assignment from his employer without moral scruples.

Once the rockets go up
Who cares where they come down
"that's not my department,"
said Wernher von Braun.

1.5.2 Technocracy

An alternative for the engineer as a "hired gun" is offered by Frederick Taylor. He proposed that engineers should take over the role of managers in the governance of companies and that of politicians in the governance of society. This proposal would lead to the establishment of a **technocracy**, that is, government by experts. Accordingly, the role of engineers would be that of technocrats who, on the basis of technological insight, do what they consider best for a company or for society. The role of technocrats is problematic for a number of reasons. First, it is not exactly clear what unique expertise engineers possess that permit them to legitimately lay claim to the role of technocrats. As we have seen, concealed behind the use of apparently neutral terms like efficiency there is a whole world

Technocracy Government by experts.

of values and conflicting interests. Admittedly engineers do have specific technological knowledge and they do know about, for example, the risks that may be involved in a technology. When it comes to the underlying goals that should be pursued through technology or the acceptable levels of risk they are not any more knowledgeable than others (the technocratic fallacy, see Chapter 4). A second objection to technocracy is that it is undemocratic and paternalistic. We speak of **paternalism** when a certain group of individuals, in this case engineers, make (moral) decisions for others on the assumption that they know better what is good for them than those others themselves. In that way paternalism denies that people have the right to shape their own lives. That clashes with the people's moral autonomy – the ability of people to decide for themselves what is good and right. Moral autonomy is often considered an important moral value.

Paternalism The making of (moral) decisions for others on the assumption that one knows better what is good for them than those others themselves.

1.5.3 Whistle-blowing

Case Inez Austin

Inez Austin was one of the few female engineers at the company Westinghouse Hanford, when in 1989 she became senior process engineer for that company at the Hanford Nuclear Site, a former plutonium production facility in the state of Washington in the United States. In June 1990, she refused for safety reasons to approve a plan to pump radioactive waste from an old underground single-shell tank to a double-shell tank. Her refusal let to several retaliatory actions by her employer. In 1990 she received the lowest employee ratings in all her 11 years at the company. Doubts were raised about the state of her mental health and she was advised to see a psychiatrist. In 1992, Austin received the Scientific Freedom and Responsibility Award from the American Association for the Advancement of Science (AAAS) "for her courageous and persistent efforts to prevent potential safety hazards involving nuclear waste contamination. Ms. Austin's stand in the face of harassment and intimidation reflects the paramount professional duty of engineers – to protect the public's health and safety – and has served as an inspiration to her co-workers." Nevertheless, after a second whistle blowing incident, relating to the safety and legality of untrained workers, her job was terminated in 1996.

Source: Based on http://www.onlineethics.org/CMS/profpractice/exempindex/austinindex.aspx (Accessed September 22, 2009).

A third role model is offered by Van Veen. Just like Boisjoly he accepted, to an important extent, his subordinate role as engineer but he did endeavor to find channels, internally and externally, to air his grievances on safety. Though he never went public

Whistle-blowing The disclosure of certain abuses in a company by an employee in which he or she is employed, without the consent of his/her superiors, and in order to remedy these abuses and/or to warn the public about these abuses.

as such his role verges on that of whistle-blower as he/she reported internal wrongs externally in order to warn society. An example of a whistle-blower is given in the boxed case on Inez Austin. The term **whistle-blowing** is used if an employee discloses certain abuses in a company in which he/she is employed without the consent of his/her superiors and in order to remedy these abuses and/or to warn the public about these abuses (cf. Martin and Schinzinger, 1996, p. 247). Abuses do not only include the endangerment of public health, safety, or the environment but also indictable offences, violation of the law and of legislation, deception of the public or the government, corruption, fraud, destroying or manipulating information, and abuse of power, including sexual harassment and discrimination. As the box shows whistle-blowing may well lead to conflicts with the employer. In fact, whistle blowers often pay a huge price possibly involving not only losing their job but also the very difficult task of getting hired again, and even the loss of friends and family.[3]

Guidelines for Whistle-Blowing

Business ethicist Richard De George has proposed the following guidelines, for when whistle-blowing is morally required:

1 The organization to which the would-be whistleblower belongs will, through its product or policy, do serious and considerable harm to the public (whether to users of its product, to innocent bystanders, or to the public at large).
2 The would-be whistleblower has identified that threat of harm, reported it to her immediate superior, making clear both the threat itself and the objection to it, and concluded that the superior will do nothing effective.
3 The would-be whistleblower has exhausted other internal procedures within the organization (for example, by going up the organizational ladder as far as allowed) – or at least made use of as many internal procedures as the danger to others and her own safety make reasonable.
4 The would-be whistleblower has (or has accessible) evidence that would convince a reasonable, impartial observer that her view of the threat is correct.
5 The would-be whistleblower has good reason to believe that revealing the threat will (probably) prevent the harm at reasonable cost (all things considered). (De George, 1990)

Whistle-blowers are often seen as people who are morally to be commended. It does not, however, seem desirable to let the professional ethics of engineers – or people of any other profession – be exclusively dependent on such practices. Although whistle-blowing may sometimes be unavoidable, as a general social framework for dealing with the potential tension between engineers and managers, it is unsatisfactory. In the first place whistle-blowing usually forces people to make big sacrifices and one may question whether it is legitimate to expect the average professional to make such sacrifices. In the second place the effectiveness of whistle-blowing is often limited because as soon as the whistle is blown the communication between managers and professionals has inevitably been disrupted. It would be much more effective if at an earlier stage the concerns of the professionals were to be addressed but in a more constructive way. This demands a role model in which the engineer as professional is not necessarily opposed to the manager. It means that engineers have to be able to recognize moral questions in their professional practice and discuss them in a constructive way with other parties.

1.6 The Social Context of Technological Development

Engineers are not the only ones who are responsible for the development and consequences of technology. Apart from managers and engineers there are other actors that influence the direction taken by technological development and the relevant social consequences. We use the term **actor** here for any person or group that can make a decision how to act and that can act on that decision. A company is an actor because it usually has a board of directors that can make decisions on behalf of that company and is able to effectuate those decisions. A mob on the other hand is usually not an actor. A variety of actors can be distinguished that usually play a role in technological development:

Actor Any person or group that can make a decision how to act and that can act on that decision.

- Developers and producers of technology. This includes engineering companies, industrial laboratories, consulting firms, universities and research centers, all of which usually employ scientists and engineers.
- **Users** who use the technology and formulate certain wishes or requirements for the functioning of the technology. The users of technologies are a very diverse group, including both companies and citizens (consumers).
- **Regulators** such as the government, who formulate rules or regulations that engineering products have to meet such as rulings concerning health and safety, but also rulings linked to relations between competitors. Regulators can also stimulate certain technological advances by means of subsidies.

Users People who use a technology and who may formulate certain wishes or requirements for the functioning of a technology.

Regulators Organizations who formulate rules or regulations that engineering products have to meet such as rulings concerning health and safety, but also rulings linked to relations between competitors.

Interests Things actors strive for because they are beneficial or advantageous for them.

Also other actors may be involved in technological development including, for example, professional associations, educational institutes, interest groups and trade unions (see Figure 1.6). All these actors have certain **interests**, – things they strive for

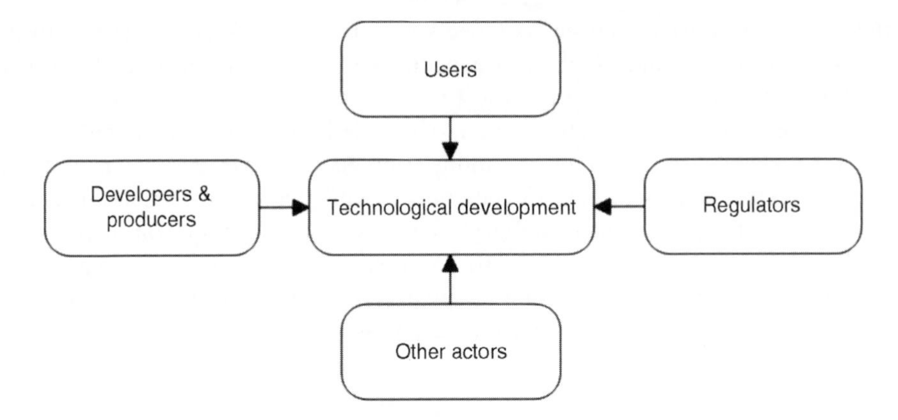

Figure 1.6 Technological development map of actors.

because they are beneficial or advantageous for them. The interests of the various actors will often conflict, so that there is no agreement on the desirable direction of technological development.

In addition to actors that influence the direction of technological development we distinguish stakeholders. **Stakeholders** are actors that have an interest ("a stake") in the development of a technology, but who cannot necessarily influence the direction of technological development. An example is people living in the vicinity of a planned construction site for a nuclear plant. Obviously these people have an interest in what type of reactor is built and how safe it is but they may not be able to influence the technology developed. Of course, such groups may organize themselves and try to get a say in technological development and they may do so more or less successfully. Stakeholders are not only relevant because they may become actors that actually influence technological development, they are also important from a moral point of view. As we have seen, stakeholders are actors whose interests are at stake in technological development. It is often assumed that morality and ethics require that we do not just neglect the interest of those actors because they are powerless but that we should somehow take them into account.[4]

Stakeholders Actors that have an interest ("a stake") in the development of a technology.

Case The Invention of Teflon

Roy Plunkett – a 28-year-old chemist at Du Pont – was requested in 1938 to develop a new, non-poisonous coolant for fridges. He therefore filled a metal tube with a little-used mixture and with tetrafluorethyleen that would perhaps possess cooling qualities. When he went to get the mixture out of the tube nothing came out but the tube was 60 grams heavier than normal. There therefore had to be something in it. After having sawn open the tube it was discovered that a pale and fatty, wax-like, white powder was stuck to the side. Nobody knew what it was so they began to experiment with the substance which turned out to be completely unique. It was given the name Teflon, after 'tef' – the nickname given by chemists to tetrafluorethyleen – followed by 'lon' – a suffix that Du Pont frequently used for its new products.

Du Pont devoted a great deal of time and money to discovering the exact characteristics of Teflon. It turned out to be complicated and expensive to produce Teflon. The first time that it was ever used was during World War II in order to reinforce the closing rings of the atom bomb. Teflon thus remained a state secret. It was not until 1946 that it was introduced to the general public.

Teflon has nowadays a wide range of uses. It is maybe best known as coating for non-stick frying pans. Although Teflon was long seen as a wonder material, it has recently come under some suspicion. In 2005, the Scientific Advisory Board of the Environmental Protection Agency (EPA) in the US found that perfluorooctanoic acid (PFOA), a chemical compound used to make Teflon, is "likely carcinogenic;" although EPA stresses (in 2010) on its website that it "has not made any definitive conclusions regarding potential risks, including cancer, at this time."[5] In 2005, scientists of US Food and Drug Administration (FDA) found small amounts of PFOA in Teflon cookware (Begley *et al.*, 2005), while DuPont scientists did not detect PFOA in such pans (Powley *et al.*, 2005). In 2006 Du Pont has committed itself to eliminating the release of PFOA to the environment (Eilperin, 2006). However, it still maintains that "evidence from 50 years of experience and extensive scientific studies supports the conclusion that PFOA does not cause adverse human health effects."[6]

Source: Based on Grauls (1993, pp. 123 ff).

The possibility of steering technological development is not only restrained by the fact that a large number of actors are involved in the development of technology but also because technological development is an unpredictable process (see Teflon box). In the course of time, a variety of methods and approaches have been developed to deal with this unpredictable character of technology development. This is done by a discipline known as **Technology Assessment (TA).** Initially TA was directed at the early detection and early warning of possible negative effects of technological development. Although such early detection and warning is important, it became increasingly clear that it is often not possible to predict the consequences of new technologies already in the early phases of technological development, as is also underscored by the Teflon example. On the other hand, it appeared that once the (negative) consequences materialize it has often become very difficult to change the direction of technological development because the technology has become deeply embedded in society and its design is more or less fixed. This problem is known as the

Technology Assessment (TA)
Systematic method for exploring future technology developments and assessing their potential societal consequences.

Collingridge dilemma This dilemma refers to a double-bind problem to control the direction of technological development. On the one hand, it is often not possible to predict the consequences of new technologies already in the early phases of technological development. On the other hand, once the (negative) consequences materialize it often has become very difficult to change the direction of technological development.

Constructive Technology Assessment (CTA) Approach to Technology Assessment (TA) in which TA-like efforts are carried out parallel to the process of technological development and are fed back to the development and design process.

Collingridge dilemma, after David Collingridge who first described it (Collingridge, 1980). Various approaches have been developed to overcome the Collingridge dilemma, one of the best known is **Constructive Technology Assessment (CTA)**. The idea behind CTA is that TA-like efforts are to be carried out parallel to the process of technological development and are fed back to the development and design process of technology (Schot, 1992; Schot and Rip, 1997). CTA aims at broadening the design process, both in terms of actors involved and in terms of interests, considerations and values taken into account in technological development. Among other things, this implies that stakeholders get a larger say in technological development.

What are the implications of the social context of technological development for the responsibility of engineers? In one sense, it diminishes the responsibility of engineers because it makes clear that engineers are just one of the many actors involved in technology development and cannot alone determine technological development and its social consequences. In another sense, however, it extends the responsibility of engineers because they have to take into account a range of stakeholders and their interests. Engineers cannot just as technocrats decide in isolation what the right thing to do is, but they need to involve other stakeholders in technological development and to engage in discussions with them.

1.7 Chapter Summary

In this chapter we have discussed the responsibility of engineers. The notion of responsibility has different meanings. One sense of responsibility, accountability, implies the obligation to render an account of your actions and the consequences of these. If you are not able to give a satisfactory account you are blameworthy. Usually four conditions need to apply in order to be blameworthy: wrong-doing, causal contribution, foreseeability, and freedom. In addition to accountability and blameworthiness, responsibility has an active component relating to preventing harm and doing good.

There are two main grounds of responsibility: the roles you play in society and moral considerations. Engineers have two main role responsibilities, one as engineers, the other as employees. As engineer you have a professional responsibility that is grounded in your role as engineer insofar as that role stays within the limits of what is morally allowed. Three professional ideals of engineers were examined as potential parts of the professional responsibility of engineers: technological enthusiasm, effectiveness and efficiency, and human welfare. The first two ideals are not always morally commendable and can in fact even become immoral when pursued in the light of immoral goals. The third ideal is morally laudable and, therefore, part of the professional responsibility of engineers.

Your professional responsibility as an engineer may sometimes conflict with your responsibility as an employee. We have discussed three models for dealing with this potential conflict: separatism, technocracy, and whistle-blowing. Separatism implies that the professional responsibility of engineers is confined to engineering matters and all decisions are made by managers and politicians. The disadvantage of this model is that engineers may end serving immoral goals and loose sight of the engineering ideal of public welfare. Technocracy means that engineers take over the decision power of managers and politicians. One disadvantage of this model is that engineers do not possess the expertise on which to decide for others what human welfare is or what is safe enough. Another disadvantage is that this model is paternalistic. Whistle-blowing means that you, as an engineer, speak out in public about certain abuses or dangerous situations in a company. Although whistle-blowing may sometimes be required it is not a very attractive model for the relation between engineers and managers. Instead of any of the three models, it might be better to work on a relation between engineers and managers that is more cooperative and mutually supportive, such as a model in which engineers think about broader issues than just engineering decisions but do not decide on these issues alone.

The responsibility of engineers is further complicated by the social context of technological development. Apart from engineers, a whole range of other actors is involved in technological development. This diminishes the responsibility of engineers as their causal contribution to technology and the foreseeability of consequences is diminished. At the same time, it introduces additional responsibilities, because engineers also need to take into account other stakeholders and their interests in the development of new technologies.

Study Questions

1 What are the five features of active responsibility according to Bovens?
2 What is the difference between passive and active responsibility?
3 What criteria (conditions) are usually applied when deciding whether someone is passively responsible (blameworthy) for a certain action and its consequences?
4 Suppose one person's actions have led to the injury of another person. What additional criteria must be satisfied in order to imply that the first person is passively responsible for the injury?
5 Do you consider Morton Thiokol responsible for the Challenger disaster? In answering this question, refer to the criteria for responsibility and use the information available.
6 Consider the following situation: An engineer who has been involved in the design of a small airplane for business travel, type XYZ, finds out that he used the wrong software to calculate the required strength of the wing. He has used a standard software package but now realizes that his package was not fit for this specific type of airplane. The very same day he finds this out, a plane of type XYZ crashes and all four passengers die. The investigation shows that the plane has crashed due to an inadequate design of the wing.

Do you consider this engineer responsible for the plane crash and the death of four people? (If you think there is not enough information to arrive at a judgment, indicate what information you would need to make a judgment and how this information would affect your judgment.)

7 In general, nobody will want to deny that engineers have an active responsibility for technologies they design and/or work with. In practice, however, many engineers find it problematic to act on this responsibility. Describe three problems for the idea that engineers should take responsibility for technologies and give a concrete example of each problem from engineering practice.
8 Explain what is meant by "separatism," and explain why the tripartite model illustrates separatism so well.
9 Why is it so difficult to steer technological development?
10 Explain why the ideal "public welfare" in professional ethics is the most important one for engineers from a moral perspective.
11 Look for an example of technological enthusiasm in your own field of study. Would you characterize this enthusiasm in this case as morally commendable, morally reprehensible, or just morally neutral? Argue your answer.

Discussion Questions

1 Do you consider Roger Boisjoly morally responsible for the Challenger disaster? And do you think his separatist argument is sound (see Section 1.5.1)?
2 Can companies, as contrasted to people, be morally responsible? In what sense are companies different from people and is this difference relevant for moral responsibility?
3 Do you think that you can ever have a moral obligation to blow the whistle in spite of the very negative consequences for you, such as dismissal or not making the grade?
4 Give an example in engineering practice, and explain what is meant by "moral responsibility" in that example and how it extends beyond role responsibility.

Notes

1 "Google: Achter het scherm" ("Google: Behind the Screen"), *Tegenlicht*, broadcast on May 7, 2006.
2 Text from the number "Wernher von Braun" by Tom Lehrer that featured in his album *That was the year that was* of 1965.
3 For more details on the legal position of whistle blowers and intiatives to protect them, see Chapter 2.
4 We will discuss the reasons for this assumption in more details in later chapters.
5 http://www.epa.gov/oppt/pfoa/pubs/pfoarisk.html (accessed April 9, 2010).
6 http://www2.dupont.com/Teflon/en_US/keyword/pfoa.html (accessed April 9, 2010).

2

Codes of Conduct

Having read this chapter and completed its associated questions, readers should be able to:

- Describe professional codes and corporate codes;
- Differentiate between three types of codes of conduct: aspirational, advisory, and disciplinary;
- Understand the role of codes of conduct with respect to the responsibility of engineers;
- Identify the strengths and weaknesses of codes of conduct;
- Evaluate the role of global codes for multinationals and for engineers.

Contents

Ethics, Technology, and Engineering: An Introduction, First Edition.
Ibo van de Poel and Lambèr Royakkers.

2.1 Introduction

Case Bay Area Rapid Transport Project

Figure 2.1 BART train. Photo: NISEE-PEER Earthquake Engineering Library, University of California, Berkeley.

In March 1972 Holger Hsortsvang, Max Blakenzee, and Robert Bruder, three engineers, working on the *Bay Area Rapid Transport Project (BART)* in California (United States) and responsible for the design and creation of an automatic guided train system, were dismissed. These engineers had been expressing their doubts about the safety of the system via internal memos since 1969 to their managers. The response was "don't make trouble." In 1971 they brought their concerns in confidence to members of the board of directors, thus bypassing their immediate superiors. That was unconventional for the BART organization and indeed for any hierarchical organization. The director they finally made contact with turned out to be very interested in their case and so he promised to raise it with the management. He furthermore promised to keep their names anonymous and do nothing to damage their interests. However, two days after the encounter the full story was published in the *Contra Costa Times*. At first the engineers denied having any involvement in the matter but once their involvement was confirmed they were immediately fired without cause or appeal. They subsequently took the matter to court.

In the wake of the affair one of the organizations to become involved was the Institute of Electrical and Electronic Engineers (IEEE). The IEEE decided to send what is known as an amicus curiae letter to the law courts. (An amicus curiae is an "friend of the court": someone, not a party to a case, who voluntarily offers information on a point of law or some other aspect of the case to assist the court.) The letter emphasized the fact that according to the IEEE's professional code, engineers are responsible for the "safety, health and welfare of the public." The IEEE also argued that the professional code is an implicit aspect of the employment contract. If this argument had been accepted by the judge then it would have meant that employees who act in accordance with what is stated in the professional code may not be simply dismissed.

After the three engineers had lost their job, their concerns were decisively confirmed on October 2, 1972, three weeks after BART began carrying passengers. There was a train system accident and several passengers were injured. Despite this, the three engineers accepted an out-of-court settlement reported to be $25,000 per person. The presumed reason for this was that they had in the first instance lied about their involvement in the matter which had weakened their case. Apart from anything else, the dismissals were very detrimental for the careers of all three engineers.

Source: Based on Anderson, Perucci, Schendel, and Trachtman (1980), Anderson, Otten, and Schendel (1983), and Unger (1994, pp. 12–17).

In this case, the three engineers acted out of a sense of professional responsibility. This professional responsibility was codified in the IEEE code of conduct and was related to the safety, health, and welfare of the public. Although their professional organization supported their behavior, it could not prevent them from being dismissed. In this chapter, we discuss the role of codes of conduct in engineering. In particular, we focus on professional codes as they have been proposed by professional engineering societies and on corporate codes, as they have been formulated by companies. In Section 2.2, we discuss these two types of codes, their structure and their content. In Section 2.3, we discuss a number of common objections that have been leveled against codes of conduct. This includes the problem that is highlighted by the case above, that acting according to the code, may nevertheless lead to dismissal. In Section 2.4, we will discuss codes of conduct in an international context.

2.2 Codes of Conduct

Codes of conduct are codes in which organizations lay down guidelines for responsible behavior of their members. Such guidelines may be detailed and prescriptive, but they can also be formulated more broadly and express the values and norms that should guide behavior and decision-making (Hummels and

Codes of conduct: A code in which organizations (like companies or professional associations) lay down guidelines for responsible behavior of their members.

Karssing, 2007). Codes of conduct are often intended as an addition to the requirements of the law. When codes of conduct are enforced this is usually done by the organization that formulated the code. For engineers, two types of codes of conduct are especially important: **professional codes** that are formulated by professional associations of engineers and, **corporate codes** of conduct that are formulated by companies in which engineers are employed.

Professional code Code of conduct that is formulated by a professional association.

Corporate code Code of conduct that is formulated by a company.

Codes of conduct are formulated for a variety of reasons, such as: increasing moral awareness; the identification and interpretation of the moral norms and values of a profession or a company; the stimulation of ethical discussion; as a way to increase accountability to the outside world; and, finally, to improve the image of a profession or company. Depending on the exact objectives of a code of conduct, a distinction can be made between three types of codes of conduct:[1]

Aspirational code A code that expresses the moral values of a profession or company.

Advisory codes A code of conduct that has the objective to help individual professionals or employees to exercise moral judgments in concrete situations.

Disciplinary code A code that has the objective to achieve that the behavior of all professionals or employees meets certain values and norms.

- An **aspirational code** expresses the moral values of a profession or company. The objective of such a code is to express to the outside world the kind of values the profession or company is committed to.
- An **advisory code** has the objective to help individual professionals or employees to exercise moral judgments in concrete situations on the basis of the more general values and norms of the profession or company.
- A **disciplinary code** has the objective that the behavior of all professionals or employees meets certain values and norms.

Most professional codes for engineers are advisory. Usually, they have the following more specific objectives: increasing awareness of and sensitivity for moral issues in the daily exercising of the profession; helping in analyzing such moral issues and in formulating key questions or issues with respect to these moral issues; and, finally, helping in coming to a judgment on these moral issues. Corporate codes of conduct are more often disciplinary. In such cases, their objective is to achieve that all employees act according to certain guidelines. The formulation of codes of conduct is only one of the activities that professional associations and companies can undertake to stimulate responsible behavior by their members. Other activities include the appointment of a confidant or committee which whom moral problems can be discussed or the organization of training sessions for dealing with moral dilemmas.

2.2.1 Professional codes

Professional codes are guidelines for the exercising of a profession and are formulated by a professional society. Professional codes have been formulated for a variety of professions like doctors, nurses, lawyers, priests, the police, and corporate managers. Also engineers have professional codes of conduct.

What is a Profession?

A **profession** is an occupation with specific characteristics. There is no agreement on what characteristics are exactly required to call an occupation a profession. The following characteristics are often mentioned (see, for example, Layton, 1971; Noble, 1977; Disco, 1990):

1 The use of specialized knowledge and skills requiring a long period of study.
2 A monopoly on the carrying out of the occupation: not everybody can call himself an engineer or do engineering work.
3 The assessment of whether the professional work is carried out in a competent way is done, and can only be done, by colleague professionals. They are the only ones who posses the knowledge and skills to apply the right standards of judgment.

Profession Often mentioned characteristics of a profession include: 1) use of specialized knowledge and skills; 2) a monopoly on the carrying out of the occupation; 3) assessment only possible by peers. In addition the following two requirements are also sometimes mentioned: 4) service orientation to society; and 5) ethical standards.

Some authors have added two further characteristics (for example, Davis, 1998; Harris, Pritchard, and Rabins, 2005):

4 A profession provides society with products, services or values that are useful or worthwhile for society, and is characterized by an ideal of serving society.
5 Ethical standards, derived from or relating to the society-serving ideal of the profession, regulate the daily practice of professional work.

These authors view professional codes as an expression of the service ideal to society and the ethical standards that regulate the profession. Authors who do not include these two additional aspects in the definition of a profession are often more skeptical about the purpose of professional codes. They stress that professions may be self-serving and that codes of conduct might primary be a means to acquire status and other privileges.

Historically, the development of professional codes for engineers began in England in 1771 with the code of the *Smeatonian Society*. More influential for the current professional codes for engineers was the formulation of a range of professional codes for different engineering professions like civil, mechanical, and electrical engineering in the first decade of the twentieth century in the United States. The early codes comprised rules for engineers that chiefly pertained to etiquette. The professional code regulated people's entry into the profession and the behavior of members towards each other and in relation to employers and clients. While the early codes did not address broader social issues raised by engineering, this changed after World War II. The gas chambers and scientific experiments that had been carried out by the Germans on people during the World War II gave science and technology a bad image. The atomic bomb also showed clearly that technology gave rise to certain moral issues.

Case The Atomic Bomb

Figure 2.2 Atomic bomb mushroom cloud. Photo: Library of Congress/United States Department of Defence.

In 1932 James Chadwick discovered the neutron, which later proved the key to nuclear fission and the discovery of the atomic bomb. The Hungarian scientist Leó Szilárd as early as October 1933 realized that "a chain reaction might be set up if an element could be found that would emit two neutrons when it swallowed one neutron" (Jungk, 1958, p. 54). This chain reaction would result in the production of large amounts of energy that might be used to produce energy but might also be put to bad purposes. In the same year, Hitler had come to power in Germany and Szilárd had fled to London to escape Nazi prosecution. Szilárd therefore started lobbying for not publishing the results of studies on this topic, as he feared they could be misused by the German government; he was however not very successful.

In 1934 the research groups of both Enrico Fermi and Irene Joliot-Curie disintegrated heavy atoms by spraying them with neutrons. At this point these scientists did not realize that they had achieved fission. It took until 1938 before the experiments were rightly interpreted, after another experiment with bombarding uranium with neutrons by the German physicist Otto Hahn, who is usually credited with discovering nuclear fission. On February 2, 1939, Szilárd wrote a letter to Joliot-Curie: "Obviously, if more than one neutron were liberated, a sort of chain reaction would be possible. In certain circumstances this may then lead to the construction of bombs which would be extremely dangerous in general and particularly in the hands of certain governments" (Jungk, 1958, p. 77), and "We all hope that there will be no or at least not sufficient neutron emissions and therefore nothing to worry about" (Jungk, 1958, p. 77). At that time, Joliot-Curie was just at the point of experimental realization of the mentioned chain reaction and her group published the results to the dismay of Szilárd.

As Szilárd feared that the Germans might be able to develop an atomic bomb, he began to look for ways to persuade the US government also to do so. In August 1939, he succeeded in convincing Einstein in signing a letter to President Roosevelt in which they warned for the developments in Germany and urged for more American studies on the subject. The letter eventually reached Roosevelt in October 1939, and contributed to the establishment of the so-called Manhattan Project, a large research project in the US that would eventually result in the production of atomic bombs. After the war, Einstein came to regret his cooperation deeply: "If I had known that the Germans would not succeed in constructing the atom bomb, I would never have lifted a finger" (Jungk, 1958, p. 87).

Towards the end of the war, a number of scientists working on the Manhattan Project became concerned about the use of the atomic bomb they had developed by the US government. In July 1945, 69 scientists signed a petition drafted by Szilárd. This petition, among other contained the following passages (www.dannen.com/decision/45-07-17.html):

> We, the undersigned scientists, have been working in the field of atomic power. Until recently, we have had to fear that the United States might be attacked by atomic bombs during this war and that her only defense might lie in a counterattack by the same means. Today, with the defeat of Germany, this danger is averted and we feel impelled to say what follows:
>
> The war has to be brought speedily to a successful conclusion and attacks by atomic bombs may very well be an effective method of warfare. We feel, however, that such attacks on Japan could not be justified, at least not unless the terms which will be imposed after the war on Japan were made public in detail and Japan were given an opportunity to surrender.
>
> The added material strength which this lead [in the development of the atomic bomb] gives to the United States brings with it the obligation of restraint and if we were to violate this obligation our moral position would be weakened in the eyes of the world and in our own eyes. It would then be more difficult for us to live up to our responsibility of bringing the unloosened forces of destruction under control.

The signed petition never reached President Truman. On August 6, 1945, the US dropped the atomic bomb "Little Boy" on the city of Hiroshima, followed on August 9 by the dropping of the "Fat Man" nuclear bomb over Nagasaki. The bombs killed as many as 140,000 people in Hiroshima and 80,000 in Nagasaki by the end of 1945. On August 15, 1945, Japan announced its surrender to the Allied Powers.

Source: Mainly based on Jungk (1958).

One of the ways of restoring the social image of science and technology after World War II was by establishing professional codes. In 1950 the German engineers' association, the Verein Deutscher Ingenieure (VDI), drew up an oath for engineers, which was clearly inspired by the dubious role of some engineers and scientists during World War II. One of the things stated in the professional code was that engineers should not work for those who fail to respect human rights.[2] Also in the United States, most of the professional codes were reformulated after World War II: the duty of the engineer to serve the public interest was especially stressed in the new codes of conduct. Organizations like the National Society of Professional Engineers (NSPE), the American Society of Civil Engineers (ASCE) and The American Society of Mechanical Engineering (ASME) formulated codes of conduct stating that engineers "should hold paramount the safety, health and welfare of the public."

In addition to national engineering societies, Europe has an overarching professional organization, the European Federation of National Engineering Associations (FEANI). FEANI was established in 1951 by a group of German and French engineers. At the moment, professional associations from 29 European countries are member of FEANI (www.feani.org accessed August 24, 2007). FEANI has formulated a universal statement regarding the conduct of professional engineers, which can be implemented by national member's societies in their code of conduct. The FEANI code thus has a quite different status than most US codes like the NSPE code which is reflected in the content of the code, in particular the FEANI code is much more general (and vague) and contains much less detail than, for example, the NSPE code.

Professional codes for engineers provide content to the responsibility of engineers. They express the moral norms and values of the profession. Most modern professional codes relate to three domains: 1) conducting a profession with integrity and honesty, and in a competent way; 2) obligations towards employers and clients; and 3) responsibility towards the public and society.

Integrity and competent professional practice

All professional codes include the obligation to practice one's profession with **integrity** and **honesty**, and in a competent way. This is the traditional core of all professional codes. To practice one's profession in a competent way means that the practitioner must be competent and the professional practice must be conducted skillfully. This implies that the practitioner must be well enough educated, must keep up to date in his field and must take only work in his field of competence. With integrity and honesty we mean that the profession must be conducted in an honest, faithful, and truthful manner. This entails, for instance, that facts may not be manipulated and agreements must be honored. Sometimes it is also stipulated that the profession must be practiced in an independent and impartial way. Usually this is meant to imply that engineers should avoid conflicts of interests. You have a **conflict of interest** if you have an interest that, when pursued,

Integrity Living by one's own (moral) values, norms and commitments.

Honesty Telling what one has good reasons to believe to be true and disclosing all relevant information.

Conflict of interest The situation in which one has an interest (personal or professional) that, when pursued, can conflict with meeting one's professional obligations to an employer or to (other) clients.

conflicts with meeting your obligations to your employer or clients. This may be a personal interest, like when you have stocks in a company that produces a certain kind of measuring apparatus and you have to advise a large client about what measuring apparatus to use. It can also be an interest that derives from another professional role, for example when you advise two competing firms. Although conflicts of interest do not necessarily lead to immoral behavior it is better to avoid them because a conflict can corrupt your professional judgment and diminishes your trustworthiness as an engineer. If a conflict of interest is unavoidable it should at least be disclosed to the interest parties.

> Engineers shall perform services only in the areas of their competence. (NSPE Code of conduct)
>
> Engineers shall issue public statements only in an objective and truthful manner. (NSPE Code of conduct)
>
> Engineers shall not be influenced in their professional duties by conflicting interests. (NSPE Code of conduct)
>
> Engineers shall maintain their relevant competences at the necessary level and only undertake tasks for which they are competent. (FEANI)

Obligations towards clients and employers

Obligations towards clients and employers are mentioned in most professional codes. In many cases, it is stipulated that engineers should serve the interests of their clients and employers and that they must keep secret the confidential information passed on by clients or employers.

> Engineers shall act for each employer or client as faithful agents or trustees. (NSPE Code of conduct)
>
> Engineers shall not disclose, without consent, confidential information concerning the business affairs or technical processes of any present or former client or employer, or public body on which they serve. (NSPE Code of conduct)
>
> Engineers shall provide impartial analysis and judgement to employer or clients, avoid conflicts of interest, and observe proper duties of confidentiality. (FEANI)

Social responsibility and obligations towards the public

Virtually all professional codes in one way or another emphasize the social responsibility of engineers. Matters frequently referred to are: safety; health; the environment; sustainable development; and the welfare of the public. According to a limited number of professional codes engineers must inform the public about the aspects of the technology in which they are involved and that are relevant to the public, such as the risks and hazards involved.

Engineers shall hold paramount the safety, health, and welfare of the public. (NSPE Code of conduct)

Engineers shall at all times strive to serve the public interest. (NSPE Code of conduct)

Engineers are encouraged to adhere to the principles of sustainable development in order to protect the environment for future generations. (NSPE Code of conduct)

Engineers shall carry out their tasks so as to prevent avoidable danger to health and safety, and prevent avoidable adverse impact on the environment. (FEANI)

2.2.2 Corporate codes

Corporate codes are voluntarily commitments made by individual companies or associations of companies setting certain values, standards, and principles for the conduct of corporations. Corporate codes are usually more recent than professional codes. They have been formulated since the 1960s and 1970s, particularly in reaction to corporate scandals (Ryan, 1991). According to a survey that was carried in 2001 and 2002, 52 percent of the 200 largest companies in the world have a corporate code (Kaptein, 2004). Below, we will discuss the main elements of the various kinds of corporate codes: the mission, the core values, the responsibilities towards stakeholders and detailed rules and norms.

Corporate Social Responsibility

Corporate Social Responsibility The responsibility of companies towards stakeholders and to society at large that extends beyond meeting the law and serving shareholders' interests.

The formulation of corporate codes is based on the assumption that companies have a **corporate social responsibility**, that is, a responsibility towards stakeholders and to society at large. This assumption has been contested by several authors who maintain that the responsibility of a company is limited to making profit within the limits of the law. This so-called classical view on corporate responsibility can be traced back towards Adam Smith, the founder of modern economics. According to Smith, the invisible hand of the market makes everyone better off if all people, producers and consumers alike, only pursue their own interests (Smith, 1776). An important contemporary defender of the classical view is the economist and Noble Prize laureate Milton Friedman. According to Friedman, companies only have responsibilities towards their shareholders and not to any other stakeholders, society, or the environment (Friedman, 1962). He considers it undesirable that companies take into account other stakeholders' interests and views. He provides two arguments for this statement. First, money spent by a corporation on social responsibility is ultimately the money of the shareholders and this expenditure conflicts with their goal to maximize profits. Second, corporations are not democratically elected. When companies formulate their own ideas about what is morally allowable or desirable they are enforcing their own particular view upon others without any democratic legitimization. If any limits on corporate behavior are desirable, they have to be formulated by the government, not by companies.

A number of objections can be raised against Friedman's view. First, although responsibilities to other stakeholders can conflict with shareholders' interests, this is not always the case. Companies are aware that corporate responsibility initiatives do not necessarily have a negative impact on their bottom line, and that they can have an extremely positive impact. In other words, the thought that "ethics is a luxury we can't afford" is replaced by "ethics pays" (Paine, 2000, p. 329). Second, laws are not always adequate or effective in preventing immoral behavior. Not everything that is morally desirable can be laid down in the law. Laws also tend to lag behind technological development and companies might be in a better position to foretell moral issues raised by new technology than the government. Hence, they have a responsibility that extends beyond what the law requires.

Mission statement

Many corporate codes contain a mission statement that concisely formulates the strategic objectives of the company and answers the question what the organization stands for.

> At Microsoft, we work to help people and businesses throughout the world realize their full potential. This is our mission. Everything we do reflects this mission and the values that make it possible. (Microsoft mission statement)

> The mission of Merck is to provide society with superior products and services by developing innovations and solutions that improve the quality of life and satisfy customer needs, and to provide employees with meaningful work and advancement opportunities, and investors with a superior rate of return. (Mission statement of Merck, a pharmaceutical company)

Core values

Core values express the qualities that a company considers desirable and which ground business conduct and outcomes. They imply an appeal to the attitudes of employees but do not contain detailed rules of conduct. Often mentioned values include teamwork, responsibility, open communication and creativity.[3] Also values like customer orientation, flexibility, efficiency, professionalism, and loyalty are regularly mentioned.

As a company, and as individuals, we value:

- Integrity and honesty.
- Passion for customers, for our partners, and for technology.
- Openness and respectfulness.
- Taking on big challenges and seeing them through.
- Constructive self-criticism, self-improvement, and personal excellence.
- Accountability to customers, shareholders, partners, and employees for commitments, results, and quality. (Microsoft)

Responsibility to stakeholders

Most corporate codes also express responsibilities to a variety of stakeholders like consumers, employees, investors, society, and the environment. Competitors and suppliers are also sometimes mentioned as stakeholders. Typically, responsibility to the environment is more often mentioned in European than in American codes. Conversely, responsibilities to competitors are far more often mentioned in American than in European or Asian codes.

With respect to customers, the supply of qualitatively good products and services is often mentioned as a responsibility. Also sustainability, and enhancing the health and safety of consumers are important topics. With respect to employees, regularly mentioned responsibilities include encouraging personal development, respect, and equal opportunity. With respect to society, the most mentioned responsibility is observing the law. Also being a good corporate citizen and contributing to society are named. Less often cited responsibilities include enhancing the quality of life, sustainability, and respecting human rights.

Stakeholder principles Principles that guide the relationship between a company and its stakeholders.

In addition to responsibilities towards stakeholders, some corporate codes also contain **stakeholder principles** that guide the relationship between company and stakeholders. The most mentioned stakeholder principles are transparency, honesty (truth), and fairness (impartiality) (Kaptein, 2004). In American codes, honesty is more often included than transparency, whereas in European and Asian codes the relation is reversed. Japanese companies relatively often cite trust as a stakeholder principle compared to American and European companies.

From Lockheed Martin's *Setting the Standard; Code of Ethics and Business Conduct*:

Our commitments:

- For our *employees*: we are committed to honesty, just management, fairness, a safe and healthy environment free from the fear of retribution, and respecting the dignity due everyone.
- For our *customers*: we are committed to produce reliable products and services, delivered on time, at a fair price.
- For the *communities in which we live and work*: we are committed to observe sound environmental business practices and to act as concerned and responsible neighbors, reflecting all aspects of good citizenship.
- For our *shareholders*: we are committed to pursuing profitable growth, without taking undue risk, to exercising financial discipline in the deployment of our assets and

resources, and to making accurate, timely, and clear disclosures in all public reports and communications.
- For our *suppliers and partners*: we are committed to fair competition and the sense of responsibility required of a good customer and teammate.

Norms and rules
Norms and rules contain guidelines for employees how to act in specific situations. This may include subjects like the acceptance of gifts, fraud, conflicts of interest, confidentiality, theft, corruption, bribery, discrimination, respect, and sexual harassment.

Some rules from Intel's *How the Corporate Business Principles Apply to You*:

- Employees must follow the law wherever they are around the world and in all circumstances. Do not engage in behavior that harms the reputation of Intel or yourself. If you wouldn't want to tell your parents or your children about your action, or would be embarrassed to read about it in a newspaper, then don't do it.
- Employees must avoid both actual and perceived conflicts of interest.
- Customers and suppliers must be dealt with fairly and at arm's length.
- Employees must never attempt to bribe or improperly influence a government official, customer or supplier.

Two examples from the IBM document *Ethics and Compliance*:

> Generally, it is not appropriate for an employee to accept a supplier's invitation to attend an entertainment or sporting event at the supplier's expense. An invitation to an entertainment or sporting event such as a golf or tennis tournament may be appropriate if it demonstrably helps to build or maintain a business relationship. Before accepting such an invitation, an employee must obtain approval from a Vice President, a Regional Sales Manager or Corporate Director of Purchasing. Sound judgment is necessary for determining when invitations to such events are appropriate.

> Paying a freight forwarder to expedite a shipment through customs is not acceptable if the agent doesn't follow applicable rules and regulations, and if the agent gives money or payment in kind to a government official for personal benefit. On the other hand, expediting by following rules and regulations and without bribing officials is acceptable.

2.3 Possibilities and Limitations of Codes of Conduct

As we have seen, codes of conduct help to express the responsibilities of engineers. They are, therefore, a useful point of departure for discussions about these responsibilities. Still, in the course of time, a number of objections against code of conduct have been leveled. Below, we discuss the main objections. In judging these objections, one should keep in mind that codes of conduct may have different objectives. Especially the difference between aspirational, advisory, and disciplinary codes is relevant here. Objections against disciplinary codes are not always sound objections against advisory codes and vice versa. Although the objections discussed below show some of the limitations of codes of conduct, none of them is strong or convincing enough to conclude that codes of conduct as such are undesirable. Much depends on the actual formulation and implementation of the code.

2.3.1 Codes of conduct and self-interest

Codes of conduct are a form of self-regulation. Sometimes, they are primarily formulated for reasons of self-interest, for example to improve one's image to the outside world, to avoid government regulation or to silence dissident voices. An example in which the latter happened is the case of Jon Tozer (see box).

Case John Tozer

In 1989 the Australian engineer John Tozer criticized the decision of the Coffs Harbour authorities to pump sewage into the sea. According to him the engineers employed by the local authority had given a misleading impression of the effects upon the environment and they had failed to properly investigate the alternatives. The engineers in question were subsequently successful in removing Tozer from the Association of Consulting Engineers Australia (ACEA). Tozer was accused of having contravened the professional code by openly criticizing the work of other (associated) engineers. Because of his disbarment Tozer, who has his own consulting engineering firm, is no longer able to fulfill any contracts for customers demanding ACEA membership.

Source: Based on Beder (1993).

The fact that self-interest plays a role in formulating codes of conduct is not necessarily objectionable as long as the content of the code is ethical and serious attempts are made to live by the code of conduct. One way to ensure this is to include a range of stakeholders in the formulation and implementation of the code of conduct to avoid the code becoming one-sided.

Window-dressing Presenting a favorable impression that is not based on the actual facts

A code of conduct serving only the interests of a company or profession may amount to **window-dressing**. We speak of window-dressing if a favorable impression is presented of what the company is doing but that impression does not represent how the company and its employees actually behave. In cases of window-dressing, it may, for example, well be the case that the existence of the code is unknown to members of the organization while at the meantime the code is used in communication with the outside world. The danger of window-dressing is especially present in the case of aspirational codes because they tend to be very vague and general.

Case Google in China: A Case of Window-Dressing?

Figure 2.3 Google offices in China, Photo: Getty Images.

> While removing search results is inconsistent with Google's mission, providing no information ... is more inconsistent with our mission. (Google statement)

Google, the leading Internet search engine company in the world, entered the Chinese market in early 2000 by creating a Chinese-language version of its home page, google.com, that was located in the United States but that could handle search requests from China. In this way, the technology was not subject to Chinese censorship laws as the facilities were not within China's physical boundaries, and Google did not need a license from the Chinese government to operate its business. In 2002, the Chinese version of Google was shut down by the Chinese government for two weeks. When reinstated, it was very slow for all Chinese users and completely inaccessible for Chinese colleges and universities. By 2005, the Chinese search engine company Baidu emerged as the leading Internet search company in China. To compete with Baidu, Google decided in 2006 to launch a Chinese website (www.google.cn) and agreed to censor its content enforced by means of filters known as "The Great Firewall of China." "Harmful" content included material concerning democracy (e.g., freedom), religious cults (e.g., Falun Gong), or antigovernment protests (e.g., Tiananmen Square). Google received much criticism from human rights advocates because it censored information such as human rights.

A moral question is here whether Google's slogan "Don't be Evil" ("It's about providing our users unbiased access to information") and their mission statement "Google's mission is to organize the world's information and make it universally accessible and useful" have been consistently followed. By censoring information, one could argue that Google has strayed from dedication to helping every user get unrestricted access to content on the Internet. Google admitted that the launching of google.cn was problematic with respect to their mission. In the words of Schrage, Google's vice president of Global Communications and Public Affairs: "[Google, Inc., faced a choice to] compromise our mission by failing to serve our users in China or compromise our mission by entering China and complying with Chinese laws that require us to censor search results. ... Self-censorship, like which we are now required to perform in China, is something that conflicts deeply with our core principles. ... This was not something we did enthusiastically or something we're proud of at all."

On March 22, 2010 after a cyber attack on Google's servers and increased demands for censoring, Google decided no longer to censor its search results. In the words of David Drummond, senior vice president of Google Corporate Development and Chief Legal Officer: "On January 12, we announced ... that Google and more than 20 other US companies had been the victims of a sophisticated cyber attack originating from China, and that during our investigation into these attacks we had uncovered evidence to suggest that the Gmail accounts of dozens of human rights activists connected with China were being routinely accessed by third parties, most likely via phishing scams or malware placed on their computers. We also made clear that these attacks and the surveillance they uncovered – combined with attempts over the last year to further limit free speech on the web in China including the persistent blocking of websites such as Facebook, Twitter, YouTube, Google Docs and Blogger – had led us to conclude that we could no longer continue censoring our results on Google.cn. So earlier today we stopped censoring our search services ... on Google.cn. Users visiting Google.cn are now being redirected to Google.com.hk, where we are offering uncensored search in simplified Chinese, specifically designed for users in mainland China and delivered via our servers in Hong Kong" (http://googleblog.blogspot.com/2010/03/new-approach-to-china-update.html (accessed April 11, 2010)). On March 30, 2010, the Chinese government blocked access to Google's search engine from Mainland China.

Source: Based on Martin (2008), Dann and Haddow (2008), and Congressional Testimony of Schrage (2006).

2.3.2 Vagueness and potential contradictions

In the application of codes of conduct to concrete situations, one is frequently confronted with rather vague concepts and rules that need interpretation. Depending on the exact interpretation of such concepts and rules, codes of conduct sometimes result in contradictory recommendations about what to do in a specific situation.

> **Uncritical loyalty** Placing the interests of the employer, as the employer defines those interests, above any other considerations.

One relevant notion from codes of conduct that is in need of further clarification and interpretation is "loyalty." The NSPE code of conduct, for example, requires that engineers "shall act for each employer or client as faithful agents or trustees." This means that engineers need to be loyal to their company (Harris, Pritchard, and Rabins, 2005, p. 191). But what does loyalty exactly amount to? Take, for example the case of the three BART engineers discussed at the beginning of this chapter. Was the engineers act disloyal because they spoke out against their organization? The answer to this question is yes if one interprets loyalty as **uncritical loyalty**. Harris, Pritchard, and Rabins (2005, p. 191) define uncritical loyalty to an employer as "placing the interests of the employer, as the employer defines those interests, above any other consideration." Such uncritical loyalty may, however, be misguided (Martin and Schinzinger, 1996, pp. 193–195). First, one might disagree about what the interests of the employer are. In the BART case, it might well be argued that it was not in the interest of the BART organization to keep silent about the technical problems. So conceived, the BART engineers acted loyally to the interests of the company. Second, it might be doubted whether the interests of the company should always override any other concerns, especially in cases when the public is put at danger. To deal with such objections, Harris, Pritchard, and Rabins propose the notion of **critical loyalty** which they define as "giving due regard to the interest of the employer, insofar as this is possible within the constraints of the employee's personal and professional ethics."

> **Critical loyalty** Giving due regard to the interest of the employer, insofar as this is possible within the constraints of the employee's personal and professional ethics.

Apart from vagueness, codes of conduct may be plagued by inconsistencies, both within codes and between codes. Let us look at the rules for confidentiality and disclosure of information contained in three different codes of conduct.

1 NSPE (National Society of Professional Engineers)

> Engineers shall not reveal facts, data, or information without the prior consent of the client or employer except as authorized or required by law or this Code. (Rule of practice 1c)
>
> Engineers having knowledge of any alleged violation of this Code shall report thereon to appropriate professional bodies and, when relevant, also to public authorities, and cooperate with the proper authorities in furnishing such information or assistance as may be required. (Rule of practice 1f)

2 FEANI (European Federation of National Engineering Associations):

> Engineers shall … observe proper duties of confidentiality.
>
> Engineers shall be prepared to contribute to public debate on matters of technical understanding in fields in which they are competent to comment.

3 IEEE (Institute for Electrical and Electronic Engineers):

> We, the members of the IEEE, ... agree to accept responsibility when making engineering decisions consistent with the safety, health and welfare of the public, and to disclose promptly factors that might endanger the public or the environment.

There are important differences between these three codes. The IEEE code does not contain a confidentiality requirement, while the other two do. Conversely, the FEANI code is silent about informing third parties when the code is violated or the public is put at risk, probably because the code is only intended as a common framework that can be further detailed by member societies in their own national codes. Note also that the NSPE Code identifies different parties that should be informed in the case of code violations than the IEEE code. Whereas the IEEE Code would encourage the BART engineers to speak out in public, the NSPE code tells them to inform the proper authorities. The prescription flowing from the FEANI code is less clear. If one interprets "contributing to public debate" as informing the public about possible hazards, one might say that engineers have a right to speak out on basis of the second rule in the box. On this interpretation, "contributing to public debate" conflicts with the rule about confidentiality. This conflict is not resolved in the code. This conflict might be avoided by an interpretation of "contributing to public debate" that excludes making public confidential information, even if this is information about the possible malfunctioning of a technical system.

As this example reveals the degree to which codes of conduct are vague and potentially contradictory is different from code to code. This means that attempts can be made to avoid vagueness and contradictions. The NSPE has gone some way in doing so in its code. In addition, the Board of Ethical Review of the NSPE has in the past published anonymous cases in which a judgment was presented whether certain behavior was in accordance with the code of conduct or not.[4]

2.3.3 Can ethics be codified?

Some authors have argued that the idea of drafting a code of conduct is misperceived because ethics cannot be codified. In a sense, this objection is the mirror of the previous one. Whereas people who criticize the vagueness and potential contradictions in codes of conduct are worried that such codes do not uniformly prescribe certain behavior, people who argue that ethics cannot be codified are often worried that codes of conduct contain strict prescriptions which conflict with what ethics is about according to them. We will consider three different arguments why ethics cannot be codified.

One argument is that ethics requires individual moral judgment, instead of blindly following a code (Ladd, 1991). In the terminology of the philosopher Immanuel Kant, following a code of conduct may be based on heteronymous motives, that is, motives originating outside the acting person like fear for sanctions while moral behavior requires autonomous decisions and behavior (see further Chapter 3). However, even if ethics requires autonomous decision-making, it does not follow that codes of conduct are necessarily objectionable. What is objectionable is a certain uncritical way of using codes of conduct. However, an advisory code need not conflict with the moral autonomy people retain in deciding whether to follow the code or not.

Nevertheless, in the case of disciplinary codes the argument may be sound because disciplinary codes suppose that the code is strictly adhered to.[5]

A second argument is that codes of conduct are not morally binding (cf. Ladd, 1991). As the box shows, a variety of arguments why codes of conduct are binding can be given. Even if one rejects the view that codes of conduct entail a contract, one might still argue that codes of conduct express already existing moral responsibilities and obligations. In that case, a code of conduct cannot create new moral obligations beyond what was already morally required. From this, however, it does not follow that a code is superfluous. It might still be helpful, for example, to remind people of their moral obligations and responsibilities.

Why are Codes of Conduct Morally Binding?

Three explanations have been offered why codes of conduct are morally binding:

1 One possible explanation is that codes of conduct entail an implicit contract between engineering as a profession and the rest of society (Harris, Pritchard, and Rabins, 2005). According to this explanation, professionals serve a moral ideal in exchange for privileges such as status, a monopoly on carrying out the occupation and good salaries. In this explanation, professionals are bound by professional codes because they have implicitly signed a contract with society. This contract creates a moral obligation to follow the code of conduct of a profession.

2 A second explanation is offered by Michael Davis. He defines a profession as follows: "A profession is a number of individuals in the same occupation voluntarily organized to earn a living by openly serving a certain moral ideal in a morally-permissible way beyond what law, market, and morality would otherwise require" (Davis, 1998, p. 417). One important feature of this definition is that being a profession is a voluntary choice. According to Davis, the existence of professional codes for engineers testifies that engineers indeed have made this choice. Such codes are binding because being a member of a profession implies an implicit contract with your colleague professionals. This contract creates a level playing field so that all professionals can pursue the moral ideal.

3 A third explanation is that the codes of conduct as such are not morally binding but that they express moral responsibilities that are grounded otherwise. Michael Pritchard, for example, has argued that engineering codes of conduct are based on common morality (Pritchard, 2009).

Similar arguments may be given for corporate codes. These can also be seen as (1) a contract between a company and society or (2) as a contract between employees of a company or (3) as an expression of the moral responsibilities and obligations a company and its employees have on other grounds.

A third argument against codes of conduct is that they presuppose that morality can be expressed in a set of universal moral rules. One reason why this is questionable is that engineering is too diverse, both in terms of disciplines (civil engineering, mechanical engineering, electrical engineering, aerospace engineering, etc.) and in terms of activities (research, design, testing, maintenance, etc.) for one code to apply. This objection can, however, be dealt with by having a variety of codes of conduct. A more fundamental objection is that sound moral judgment always requires taking into account the particularities of a situation (e.g., Dancy, 1993). According to this line of reasoning, it is not surprising that codes of conduct always require interpretation in particular situations.

Two points are worth noting about these three arguments. First, the arguments are merely directed against disciplinary codes. Such codes are strictly prescriptive and are enforced. Enforcement usually requires that the room for interpretation of the code is limited. Moreover, enforcement makes it desirable that the code is morally, or at least legally, binding. The arguments are less, if at all, convincing in the case of advisory and aspirational codes. Second, in as far as especially the first and third argument are sound, they imply that it is neither possible nor desirable to try to avoid all room for interpretation in the formulation of a code of conduct. This suggests that one needs to accept some degree of vagueness and some potential conflicts in codes of conduct.

2.3.4 Can codes of conduct be lived by?

Codes of conduct sometimes contain provisions that are very difficult or impossible to follow in practice. Professional codes can, for example, justify or require actions that go against the interest of the employer. The BART case, which with this chapter started, is an example. More generally, professional codes sometimes require that engineers inform the public timely and completely if the safety, health, or welfare of the public is put at stake in a technological project. This duty to inform the public can conflict with the confidentiality duty that engineers also have according to the law in many countries. If engineers in such situations release information outside the company in which they are working, they are blowing the whistle (see Section 1.5.3).

Engineers, and other employees, who blow the whistle are usually in a weak position from a legal point of view (Malin, 1983). The situation is different from country to country, but the laws that regulate employment contracts in most countries either impose certain **confidentiality duties** on employees or they allow the employer to order the employee to keep silent certain specific information, or they do both. The reason for this is twofold. First, confidentiality may be required to protect the competitive position of one company versus another. Second, such laws are intended to avoid employees disproportionately damaging the company for which they are working by making certain information public. Breaching confidentiality duties may be a ground for dismissal in some countries. In other countries, like the United States, employees can be dismissed at will by the company.[6] However, the employee can hold the company liable for the damage of dismissal on unjust grounds.

Confidentiality duties Duties on employees to keep silent certain information.

Limits to confidentiality duties

There are limits to the confidentiality duties that companies can impose upon their employees. First, in many countries freedom of speech is legally protected. Historically, freedom of speech is understood to apply to the relation between the state and an individual citizen and not to the relation between a company and an individual employee, which is basically a relation between citizens, according to the law. There is, however, a tendency in law also to apply fundamental rights like the freedom of speech to relations between organizations and individuals. This does not mean that employees have complete freedom of speech, but it might mean that confidentiality duties should be weighed against, or be proportional to the freedom of speech of an employee and the legitimate interests of an employer. Second, in some cases there are legal requirements to make public certain information, or to inform the government or the public prosecutor about certain abuses. These legal requirements may override confidentiality duties. Third, engineers might argue that they have a professional duty, based on their professional code of conduct, to make public certain information. This happened in the BART case and was supported by the professional association of electrical engineers, the IEEE, but to no avail. Fourth, employees can argue that it is in the public interest that certain information is made public. Again, the success of this strategy in court seems limited. In response, several governments have formulated special laws to protect whistleblowers (see box). In the US there has been legislation protecting whistle blowers for 20 years. In recent times this has been adapted to give whistle blowers greater protection. Recently large financial rewards have been paid to whistle blowers who brought to light fraud or tax abuse. Nevertheless, also in these cases whistle blowers usually only have a limited amount of legal leverage in the first place and they almost always eventually lose their jobs.

Protection of Whistle Blowers

In several countries, attempts have been undertaken to protect whistle blowers legally. The main initiatives have been undertaken in the United States and the United Kingdom (Hassink, de Vries, and Bollen, 2007).

In the United States, the Sarbanes-Oxley Act (SOX) came in force in 2002. This act requires companies to adopt policies for internal whistle blowing with respect to accounting and auditing. Companies can also apply such procedures to other kinds of violations covered by their code of conduct. Prior to SOX, federal whistleblower statutes only covered the public sector, or related to more specific areas like safety and the environment.

In the United Kingdom, the Public Interest Disclosure Act of 1998 protects both internal and external whistle blowers from retaliation, but does not have provisions with respect to whistle blowing policies of companies. The Combined Code on Corporate Governance of 2003, issued by the Financial Services Authority, encourages the institutionalization of whistle blowing policies by companies. Corporations should follow this code or explain why they did not.

A code of conduct is hardly credible if living by it requires engineers to accept dismissal on a regular base. This is especially a problem for professional codes that require engineers to blow the whistle. Nevertheless, there are a number of initiatives that can be undertaken to improve the degree to which such codes can be lived by. First, the law may be changed to provide better protection for whistle blowers. Second, companies can include a right to inform the public in certain well-circumscribed cases in their corporate code and can formulate policies so that employees can indeed live by such codes. Some companies, like the chemical concern DSM, have formulated policies or procedures for whistle blowing.[7] Also professional associations can undertake initiatives, like providing legal support to individual engineers in cases where adhering to the professional code creates conflict with the employer. The IEEE has done that in the past. Some professional organizations like the NSPE have also published lists of companies that live by the professional code.

2.3.5 Enforcement

Enforcement is only an objective in the case of disciplinary codes. Active enforcement of codes of conduct seems to be an exception, especially for professional codes. Below, we will elaborate on the reasons for this and discuss what possibilities for enforcement exist.

Professional codes

One obvious reason why professional codes are often not enforced is that they are often advisory and that enforcement is not an objective of advisory codes. An underlying reason for the lack of enforcement, and for the choice to formulate advisory rather than disciplinary codes, is that professional codes do not have a legal status. Moreover, the possibilities for professional associations to enforce professional codes are limited. Enforcement requires sanctions and the most severe sanction that professional societies can exercise with respect to their members is usually loss of membership. The effect of that sanction is limited because in most countries, membership of a professional association is voluntary and is not required to exercise the profession of an engineer. A notable exception is consulting engineering in the United States and Australia. Consulting engineers in these countries have to be registered as engineers in order to carry out their profession if they are not employed by a company but have their own firm. Such registration is also sometimes required for specific groups of engineers in other countries. If registration is required, loss of registration and thus loss of the ability to work as a professional engineer can be the consequence of an engineer breaching his or her professional code. The case of John Tozer, discussed earlier, is an example. In most cases, no attempts are made by professional associations to enforce their code of conduct.

Corporate codes

Corporate codes also usually lack a legal status. Nevertheless, enforcement or at least monitoring of the code is more common than in the case of professional codes. Of the world largest companies that have a code, 52 percent report monitoring of compliance with the code (Kaptein, 2004). Generally speaking, corporate codes offer more possibilities for enforcement than professional codes. The reason for this is that companies usually influence the daily practice of individual engineers to a much larger extent than do professional associations. Companies have more possibilities to stimulate or discourage the behavior of individual engineers than professional associations. Ultimately, they can dismiss engineers if in breach of the code of conduct; a sanction that is much more severe than loss of professional membership.

Corporate codes can also be enforced externally, that is, through an external organization assessing the company in terms of its code of conduct. This is called **external auditing**. An increasing number of companies are voluntary audited by accountancy or consultancy firms with respect to, for example, safety, environment, social issues, and integrity (Hummels and Karssing, 2007). An advantage of such external assessment is that it helps to stop the corporate code of conduct being interpreted and enforced at will. In the absence of external audits, it is conceivable that those on the work floor are punished severely for not obeying the corporate code of conduct while people at higher levels in the organizations, that is, those persons who also interpret and enforce the code, are judged more mildly. External auditing also increases the credibility, and so the image, of a company. External auditing may also be required for the acquisition of a hallmark that guarantees customers of the company that certain standards are met. External auditing or enforcement can also be carried out by branch organizations. This requires a code of conduct on the level of an entire business branch. In several countries, the chemical industry has established such codes of conduct ("responsible care" – see: www.responsiblecare.org [accessed November 2, 2009]). Such branch codes have the additional advantage that companies who want to live by certain moral standards are not punished for that financially or commercially.

External auditing Assessing of a company in terms of its code of conduct by an external organization.

Even if corporate codes are not enforced, they offer better possibilities for stimulating responsible behavior than many professional codes. One reason is that external parties can criticize a company for not living by its own code of conduct. This is of course also the case with professional associations but companies are often more sensitive to external criticism than professional associations.

Case Brent Spar

According to its code of conduct, Shell is committed to contributing to sustainable development (see also Appendix 4). In 1999 Shell decided to sink the oil platform Brent Spar instead of dismantling it. The British government gave Shell permission to carry out this option. However, subsequently Shell was put under great pressure by environmental organizations, in particular by Greenpeace. Greenpeace argued that dismantling was more environmentally friendly and, moreover, saw the sinking of a platform as an undesirable precedent for the discarding of oil platforms. Because Greenpeace was able to mobilize the public and consumers of Shell products, among others through an occupation of the Brent Spar, Shell eventually felt forced not to sink the Brent Spar.

2.4 Codes of Conduct in an International Context

2.4.1 Global codes for multinationals

The 1990s witnessed a proliferation of corporate codes of conduct and an increased emphasis on corporate social responsibility. These codes emerged in the aftermath of a period that witnessed a major shift in the economic role of the state, and in policies toward multinational corporations and foreign direct investment. In the 1970s many national governments had sought to regulate the activities of multinational companies, since these companies were widely criticized for their behavior in developing countries. Host governments and labor organizations claimed that multinational companies failed to operate in harmony with local economic, social, and political objectives. The 1980s was a decade of deregulation, since efforts at regulation had been unsuccessful, and increased efforts were undertaken to attract foreign investment. Foreign direct investment in the global economy began to reach unprecedented levels, significantly increasing the influence of multinational companies on the prospects of developing countries. Many governments of lesser-developed nations saw foreign capital as key to economic growth and actively encouraged foreign investment. However, few such nations had the power to enforce corporate regulation. As a consequence, this allowed some multinationals to degrade the environment, abuse human rights, and provide little benefit to local or national development. The view that the best way of companies to promote social development in a developing nation is simply by increasing the overall level of economic activity through trade and investment, however, was changing. The new phrase became the "triple bottom line" (3BL or "People, Planet, and Profit") of economic, social, and environmental outcomes (Elkington, 1994). It is in this context that the recent wave of voluntary codes must be understood, which go beyond simple business or labor matters, to demonstrate that they are motivated by a sense of social responsibility, particularly in light of the increased liberalization of markets (cf. Sethi and Williams, 2000; Cottril, 2000). These codes of conduct have been seen as pivotal in the global marketplace (cf. Radin, 2004). US companies began introducing such codes in the early 1990s, and the practice spread to Europe in the mid-1990s. The codes tend to focus on the impact of multinational companies in two main areas: social conditions and the environment.

However, many voluntary codes of conduct of multinational companies were vague declarations of business principles applicable to international operations. A number of organizations have anticipated this by developing a **global code of conduct** that multinational companies can use as a guide to develop and/or revise their codes of conduct, especially related to investments in developing countries. Three major global codes of conduct are the Caux Pound Table principles (www.cauxroundtable.org/documents/Principles%20for%20Business.PFD), the Organization of Economic Cooperation and Development guidelines for multinational companies (www.oecd.org/daf/investment/

Global code of conduct A code of conduct that is believed to apply worldwide.

guidelines), and the United Nations Global Compact (www.unglobalcompact.org/AboutTheGC/TheTenprinciples/index.html). The United Nations Global Compact (UNGC) is the world's largest, global corporate citizenship initiative. It is concerned with exhibiting and building the social legitimacy of business and markets by offering a framework for businesses that are committed to aligning their operations and strategies with ten principles in the areas of human rights, labor, the environment, and anti-corruption (see box). The principles are derived from *The Universal Declaration of Human Rights*, *The International Labour Organization's Declaration on Fundamental Principles and Rights at Work*, *The Rio Declaration on Environment and Development*, and *The United Nations Convention Against Corruption*. Many multinationals are involved in this voluntary initiative of the UNGC. This initiative of the United Nations is meant to stimulate corporate responsibility. Although the guidelines are not directly binding on companies, adhering companies are expected to promote them and to follow procedures for resolving alleged violations.

The UNGC states that business, trade and investment are essential pillars for prosperity and peace. But in many areas, business is too often linked to serious issues – for example, exploitative practices, corruption, income equality, and barriers that discourage innovation and entrepreneurship. Following the ten principles can in many ways build trust and social capital, contributing to broad-based development and sustainable markets.

United Nations Global Compact Principles

Human Rights

Principle 1: Businesses should support and respect the protection of internationally proclaimed human rights; and
Principle 2: make sure that they are not complicit in human rights abuses.

Labour Standards

Principle 3: Businesses should uphold the freedom of association and the effective recognition of the right to collective bargaining;
Principle 4: the elimination of all forms of forced and compulsory labour;
Principle 5: the effective abolition of child labour; and
Principle 6: the elimination of discrimination in respect of employment and occupation.

Environment

Principle 7: Businesses should support a precautionary approach to environmental challenges;
Principle 8: undertake initiatives to promote greater environmental responsibility; and

Principle 9: encourage the development and diffusion of environmentally friendly technologies.

Anti-Corruption

Principle 10: Businesses should work against corruption in all its forms, including extortion and bribery.

Case Shell, Nigeria and the Ogoni: A Study in Unsustainable Development

Figure 2.4 Gas flaring (Shell): Woman tending her plot at Shell gas flare site, Rumuekpe, Nigeria. Photo: © Elaine Gilligan/Friends of the Earth, June 2004.

"Shell is a global group of energy and petrochemical companies. Our aim is to meet the energy needs of society, in ways that are economically, socially and environmentally viable, now and in the future" (www.shell.com). The company is involved is several voluntary social and environmental initiatives, such as the United Nations Global Compact.

The Nigerian government's June 4, 2008 decision to replace the Shell Petroleum Development Company (SPDC) – Shell's Nigerian subsidiary – as operator of oil concessions in Ogoni areas offers an opportunity for ending one of the longest running conflicts between a multinational oil company and a local community in the Niger Delta. The Niger Delta was once considered the breadbasket of Nigeria because of its rich ecosystem, a place where people cultivated fertile farmlands and benefited from abundant fisheries.

The origins of the conflict between the Ogoni and SPDC date back to the company's discovery of oil in this part of the Niger Delta in 1958. Nigeria was still under British colonial rule, and the Ogoni, like all other minority ethnic groups in the Delta, had no say in the exploitation agreements. Even after independence in 1960, they were not accorded a real stake in oil production.

There were more than 100 oil wells, mostly operated by SPDC. As elsewhere in the Delta, the environmental effects of oil exploration and production in Ogoni territory were severe. Land and water pollution from spills played havoc with the ecosystem. Villagers lived with gas flares burning 24 hours a day (some for over 30 years) and air pollution that produced acid rain and respiratory problems. Above-ground pipelines cut through many villages and former farmland.

SPDC refused to accept responsibility for environmental repercussions and largely denied there was an issue. As late as 1995, for example, an SPDC document insisted that: "Allegations of environmental devastation in Ogoni, and elsewhere in our operating area, are simply not true. We do have environmental problems, but these do not add up to anything like devastation." In response to criticism of its community relations practices, SPDC insisted that most of the Ogoni demands for social benefits and infrastructural development were the responsibility of the government, not an oil company. It maintains that it has responded "promptly, fairly, and completely" to community complaints in Ogoni land but that many, such as those articulated in the Ogoni Bill of Rights, are of a political nature and thus beyond its competence.

In response the Ogoni founded in 1992 the Movement for the Survival of Ogoni People (MOSOP), led by Ken Saro-Wiwa. From the start it adopted a policy of non-violence. MOSOP demanded that SPDC take responsibility for its massive environmental devastation of their homeland and denounced the injustices that Shell has inflicted on the Ogoni and other peoples in the Niger Delta. In 1995, Ken Saro-Wiwa and 13 other MOSOP leaders were subjected to a secret tribunal that, based on unsubstantiated allegations, sentenced nine of the men to death by hanging. They were accused of incitement to murder. All nine were summarily executed without any opportunity for appeal.

Most Ogoni saw Shell as the architect of the events. The company strongly denied any complicity in the military repression of the Ogoni. However, the impression persisted that it had a hand in the repression. The Ogoni resolved never to allow SPDC to resume operations on their land. Many regarded its pledge not to use armed escorts and only to resume operations with host communities' consent as mere posturing. Relations between SPDC and the Ogoni have remained tense ever since.

A major issue that has to be dealt with in the context of the exit of SPDC is environmental clean-up. No significant study has been conducted to determine reliably the precise impact of oil industry-induced environmental degradation on human livelihoods in the area, but there are indications of severe damage. The African Commission on Human and Peoples' Rights held that "the pollution and environmental degradation in Ogoni was to a level unacceptable and has made living in Ogoni land a nightmare." SPDC policy, according to the company, is to clean up environmentally-damaging incidents related to its operations regardless of cause, but only to pay compensation if the incident occurred as a result of its own operational failure. When environmental damage occurs as a result of sabotage (a common occurrence according to SPDC), the company is forbidden by Nigerian law from paying compensation. SPDC continues to pledge cooperation with the proposed United Nations Environment Programme (UNEP) environmental assessment, though it has not promised that it will pay any damages related to UNEP findings.

Source: Based on International Crisis Group (2008) and Boele, Fabig, and Wheeler (2001).

2.4.2 Global codes for engineers

The globalization of the world's economies has also increased the working space of engineers. Engineering products and production facilities often transcend national boundaries. Engineers travel across the world and meet other cultures by interacting with foreign engineers. Multinational companies employ engineers from different cultural backgrounds in the same corporate environment. So, engineering has become a global activity and increasingly requires a global approach and acceptable global guidance.

The engineering profession in the United States has been a world leader in promoting engineering ethics code development and associated educational activities. Due to their leadership other nations have followed the American lead and have adopted US codes. The Nation Society for Professional Engineers (NSPE), for example, reports that its code is used by the Japan Consulting Engineers Council. It is also expected that a code very similar to American ones will soon be adopted by the Japan Accreditation Board for Engineering Education (JABEE) that was established in 1999 (Luegenbiehl, 2004). However, this approach may well be counterproductive, since it neglects the cultural differences between Japan and the United States. The US codes are based on the notion of **professional autonomy**: "empowering individuals to reason more clearly and carefully concerning moral questions, rather than to inculcate any particular beliefs" (Schinzinger and Martin, 2000, p. 14). However, not all nations value autonomy to the same degree as the United States. For example, Japanese society emphasizes group values in educational and socialization practices, instead of individualism as in the United States.

Professional autonomy The ideal that individual professionals achieve themselves moral conclusions by reasoning clearly and carefully.

Whereas many professionals in the United States focus on individual career development, the Japanese professionals are more devoted to the company's goals. Most Japanese people have a strong sense of loyalty, so whistle blowers would probably not be accepted by Japanese society. As engineering ethicist Heinz Luegenbiehl writes:

> The ideal [American] professional model requires that the engineer and the engineering profession be autonomous so as to protect the public in the face of corporate self-interest. The ideal Japanese model, on the other hand, requires the engineer to function harmoniously as an integral part of the group in a system where the corporation serves the needs of society. The potential for professional autonomy is very limited in the Japanese model. In the Western model the profession guarantees the quality of the engineer's work through its contract with the larger society. In the Japanese model the corporation serves the same function. … Seen in terms of engineering, it is therefore the corporation which takes responsibility for, and guarantees, the engineer's work. The engineers, for their part, are an integral part of the larger group and, knowing that their fate is tied to that of the corporation, would be aware that they would not profit from individual actions. The corporation, in turn, sees its interest tied to those of the nation. The core demand for "safety, health, and welfare of the public," the primary goal of an engineering ethics, can then be achieved through the corporation, since it is not expected to act based solely on the interests of its owners. (Luegenbiehl, 2004, pp. 71–72)

Other commentators have shown some more cultural differences between nations, and have argued that drafting a global code for engineers is not a straightforward process (Downey, Lucena, and Mitcham, 2007). It requires continuing efforts to understand and appreciate cultural differences (Weil, 1998). An example of a rather successful effort from which we can learn is a recent project to devise a common code of conduct for American, Canadian, and Mexican engineers under the North American Free Trade Agreement (NAFTA). The objectives of this project were 1) to study the aspects of conduct and ethics related to engineering practice under the provisions of the NAFTA, and 2) to develop a mutually agreed upon set of ethical principles.

The main challenge of a global code for engineers is to create consistency in spite of cultural differences. As we have seen, autonomy cannot serve as an uncontested universal foundational assumption for building a global code for engineers. Heinz Luegenbiehl (2010) proposes some principles for a global code for engineers based on the nature of engineering activity and the universal use of reason in engineering (see box). The universal foundational assumption is that all engineers, independent of their cultural background, must accept the premise that the use of reason is a valid decision-making instrument.

Ethical Principles for Engineers in a Global Environment

- The Principle of Public Safety: Engineers should endeavor, based on their expertise, to keep members of the public safe from serious negative physical consequences resulting from their development and implementation of technology.

- The Principle of Human Rights: Engineers should endeavor to ensure that fundamental rights of human beings will not be negatively impacted as a result of their work with technology.
- The Principle of Environment and Animal Preservation: Engineers should endeavor to avoid damage to the animal kingdom and the natural environment which would result in serious negative consequences, including long-term ones, to human life.
- The Principle of Engineering Competence: Engineers should endeavor to engage only in engineering activities which they are competent to carry out.
- The Principle of Scientifically Founded Judgment: Engineers should endeavor to base their engineering decisions on scientific principles and mathematical analysis, and seek to avoid influence of extraneous factors.
- The Principle of Openness and Honesty: Engineers should endeavor to keep the public informed of their decisions which have the potential to seriously affect the public, and to be truthful and complete in their disclosures. (Luegenbiehl, 2010)

Charles Harris (1998) has proposed some principles for a global code that apply to engineers operating in developing countries, based on Richard De George's (1993) guidelines for multinational corporations in the international environment. De George's guidelines, however, apply to multinational companies or to their managers. They cannot be simply applied to engineers. First, engineers have a lesser scope of responsibility than managers. Engineers are responsible primarily for the design, production, and implementation of technology, and are therefore more narrowly focused than managers, who are responsible for the total well-being of the enterprise (Harris, 1998, p. 324). Second, engineers do not make management decisions, and have relatively little decision-making power within the corporate hierarchy (Harris, 1998, pp. 324–325). Nevertheless it is not very difficult to adapt some of these guidelines for the engineering practice. Engineers, Harris claims, have a responsibility:

1 to refuse to engage in direct, intentional harm;
2 to refrain from participating in the design, production, or implementation of technology that produces more harm than good, all things considered;
3 to participate only in technology that promotes the country's development;
4 not to participate in the violation of human rights; and
5 to respect host-country (lesser-developed country) culture in their professional work.

In combination with Luegenbiehl's Ethical Principles for Engineers in a Global Environment (see box), these principle could function as a starting point to develop and/or revise international professional codes for engineers.

2.5 Chapter Summary

Codes of conduct are codes in which organizations lay down guidelines for responsible behavior of their members. Codes of conduct can be aspirational (mentioning the main values), advisory (assisting individuals in moral judgment) and disciplinary (enforcing rules of behavior). Professional codes are formulated by professional associations of engineers, and corporate codes are formulated by companies in which engineers are employed. Professional codes describe the professional responsibility of engineers, and corporate codes the responsibility of engineers as employees. Most professional codes relate to three domains: 1) conducting a profession with integrity and honesty, and in a competent way; 2) obligations towards employers and clients; and 3) responsibility towards the public and society. Corporate codes usually contain a mission statement (the overall objectives of the company), core values, stakeholder principles and more detailed rules and norms.

A number of objections have been raised against codes of conduct:

1 Code of conduct sometimes amount to window-dressing;
2 Codes of conduct are often vague and are potentially contradictory;
3 Ethics cannot be codified;
4 Codes of conduct cannot be lived by;
5 Codes of conduct are not enforced.

We have seen that the second and third objection mirror each other. According to the objection that ethics cannot be codified, ethics always remains a matter of judgment. This is exactly the reason why codes of ethics cannot avoid all vagueness and potential contradictions. This is not to say that vagueness and contradictions should not be avoided when possible, but the code is maybe better considered as a set of guidelines that is helpful in judging cases than as a set of strict prescriptive rules. Objections 2 and 3, then, do not really apply to aspirational and advisory codes, although they may be a problem for disciplinary codes. The same applies to objection 5 because enforcement is only an objective for disciplinary codes and not for advisory and aspirational codes. Objection 4 is serious and may be especially a problem in cases of whistle blowing, or more generally, tensions between your responsibility as professional engineer and as employee. Partly it can be solved by better attenuating the responsibility of engineers as professionals with the responsibility of engineers as employees, and thus better attenuating professional codes and corporate codes. Some companies have tried to do this.

As engineering increasingly becomes an international activity, codes of conduct increasingly become global in nature. This raises difficult questions about how to deal with cultural differences and about whether the professional autonomy model on which most US professional codes are based can be exported. Nevertheless it seems possible to formulate a global professional code for engineers that contains at least some more or less commonly accepted principles.

Study Questions

1 The Software Engineering Code of Ethics and Professional Practice of the Association for Computing Machinery states that "The dynamic and demanding context of software engineering requires a code that is adaptable and relevant to new situations as they occur. However, even in this generality, the Code provides support for software engineers and managers of software engineers who need to take positive action in a specific case by documenting the ethical stance of the profession. The Code provides an ethical foundation to which individuals within teams and the team as a whole can appeal. The Code helps to define those actions that are ethically improper to request of a software engineer or teams of software engineers. The Code is not simply for adjudicating the nature of questionable acts; it also has an important educational function. As this Code expresses the consensus of the profession on ethical issues, it is a means to educate both the public and aspiring professionals about the ethical obligations of all software engineers" (http://www.acm.org/about/se-code (accessed November 2, 2009)).

Is this code aspirational, advisory, or disciplinary? Explain your answer.

2 Give an example of a situation in which you have a professional responsibility to do something *but* not a legal responsibility.

3 What is meant by "a code is nothing, coding is everything?"

4 What are the most important objectives of professional codes of conduct?

5 Why is enforcement an explicit objective for disciplinary codes? Why is enforcement often difficult to obtain for professional engineering codes of conduct?

6 What are corporate codes? Discuss three objections to and/or shortcomings of corporate codes.

7 What are the two arguments of Milton Friedman's criticism of corporate social responsibility? Give some objections against these arguments.

8 Like engineers, medical doctors and lawyers also have professional codes. Unlike engineering codes, however, these codes typically are accompanied by disciplinary law, so that doctors or lawyers who violate the code can be excluded from practicing the profession. Provide an argument *for* and an argument *against* the adoption of similar disciplinary law for engineers.

9 What is valuable about loyalty? What is problematic about loyalty? Be careful to indicate what concept of loyalty you are using in answering this question.

10 To gain the protection of the UK's Public Interest Disclosure Act, those who reveal organizational malpractices have to satisfy a number of conditions that witnesses in other criminal investigations do not have to satisfy, for example, deriving no financial gain from the case and not having been involved in the crime at any stage. Critically evaluate the merits of these conditions. Compare them also with the guidelines for whistle-blowing mentioned in Section 1.5.3.

11 Look for a professional code of conduct in your own area:

a. Do you recognize the three general content areas mentioned in the text in this code?

b. Is the code vague at some points? Where?

c. Are their potential contradictions between the provisions of the code? Does the code contain provisions to deal with these contradictions?

d. Are there any provisions in the code that are impossible to live by? Which ones?

e. Do you agree which the professional responsibility set out in the code? Are you missing anything?

12 Look for a corporate code of an engineering company. In what respects are the responsibilities of engineers that are articulated in this code different from the responsibilities articulated in professional codes (like the code of the NSPE)? Is this code conflicting at certain points with, for example, the professional code of the NSPE? If there is a conflict what code should, in your view, take precedence and why?
13 Do you agree that engineers have a responsibility for human rights as some global codes of conduct suggest? Is this responsibility restricted to not engaging in violations of human rights or do engineers also have a responsibility to enhance human rights through their engineering projects?
14 Draft a code of conduct to cover e-communications (email, Web use and so on). Explain and justify your proposed code.
15 One of the principles for a global code of conduct for engineers mentioned by Luegenbiehl is the principle of scientific-founded judgment. What do you think that Luegenbiehl means with extraneous factors? Would considerations of safety or human welfare count as extraneous factors that should not influence engineering decisions?
16 The US government allows employees of aircraft manufacturers like Boeing to serve as inspectors for the Federal Aviation Agency (FAA) that is responsible for regulating the aircraft industry and doing safety and quality inspections. What would be the reasons for the US government to allow this? Is this a conflict of interest? Would it be unethical for an engineer employed by Boeing also to act as inspector for the FAA?

Discussion Questions

1 If you were to give ethical training to engineers, would you stress knowing the law, company rules and codes of conduct, or would you instead focus, on explaining the principles behind these rules. Are there any common principles behind these rules? Which ones?
2 Loyalty or integrity: which should be the most important to engineers?
3 What do you see as the main ethical issues arising from globalization?
4 Cases like Shell in Nigeria and Google in China that were discussed in this chapter seem to suggest that codes of conduct are a dead letter when it comes to moral decision-making in practice. Discuss whether codes of conduct are indeed just window-dressing in cases like this or whether they have any positive effect. Can you think of ways to bridge the gap between what companies like Shell and Google say in their codes and what they do in practice? Should multinational companies maybe avoid undemocratic countries like Nigeria and China to avoid tough ethical decisions?
5 Choose any Fortune 500 company. Locate the company's code of ethics published on the company's Web page. Evaluate the code in terms of the United Nations Global Compact Principles.

Notes

1 For a comparable distinction, see Frankel (1989).
2 VDI, "Bekentennis der Ingenieurs" [The Confession of Engineers] (1950), included in Lenk and Ropohl (1987, p. 280).
3 The description of the content of corporate codes of conduct here and below is based on Kaptein (2004).

4 The cases can be found at www.niee.org/cases/index.htm (accessed August 24, 2007).
5 There may be, however, non-moral arguments for having a disciplinary code.
6 Convention 158 of the International Labour Organization states that an employee "can't be fired without any legitimate motive" and "before offering him the possibility to defend himself." The US has not ratified this convention.
7 DSM Alert: Whistle blowing policy and procedure for expressing concerns about expected serious misconduct at DSM, 2004.

3

Normative Ethics

Having read this chapter and completed its associated questions, readers should be able to:

- Describe normative judgments, and distinguish them from descriptive judgments;
- Describe norms, values and virtues;
- Describe the four ethical theories: utilitarianism, Kantian theory, virtue ethics, and care ethics;
- Identify the criticisms of the four ethical theories;
- Apply the ethical theories to moral issues in engineering practice;
- Reflect upon how ethical theories may impact on making moral decisions.

Contents

Ethics, Technology, and Engineering: An Introduction, First Edition.
Ibo van de Poel and Lambèr Royakkers.

3.1 Introduction

Case The Ford Pinto

Figure 3.1 Ford Pinto. Photo: Bettmann Archive/Corbis.

On August 10, 1978, on Highway 33 in the neighborhood of Goshen, Indiana (United States), a tragic accident occurred. A truck rear-ended a five-year-old Ford Pinto carrying three teenagers: sisters Judy and Lynn Ulrich (ages 18 and 16, respectively) and their cousin Donna Ulrich (age 18). The collision caused the gas [petrol] tank to rupture and explode, killing all three teens.

Subsequently an Elkart County grand jury returned a criminal homicide charge against Ford, the first ever against an American company. During the following 20-week trial, the judge advised the jury that Ford should be convicted of reckless homicide if it were shown that the company had engaged in "plain, conscious and unjustifiable disregard of harm that might result (from its actions) and the disregard involved a substantial deviation from acceptable standards of conduct." The key phrase around which the trial hinged was "acceptable standards." Towards the end of the 1960s, Ford Motor Company, one of the world's largest car manufacturers, was gradually losing market share. Ford was losing ground to the smaller and cheaper European cars. In 1968, President Lee Iacocca decided a small cheap car had to be designed quickly. This was to become the Ford Pinto. The decision was made to put it onto the market for less than $2000 in 1970. This was a very competitive price but the time schedule for the car's development was rushed. At the time, car development normally required around 43 months. Only 24 months were reserved for the Ford Pinto. Because the Pinto had to cost a maximum of $2000, a radical design was selected in which styling took precedence over engineering design. The safety aspect of the design did not receive sufficient priority. There was no experience with small cars within the company at all. Among other things, this led to the positioning of the petrol tank just behind the rear axle. Later it was found that the gear construction in the rear axles (the differential) was situated such that it would puncture the petrol tank in the event of a collision. In Ford's tests of the Pinto prototype, this problem occurred at speeds as low as 35 km per hour. The puncture of the tank caused an extremely hazardous situation. These test results were passed on to the highest management level within Ford. From other tests it was shown that there were two simple ways to considerably reduce the risk that the petrol tank would be ruptured. It was possible to alter the design to allow the petrol tank to be situated above the axle. It was estimated that the change in the design would raise the price of the car by $11. A second option was to protect the tank with a rubber layer, which was probably a cheaper option. However, because the design met the safety requirements of the government, the Pinto was taken into production without any alterations.

To justify its actions, Ford made a cost-benefit analysis. In this cost-benefit analysis, which was published under the heading "Fatalities Associated with Crash-Induced Fuel Leakage and Fires," it was asserted that the extra costs of $11 did not weigh against the benefit that society would derive from a smaller number of wounded passengers and fatalities. This statement was argued as follows:

The societal benefits of the riskier design that costs $11 less was estimated at nearly $50 million: 180 lives lost, 180 wounded and 2100 cars burnt out. The calculation for this was 180 lives × $200,000 + 180 seriously wounded × $67,000 + 2100 burnt out cars × $700 = $49.53 million. This was considered to be the total societal benefit.

Against this there was the cost of improving the cars: 11 million cars and 1.5 million trucks had to be called back and retrofitted against an estimated costs per unit of $11, amounting to a total cost of 137 million dollar (12.5 million × $11). A memorandum attached to the report described the costs and benefits as in Table 3.1.

Table 3.1 Benefits and costs

Benefits		
Savings	*Unit cost (US$)*	*Total (US$)*
180 burn deaths	200 000	36 000 000
180 serious burn injuries	67 000	12 060 000
2,100 burned vehicles	700	1 470 000
Total		*49 530 000*
Costs		
Sales	*Unit cost (US$)*	*Total (US$)*
11 million cars	11 per car	121 000 000
1.5 million light trucks	11 per truck	16 500 000
Total		*137 500 000*

The estimation by Ford of the number of lives lost and wounded incurred was based on statistical data. The estimation Ford made that a human life is worth $200,000 was based on a report of the National Highway Traffic Safety Administration (see Table 3.2).

Table 3.2 Component costs

Component		*1971 costs (US$)*
Future productivity losses	Direct	132 000
	Indirect	41 300
Medical costs	Hospital	700
	Other	425
Property damage		1 500
Insurance administration		4 700
Legal and Court		3 000
Employer losses		1 000
Victim's pain and suffering		10 000
Funeral		900
Assets (lost consumption)		5 000
Miscellaneous accident cost		200
Total per fatality		*200 725*

The conclusion that Ford drew was clear: a technical improvement costing $11 per car which would have prevented gas tanks from rupturing so easily was not cost-effective for society. The $137 million cost of the safer model clearly outweighed the benefits of $49.53 million. Altering the Pinto for $11 a car would cost society ($137 million – $49.53 million) $87.47 million.

On March 13, 1980, the Elkhart County jury found Ford not guilty of criminal homicide. However, under pressure from various institutions, Ford recalled 1.5 million cars for refitting, and this case and many other similar Pinto accidents cost Ford millions of dollars in legal settlements to accident victims. Ford also suffered a great deal of damage to its reputation.

Source: Based on Birch and Fielder (1994).

The argument of Ford is controversial and has evoked a lot of debate. It painfully illustrates that expressing the value of human life in monetary terms involves the danger of neglecting fundamental human rights, such as the right to life. The Ford Pinto case has become one of the most well-known cases in applied ethics, since it raises many questions of ethical importance. Some people conclude that Ford was definitely wrong in designing and marketing the Pinto, and others believe that Ford was neither legally nor morally blameworthy, and acted right in producing the Pinto. Reflecting on this case, several ethical questions emerge: Was Ford acting wrong in rushing the production of the Pinto? Even though Ford violated no federal safety standards or laws, should the company have made the Pinto safer in terms of rear-end collisions, especially regarding the placement of the gas tank? Was it acceptable that Ford used cost-benefit analysis to make a decision relating to safety, specifically placing dollar values on human life and suffering? Should companies like Ford play a role in setting safety standards? What were the responsibilities of the Ford design engineers and crash-test engineers?

Different arguments can be used to answer these questions. Despite the apparent differences, some recurrent patterns can be found in the moral arguments that are used in cases like this, and the cases we have already seen: the Challenger case (Section 1.2) and the BART case (Section 2.1). These patterns are related to ethical theories that have been developed by various philosophers. Ethical theories help us to sort out our thinking and to develop a coherent and justifiable basis for dealing with moral questions. The role of ethical theories is to provide certain arguments or reasons for a moral judgment. They provide a normative framework for understanding and responding to moral problems, so improving ethical decision-making or, at least, avoiding certain shortcuts, such as neglecting certain relevant features of the problem or just stating an opinion without any justification. In this chapter we shall therefore introduce three of the best-known ethical theories: consequentialism, duty ethics, and virtue ethics. These theories each have their own criteria with which they determine whether an action is right or wrong. Before we go into these three theories we shall discuss what we mean by morality and ethics (Section 3.2) and distinguish between descriptive and normative judgments (Section 3.3). In Section 3.4 we shall look into the points of departure of ethics: values, norms, and virtues. These points of departure often recur in ethical theories. Before we discuss the three most important ethical theories, we shall first consider the two most extreme approaches to ethics: normative relativism and absolutism (Section 3.5). In Section 3.6 we shall indicate how the three best-known ethical theories – consequentialism, duty ethics, and virtue ethics – are related to each other. These three ethical theories will be discussed at length in the sections that follow (Sections 3.7, 3.8 and 3.9). In Section 3.10 we will discuss a relatively recent approach to ethics as an alternative to the familiar moral theories: care ethics. Finally, in Section 3.11, we clarify our position with respect to applied ethics.

3.2 Ethics and Morality

Defining the term ethics is not easy. Ethics has had many meanings over the centuries. The term is derived from the Greek word ethos, which can be translated as "custom" or "morals," but also as "conviction." "Ethica" stood for the science that considered what was good or bad, wise or unwise, about people's deeds. The Romans translated "ethos" into the Latin "mos" (plural "mores"), which is the root of the word "moral." Ethics and moral stem from the same source. Over the centuries "moral" has taken on the meaning of the totality of accepted rules of behavior (of a group or culture). In this text, we will distinguish ethics from morality. The term ethics will be reserved for a further consideration of what is moral.

Ethics is the systematic reflection on what is moral.

Morality is the whole of opinions, decisions, and actions with which people, individually or collectively, express what they think is good or right.

Ethics The systematic reflection on morality.

Morality The totality of opinions, decisions, and actions with which people express, individually or collectively, what they think is good or right.

In this book we define ethics as the systematic reflection on morality. Morality is defined here as the totality of opinions, decisions, and actions with which people express what they think is good or right. This roughly agrees with the often used definition of morality as the totality of norms and values that actually exist in society.

Systematic reflection on morality increases our ability to cope with moral problems, and thus moral problems that are related to technology as well. Ethics, however, is not a manual with answers; it reflects on questions and arguments concerning the moral choices people can make. Ethics is a process of searching for the right kind of morality.

The study of ethics can be both of a descriptive or prescriptive nature. **Descriptive ethics** is involved with the description of the existing morality, including the description of customs and habits, opinions about good and evil, responsible and irresponsible behavior, and acceptable and unacceptable action. It studies the morality found in certain subcultures or during certain periods of history. Prescriptive or **normative ethics** takes matters a step further. Descriptive ethics can discuss the morality of foreign nations or monthly magazines for men without passing judgment. Normative ethics, which is central to this book, moves away from this detachment. By definition normative ethics is not value-free; it judges morality. It considers the following main question: do the norms and values actually used conform to our ideas about how people should behave? Normative ethics does not give an unambiguous answer to this question, but in its moral judgment various arguments are given based on various ethical theories. These ethical theories provide viewpoints from which we can critically discuss moral issues.

Descriptive ethics The branch of ethics that describes existing morality, including customs and habits, opinions about good and evil, responsible and irresponsible behavior, and acceptable and unacceptable action.

Normative ethics The branch of ethics that judges morality and tries to formulate normative recommendations about how to act or live.

3.3 Descriptive and Normative Judgments

One central question in normative ethics is "what is a right opinion, decision, or action?" To answer this question a judgment has to be made about the opinion, decision, or action in question. This is a normative judgment, because it says something about what "correct behavior" or a "right way of living" is. Normative judgments are value judgments but not descriptive judgments. **Descriptive judgments** are related to what is actually the case (the present), what was the

Descriptive judgment A judgment that describes what is actually the case (the present), what was the case (the past), or what will be the case (the future).

case (the past), or what will be the case (the future). Descriptive judgments are true or false. The assertion "the Challenger met all safety standards of the time" is a descriptive judgment: the assertion is true or false. Sometimes the truth of a descriptive statement has not yet been determined because testing is impossible. Take for example the statement "God exists." Science plays an important role in determining the truth of descriptive judgments. A **normative judgment** is a value judgment. Value judgments indicate whether something is good or bad, desirable or undesirable; they often refer to how the world should be instead of how it is. Such kinds of value judgments often refer to moral norms and values. This can give rise to meaningful discussions, which is not the case for judgments of taste, such as "I do not like Brussels sprouts." Examples of moral judgments are "the Challenger should never have been launched," "Engineers should faithfully provide measurements.", and "stealing is bad."

Normative judgment Judgment about whether something is good or bad, desirable or undesirable, right or wrong.

The distinction between descriptive and normative judgments is not always that easy. The statement "taking bribes is not allowed" can be both a normative and a descriptive judgment. If the statement means that the law declares that taking bribes is illegal then it is a descriptive judgment. If however the statement means that bribery should be forbidden, then it is a normative judgment.

3.4 Points of Departure: Values, Norms, and Virtues

Norms, values, and virtues are the points of departure, respectively, for the three primary normative theories that we will discuss in Sections 3.7, 3.8, and 3.9. We shall discuss them in detail below.

3.4.1 Values

Values help us determine which goals or states of affairs are worth striving for. Moral values are related to a good life and a just society. They have to be distinguished from the preferences or interests of individual people. Preferences or interests are matters people feel they should strive for, for themselves. Moral values are lasting convictions or matters that people feel should be strived for in general and not just for themselves to be able to lead a good life or to realize a just society. A typical example of this is the slogan of the French Revolution: "liberté, égalité, fraternité" (freedom, equality and brotherhood). This slogan did not express a personal preference – such as "I want to be rich" – but expressed values that were felt to be of importance for everyone. Other examples of moral values include justice, health, happiness, and charity. Values are not limited to people; companies have them too. They often formulate their most important moral values (core values) in their mission statement (see Section 2.2.2).

Values Lasting convictions or matters that people feel should be strived for in general and not just for themselves to be able to lead a good life or to realize a just society.

A distinction can be made between intrinsic and instrumental values. An **intrinsic value** is an objective in and of itself. An **instrumental value** is a means to realizing an intrinsic value. The value of money for Scrooge McDuck is intrinsic. He values money independently of what you can do with money. For Mother Theresa, however, money was an instrumental value to realize a higher end: helping the poor. A person can consider his work to be both of intrinsic and instrumental value. If work is meant to support the value of becoming rich, it is an instrumental value. If a person has much job satisfaction, then work is an intrinsic value.

Intrinsic value Value in and of itself.

Instrumental value Something that is valuable in as far as it is a means to, or contributes to something else that is intrinsically good or valuable.

Case Biometric Technology and Data Matching at Super Bowl XXXV

A large spectator event like the Super Bowl presents a prime target for terrorists. Fearing the potential for such an attack or other serious criminal incident, law enforcement agencies in Florida turned for help to biometrics: the use of a person's physical characteristics or personal traits for human recognition. At Super Bowl XXXV in January 2001, a biometric system relying on facial recognition was used. This technology scanned the faces of individuals entering the stadium. The digitized facial images were then instantly matched against images in a centralized database of suspected criminals and terrorists. At the time, this practice was criticized by civil-liberty proponents and privacy advocates. In the post-September 11 (2001) world, however, practices that employ technologies such as face-recognition devices have received overwhelming support from the American public.

Source: Based on Tavani (2004).

In the literature on computer ethics, the threat to personal privacy is one of the most debated ethical problems. The distinction between instrumental and intrinsic values suggests two common ways to attempt to justify privacy. The most common justification is that privacy has instrumental value. It offers us protection against harm. For example, if a person is tested HIV+ and this is publicly known, then an employer might be reluctant to hire him and an insurance company might be reluctant to insure him. The justification of privacy, however, would be more secure if we could show that it has intrinsic value. While few authors argue that privacy is only an intrinsic value, others argue that while privacy is instrumental, it is not merely instrumental. For example, computer ethicist Deborah Johnson proposes that we regard privacy as an essential aspect of autonomy (Johnson, 2001). Autonomy is fundamental to what it means to be human, to our values as human beings (see also Section 3.8).

So, privacy is a necessary condition for an intrinsic value: autonomy. Johnson argues that the loss of privacy would therefore be a threat to our most fundamental values. For example, if a person is being watched by constant surveillance, this has an enormous effect on how the person behaves and how he or she sees himself or herself.

3.4.2 Norms

Norms Rules that prescribe what actions are required, permitted, or forbidden.

Norms are rules that prescribe what concrete actions are required, permitted or forbidden. These are rules and agreements about how people are supposed to treat each other. Values are often translated into rules, so that it is clear in everyday life how we should act to achieve certain values.

One example of a value within our traffic system is safety. However, the value alone is not enough to guarantee safety on the road. To this purpose, we need rules of behavior or norms: prescribed actions that indicate what we must do or must not do in a given situation. The value "safety" in a traffic system is mainly specified by the legal norms from the traffic regulations. In the Dutch regulations, for example, we have the rule that drivers coming from the right must always be given way.

Moral norms are indications for responsible action. Next to moral norms there are other kinds of norms, such as legal norms (for example, traffic rules), precepts of decorum (for example. "you should not talk when your mouth is full"), and rules of play (for example, in Ludo you can only place a counter on the playing board once you have thrown a six with the die). Some moral norms, like "Thou shalt not kill" and "Thou shalt not steal," have been turned into laws.

The difference between values and norms can be described as follows. Values are abstract or global ideas or objectives that are strived for through certain types of behavior; it is what people eventually wish to achieve. Norms, however, are the means to realize values. They are concrete, specific rules that limit action. Without an interpretation, the objective cannot be achieved. Take for example the need for traffic regulations to guarantee traffic safety. In addition, norms have no meaning or are ineffective if the underlying value is unclear or is lacking. So one can imagine that the norm "all bicycle bells must be blue" will be largely ignored. The norm has no meaning – there is no underlying value. These differences are summarized in Table 3.3.

Table 3.3 Differences between values and norms

Values	*Norms*
Ends	Means
Global	Specific
Hard to achieve without norms	Ineffective without values

Source: Based on Jeurissen and Van de Ven (2007, p. 57).

3.4.3 Virtues

Next to values and norms we have another moral point of departure: **virtues**. The philosopher Alasdair MacIntyre describes virtues as a certain type of human characteristics or qualities that has the following five features:

Virtues A certain type of human characteristics or qualities.

1 They are desired characteristics and they express a value that is worth striving for.
2 They are expressed in action.
3 They are lasting and permanent – they form a lasting structural foundation for action.
4 They are always present, but are only used when necessary.
5 They can be influenced by the individual (MacIntyre, 1984a).

The last statement suggests that people can learn virtues. It is a matter of the shaping of a person's character or personality. This occurs during our upbringing or our learning process within an organization. Examples of virtues are justice, honesty, courage, loyalty, creativity, and humor.

We can distinguish moral virtues (or character virtues) from intellectual virtues. Intellectual virtues focus on knowledge and skills. Moral virtues are the desirable characteristics of people – the characteristics that make people good.

On the basis of the preceding description of moral virtues, they seem to be similar to values. Many of the characteristics that we qualify as moral virtues are also values, such as integrity and being just. The difference is that the notion of virtue mainly refers to the character development someone has to have gone through to truly realize those values.

Moral virtues are indispensable in a responsible organization. An organization can formulate nice values like integrity, respect, and responsibility as much as it likes, but without the moral virtues being present in the character of its employees little will be accomplished. The values indicate which characteristics (virtues) an organization prizes or expects of its employees – what kind of people it expects its employees to be.

> Moral values help us determine which goals or states of affairs are worth striving for in life, to lead a good life or to realize a just society.
>
> Moral norms are rules that prescribe what action is required, permitted, or forbidden.
>
> Moral virtues are character traits that make someone a good person or that allow people to lead good lives.

3.5 Relativism and Absolutism

Before discussing the three most important theories in normative ethics, we shall look at two extreme theories that seem to be very tempting at first when it comes to forming a moral judgment: normative relativism and absolutism.

3.5.1 Normative relativism

Normative relativism An ethical theory that argues that all moral points of view – all values, norms, and virtues – are equally valid.

Normative relativism argues that all moral points of view – all values, norms and virtues – are relative. What is good or responsible for one person is not necessarily so for another. A moral judgment or choice is simply a personal opinion: "If I think it is good (or bad) to do A, then it *is* good (or bad) to do A." So the defense of such a claim is subjective and random: there are no guidelines about behavior that are objective and independent of time, place, and culture. In other words, there are no universal norms according to this theory, that is, norms that are universally applicable and should be respected by all. Furthermore, normative relativism states that the various values and norm systems for each culture are equal, so that it is impossible to say that certain norms and values are better than others. This means that we have to respect all value and norm systems.

There are three problems with this theory. First, it seems to involve an inherent contradiction. The theory states that there are no universal norms, but at the same time it uses a universal norm: "Everybody has to respect the moral opinions of others." Second, it makes any meaningful moral discussion totally impossible, because you can always appeal to your freedom of opinion, which by definition is neither better nor worse than other opinions. The question is whether this is a valid standpoint. Should the torture of political prisoners be tolerated because this is customary within a given culture? Are there no moral limits to such tolerance? Do we not all object to this kind of relativistic argument to defend the torture of political prisoners? Finally, normative relativism can lead to unworkable or intolerable situations. Engineers work in teams or are employed within a company where there are written – and unwritten – rules to promote cooperation (for example, attending meetings on time). A system that allowed engineers to disregard these rules based on his or her personal values (which other people have to respect) would create an unworkable situation.

3.5.2 Absolutism

Universalism An ethical theory that states that there is a system of norms and values that is universally applicable to everyone, independent of time, place, or culture.

Absolutism A rigid form of universalism in which no exceptions to rules are possible.

The other extreme position is *absolutism*: a rigid form of universalism. **Universalism** states that there is a system of norms and values that is universally applicable to everyone, independent of time, place, or culture. **Absolutism** can have a religious nature, where a god determines the universal norms (also known as a dogmatic schema). In many types of universalism room is left to transgress a universal norm in specific exceptional circumstances. In contrast with absolutism, most types of universalism allow for the possibility that not all norms and values are universal. In absolutism a norm like "Thou shalt not kill" would be considered to be universally applicable, but one can imagine situations where killing a person may be the most morally responsible thing to do. Absolutism does not make any exceptions: a rule is a rule.

Absolutism has three main problems. First, we cannot work with the notion that a universal norm prescribes the best action in all situations. Killing someone out of self-defense is justifiable, despite the universal norm "Thou shalt not kill." Second, absolutism gives no answers for conflicting norms. This occurs, for example, in the case of a whistleblower (see Section 1.5.3): on the one hand you have an obligation to maintain confidentiality but on the other you have the obligation to warn society about risks. According to which generally applicable norm should you act? Third, absolutism offers no room for an independent moral judgment, since it often stems from dogmatism. Independent moral judgment was central to the philosophy of the Enlightenment (in the seventeenth and eighteenth centuries). Enlightenment philosophers like Bentham, Mill, and Kant encouraged people to move away from prejudices and dogmatic schemas. The idea was that reason allows man to design his own rules of behavior. Humans, as rational agents, were not supposed to blindly follow traditional moral guidelines, such as the morality dictated by God.

Considering the discussion above, we can state that a choice based only on normative relativism or absolutism is at the very least *ethically* suspect, since ethics reflects on morality, and calls us to make reasoned judgments about it. The ethical theories we shall discuss in the following section are more rational theories than normative relativism or absolutism. Two of them originate from the tradition of philosophy of the Enlightenment.

Immanuel Kant summarized the essence of Enlightenment as follows: "Enlightenment is man's release from his self-incurred tutelage. Tutelage is man's inability to make use of his understanding without direction from another. Self-incurred is this tutelage when its cause lies not in lack of reason but in lack of resolution and courage to use it without direction from another. *Sapere aude!* 'Have courage to use your own reason!' – that is the motto of enlightenment." (Kant, 1990 [1784])

3.6 Ethical Theories

We will now discuss three primary ethical theories and attempt to synthesize their applications. These three are consequentialism, deontology, and virtue ethics. We can distinguish these theories from each other by their approach to the structure of human action and the primary focus or point of departure they use to theorize ethics (see Table 3.4)

The structure of human action means that an action is carried out by a certain actor (person or institution) with a certain intention, which then leads to certain consequences. So, we can evaluate each moral action from three perspectives: the actor, the action and the consequences.

If we evaluate the action from the perspective of the *action* itself, we make use of deontological ethics or deontology (Greek: δέον *(deon)* meaning *obligation* or *duty*): duty ethics. Here, the point of departure is *norms.* It is your moral obligation to

Table 3.4 Differences between the ethical theories

	Actor	*Action*	*Consequences*
Theory	Virtue ethics	Deontology	Utilitarianism
Points of departure	Virtues	Norms	Values

ensure that your actions agree with an applicable norm (rule or principle). One example of such an applicable norm is the "Golden Rule," which can be found in the texts of various religions: "Do unto others as you would have them do unto you."

If we look at the *actor* and his/her characteristics to pass moral judgment on an action, then we make use of virtue ethics. It is neither the incidental action that counts nor the consequences of the action, but it is the quality of the person acting that makes the action morally right or not. Here, the moral point of departure is *virtues*, which allow people to realize a good life.

If we disregard both the actor and the action in the moral judgment of a certain action, but only consider the consequences, then we apply consequentialism. You ought to choose the action with the best outcomes. The moral point of departure is *values*. Consequentialists focus on realizing certain goals or states of affairs they feel should be strived for, for example, promoting pleasure, avoiding pain, or realizing ambitions.

There are different variants on the ethical theories mentioned above. In the following three sections we shall discuss the best-known variant for each theory: utilitarianism as a representative of consequentialism (Section 3.7), Kant's theory as a representative of duty ethics (Section 3.8), and Aristotle's virtues doctrine as a representative of virtue ethics (Section 3.9)

3.7 Utilitarianism

In **consequentialism**, the *consequences* of actions are central to the moral judgment of those actions. An action in itself is not right or wrong; it is only the consequence of action that is morally relevant. We shall limit ourselves to one type of consequentialism: **utilitarianism**. Utilitarianism is characterized by the fact that it measures the consequences of actions against one value: human pleasure, happiness, or welfare. Utilitarianism therefore is a monistic type of consequentialism. There are pluralistic types of consequentialism too, where various values must be weighed against each other in the assessment of actions.

Consequentalism The class of ethical theories which hold that the consequences of actions are central to the moral judgment of those actions.

Utilitarianism A type of consequentialism based on the utility principle. In utilitarianism, actions are judged by the amount of pleasure and pain they bring about. The action that brings the greatest happiness for the greatest number should be chosen.

3.7.1 Jeremy Bentham

Jeremy Bentham was the founder of *utilitarianism*, a word derived from the Latin *utilis* meaning useful. Utilitarianism makes the consequence of an action central to its moral judgment: an action is right if it is useful and wrong if it is damaging. The next question of course is "useful for what?" In other words, what is the purpose for which the action is a means? This purpose has to be something that has *intrinsic* value. So it has to be good in itself. This means that the utilitarian is primarily concerned with values; he first has a notion of what is intrinsically good and subsequently considers the moral rightness dependent on this notion. The value theory that Bentham connects to his ethics is **hedonism**: the idea that "pleasure" is the only thing that is good in itself and for which all other things are instrumental.

Hedonism The idea that pleasure is the only thing that is good in itself and to which all other things are instrumental.

Jeremy Bentham (1748–1832)

> By the principle of utility is meant that principle which approves or disapproves of every action whatsoever, according to the tendency it appears to have to augment or diminish the happiness of the party whose interest is in question: or, what is the same thing in other words to promote or to oppose that happiness. I say of every action whatsoever, and therefore not only of every action of a private individual, but of every measure of government. (Bentham, 1948 [1789])

Figure 3.2 Jeremy Bentham. Photo: © Classic Image / Alamy.

Jeremy Bentham was born in London on February 15, 1748. At pre-school age his father taught him Latin, Greek and music. A private teacher also taught him French language and literature. His private teacher had him read Télémaque by Fénelon. The book had a huge impact on Bentham, who identified strongly with the hero Telemachus. His dedication to the welfare of humanity was an ideal he held to throughout his life. When he was 12 he was enrolled at Queen's College in Oxford, where he took classical languages and philosophy. As a small and shy but intelligent child, he soon was given the nickname "the philosopher." Bentham looked back on this period in horror. He considered the lectures in Oxford to be useless and a waste of time – the only things he felt had been useful were lessons on logic. As a student he trained to become a lawyer, but after a few years of running a law practice he focused more and more on developing a philosophical and scientific theory of legislation and justice. He fiercely criticized the legal system, because it did nothing to improve the welfare of people. Courts of law could condemn people for "sexual crimes" even if neither party had objections to the sexual act. Bentham thought this was nonsense: if both parties agreed to an act then there could be no crime. As an alternative Bentham wanted to build a new legal system that was rational, clear, and consistent. It was to be based on ethical knowledge and not on tradition or custom. His ethical opinions for which he chose the name "utilitarianism" were set out in *An Introduction to the Principles of Morals and Legislation* (1948 [1789]). Due to the clash between his ethical opinion and conventional Christian thought, Bentham was greatly opposed to Christianity, which he considered to be a form of ascetism where pleasure was condemned. According to him, Christianity was a major obstacle to human happiness and a hindrance in the realization of utilitarianism.

Bentham was one of the earliest philosophers to argue for a complete equality between sexes, and for decriminalization of homosexuality and equal rights for homosexually inclined people. Furthermore, he is widely recognized as one of the earliest proponents of animal rights. Bentham argued that the ability to suffer, not the ability to reason, must be the benchmark of how to treat other beings. If the ability to reason were the criterion, many human beings, including babies and disabled people, would also have to be treated as though they were things.

Figure 3.3 Panopticon. Photo: Bettmann Archive/Corbis..

Bentham is probably best known in popular society for his design of the "panopticon" (which means all-seeing): it is a dome-shaped prison in which a prison warder can see all prisoners. They are kept in cell rings with windows facing inwards (Figure 3.3). The warder can observe all prisoners, but the prisoners cannot see the warder. The idea behind this is simple: if individuals are checked by an all-seeing eye (without the eye being seen), they will allow themselves to be disciplined and be controllable. The panopticon remained an obsession of Bentham's for more than 20 years.

Bentham died on June 6, 1832 in the town he was born aged 85. The real body of Bentham together with a wax head (something went wrong preserving the head) can still be admired in the University College of London in a cabinet with a glass door. During board meetings of the university, he is removed from the cabinet so that he can attend these meetings. Bentham left his fortune to the university with the condition that he would be allowed to attend all meetings of the board.

Bentham calls pleasure and pain the sovereign masters of man. That which provides pleasure or avoids pain is good, and that which provides pain or reduces pleasure is bad. Bentham places experience at the heart of his ethics. According to him, it is an elementary fact of experience that people strive by nature for pleasure and avoid pain. Moreover, people know what provides pleasure and what results in pain, and also how pleasure can be realized. Based on this experience people can form a moral judgment without the intervention of an authority such as a legislator or God.

The only moral criterion for good and bad lies in what Bentham calls the **utility principle**: the greatest happiness for the greatest number (of the members of the community). This principle is the only and sufficient ground for any action – both for individuals and collectives (e.g., companies or government). It gives us a reason to act morally. Moral terms like "proper," "responsible," and "correct" only are meaningful if they are used for actions that are in agreement with the utility principle. The greatest happiness can be determined quantitatively according to Bentham. He believed that we can calculate the expected pleasure or pain and can even indicate quite accurately how much will be produced by a given action. Here, pleasure and pain are given in terms of a measurable result, which can be made suitable for calculation. In this context he referred to a **moral balance sheet** and even drew up extensive tables. He made use of a number of circumstances, such as intensity, duration, certainty and extent of an action (see box). Applying this theory to a moral problem means drawing up a moral balance sheet. Here, the costs and benefits for each possible action must be weighed against each other. The action with the best result (providing the most utility) is the one to be preferred. According to Bentham, money can even be used to express quantities of pleasure or pain, because these experiences can (almost) always be bought and sold.

Utility principle The principle that one should choose those actions that result in the greatest happiness for the greatest number.

Moral balance sheet A balance sheet in which the costs and benefits (pleasures and pains) for each possible action are weighed against each other. Bentham proposed the drawing up of such balance sheets to determine the utility of actions. Cost-benefit analysis is a more modern variety of such balance sheets.

Value of a Lot of Pleasure or Pain, How to be Measured

Pleasures then, and the avoidance of pains, are the *ends* that the legislator has in view; it behoves him therefore to understand their *value*. Pleasures and pains are the instruments he has to work with: it behoves him therefore to understand their force, which is again, in other words, their value. To a person considered by *himself*, the value of a pleasure or pain considered *by itself*, will be greater or less, according to the four following circumstances:

- its *intensity*;
- its *duration*;
- its *certainty* or *uncertainty*; and.
- its *propinquity* or *remoteness*.

These are the circumstances which are to be considered in estimating a pleasure or a pain considered each of them by itself. But when the value of any pleasure or pain is considered for the purpose of estimating the tendency of any *act* by which it is produced, there are two other circumstances to be taken into the account; these are,

- Its *fecundity*, or the chance it has of being followed by sensations of the *same* kind: that is, pleasures, if it be a pleasure: pains, if it be a pain.
- Its *purity*, or the chance it has of not being followed by sensations of the *opposite* kind: that is, pains, if it be a pleasure: pleasures, if it be a pain.

These two last, however, are in strictness scarcely to be deemed properties of the pleasure or the pain itself; they are not, therefore, in strictness to be taken into the account of the value of that pleasure or that pain. (…) And one other; to wit:

- Its *extent*; that is, the number of persons to whom it *extends*; or (in other words) who are affected by it.

To take an exact account then of the general tendency of any act, by which the interests of a community are affected, proceed as follows. Begin with any one person of those whose interests seem most immediately to be affected by it: and take an account,

1 Of the value of each distinguishable *pleasure* which appears to be produced by it in the *first* instance.
2 Of the value of each *pain* which appears to be produced by it in the *first* instance.
3 Of the value of each pleasure which appears to be produced by it *after* the first. This constitutes the *fecundity* of the first *pleasure* and the *impurity* of the first *pain*.

4 Of the value of each *pain* which appears to be produced by it after the first. This constitutes the *fecundity* of the first *pain*, and the *impurity* of the first pleasure.
5 Sum up all the values of all the *pleasures* on the one side, and those of all the pains on the other. The balance, if it be on the side of pleasure, will give the *good* tendency of the act upon the whole, with respect to the interests of that *individual* person; if on the side of pain, the *bad* tendency of it upon the whole.
6 Take an account of the *number* of persons whose interests appear to be concerned; and repeat the above process with respect to each. *Sum up* the numbers expressive of the degrees of *good* tendency, which the act has, with respect to each individual, in regard to whom the tendency of it is *good* upon the whole: do this again with respect to each individual, in regard to whom the tendency of it is *good* upon the whole: do this again with respect to each individual, in regard to whom the tendency of it is *bad* upon the whole. Take the *balance* which if on the side of *pleasure*, will give the general *good tendency* of the act, with respect to the total number or community of individuals concerned; if on the side of pain, the general *evil tendency*, with respect to the same community.

It is not to be expected that this process should be strictly pursued previously to every moral judgment, or to every legislative or judicial operation. It may, however, be always kept in view: and as near as the process actually pursued on these occasions approaches to it, so near will such process approach to the character of an exact one." (Bentham, 1948 [1789])

The idea behind the calculation above is quite simple: an action is morally right if it results in pleasure, and it is morally wrong is it gives rise to pain. To find out which action leads to the most happiness for the greatest number of people, we need to count the pleasure and pain of all individuals. This is no simple matter, because pleasure cannot be measured objectively. First, the pleasure of different people cannot be compared; pleasure is a rather subjective term. A person can enjoy a composition by Mozart, while someone else experiences this quite differently. Second, it is not easy to compare actions: is reading a good book worth more than eating an ice cream? While applying this hedonistic calculus this will often lead to problems, because it is not clear how much pleasure a given experience produces for each person. How much pleasure do social contacts, our health, or our privacy give us? Since this is not clear, making moral judgments about human actions becomes hard. Take, for example, a company that pollutes the environment. If the company were to work in a more environmentally friendly way this would reduce the profits and the numbers of people employed. However, if the company does not become environmentally friendly then the damage to the environment will have repercussions for public health. It seems nearly impossible to draw up a quantitative moral balance sheet for these two options: continuing along the status quo or changing to environmentally friendly production.

3.7.2 Mill and the freedom principle

John Stuart Mill (1806–1873) extended and revised Bentham's thinking. There are two main respects in which Mill's thinking differs from that of his predecessor. According to Mill, qualities must be taken into account when applying the utilitarian calculus: forms of pleasure can be qualitatively compared, in which it is possible that a quantitatively smaller pleasure is preferred over a quantitatively larger one because the former pleasure is by nature more valuable than the latter. According to Mill, "[i]t is better to be a human being dissatisfied than a pig satisfied; better to be Socrates dissatisfied than a fool satisfied" (Mill, 1979 [1863]). Unfortunately, Mill does not answer the question what makes one pleasure more valuable than another. He only gives indications: "higher" desires, like intellectual ones, are to be preferred above "lower" desires, like physical or animal desires. Satisfying the desire to complete a study is more rewarding than watching "As the World Turns" every evening or to be able to eat as much as you want at every meal. The second distinction was a response to the criticism that the position of individuals cannot always be protected if the calculation indicates that the pleasure of the majority outweighs the unhappiness of a few individuals. This could result in the exploitation and abuse of minorities, because Bentham's utilitarianism does not say anything about the division of pleasure and pain among people. According to Mill we must choose the action that provides the most pleasure but does not conflict with human nature and dignity. For the latter point he introduces the **freedom principle**: everyone is free to strive for his/her own pleasure, as long as they do not deny or hinder the pleasure of others. Mill illustrates this principle using the example of drunkenness. The right to interfere with someone who is drunk only arises when the person who is drunk starts to do harm to others. Mill's principle also provides a foundation for the discussion nowadays about legalizing soft drugs (or even heroin). According to Mill, the sale and use of soft drugs should not be a matter for penal law, as this would be a violation of freedom. The fact using soft drugs is bad for your health cannot be a consideration for the legislator to intervene, because the legislator has no right to be involved with personal decisions in Mill's view. Mill illustrates this principle on the basis of drunkenness.

Freedom principle The moral principle that everyone is free to strive for his/her own pleasure, as long as they do not deny or hinder the pleasure of others.

> Drunkenness, for example, in ordinary cases, is not a fit subject for legislative interference; but I should deem it perfectly legitimate that a person, who had once been convicted of any act of violence to others under the influence of drink, should be placed under a special legal restriction, personal to himself; that if he were afterwards found drunk, he should be liable to a penalty, and that if when in that state he committed another offence, the punishment to which he would be liable for that other offence should be increased in severity. (Mill, 1859, chapter 5)

The freedom principle is also known as the **no harm principle**: "one is free to do what one wishes, but only to the extent that no harm is done to others." However, the principle can hardly ever be applied in full, since any moral problem involves possible harm to others, or at least the risk of harm.

No harm principle The principle that one is free to do what one wishes, as long as no harm is done to others. Also known as the freedom principle.

John Stuart Mill (1806–1873)

> The only freedom which deserves the name, is that of pursuing our own good in our own way, so long as we do not attempt to deprive others of theirs, or impede their efforts to obtain it. Each is the proper guardian of his own health, whether bodily, or mental and spiritual. Mankind are greater gainers by suffering each other to live as seems good to themselves, than by compelling each to live as seems good to the rest. (Mill, 1859)

Figure 3.4 John Stuart Mill. Photo: Hulton Archive/Getty Images.

John Stuart Mill was born in 1806; he was the oldest son of James Mill and proved to be a prodigy. James Mill had special ideas about raising children. At the age of three he taught his son Greek, at age four he taught him Latin, and shortly after he taught him mathematics. At age twelve John Stuart Mill wrote a book about Roman history. Mill was a proponent of utilitarianism as proposed by his godfather Jeremy Bentham. When he was 18, Mill founded a utilitarian society for youths, where lectures and discussions were held about the utility principle. When he was 20, Mill had a nervous breakdown and he suffered from severe depressions. He found that the utilitarianism of Bentham was not making him happy. Following this, he distanced himself from Bentham's ideas.

In 1823 he started to work for the East India Company under his father's authority. This work provided him with much opportunity to study and write. In 1830 he met the 23 year-old Harriet Taylor. They were highly impressed by each other. However, Harriet was married to the businessman John Taylor and she decided not to sacrifice her family because of her feelings. Her husband eventually allowed her to meet with Mill on a regular basis. According to Mill's testimony their love for each other was purely platonic. After John Taylor's death in 1849 there was no more reason not to marry, which they did in 1851. In Mill's view, Harriet's opinions had a major influence on him and especially his socio-philosophical work. Together with her, Mill called for the emancipation of women and also argued for women's right to vote. In 1869 he published *The Subjection of Women*, which is now the classical theoretical statement of the case for woman suffrage. Harriet died in 1858 in Avignon. Between 1866 and 1868, Mill was a Member of Parliament. He was considered a radical, because he supported the public ownership of natural resources, the development of labor organizations, compulsory education, birth control, an end to slavery, and equality of women. His advocacy of women's suffrage in the Reform Bill of 1867 led to the creation of the suffrage movement. He died in 1873.

John Stuart Mill was the most influential British thinker of the nineteenth century. Mill's essay *On Liberty* (1859) remains his major contribution to political thought. He proposed that self-protection is the only reason an individual or the government can interfere with a person's liberty of action. Outside of preventing harm to others, the state has no legitimate reason to compel a person to act in the way the government wishes.

3.7.3 Criticism of utilitarianism

Although utilitarianism has a strong intuitive attraction because of its simplicity, it has nevertheless received much criticism. Two important points of criticism were discussed above: happiness cannot be measured objectively and utilitarianism can lead to exploitation. Four other points of criticism are discussed below. In many cases the criticism was incorporated by utilitarians to improve utilitarianism.

The first criticism is that the consequences cannot be foreseen objectively and often are unpredictable, unknown, or uncertain. An obvious solution is to work with expected consequences and the accompanying pleasure. In the twentieth century this notion was even given a mathematical foundation using statistics.

Distributive justice The value of having a just distribution of certain important goods, like income, happiness, and career.

Next to this there is the problem of **distributive justice**. Distributive justice refers to the value of having a just distribution of certain important goods, like income, happiness, and career. Utilitarianism can lead to an unjust division of costs and benefits. According to the political philosopher John Rawls utilitarianism suffers from this problem because it does not recognize the fundamental separateness of persons (Rawls, 1971). Instead of that utilitarianism treats society as a whole in which pleasure must be increased via the criterion "the greatest happiness for the greatest number." The question concerning the distribution of happiness is neglected, even under Mill's formulation of utilitarianism. It is a tricky question because numerous issues in technology are concerned with this problem, such as how the risks and benefits of technology should be justly distributed (see Chapter 8). Despite Rawls criticism, utilitarians have tried different ways to pay attention to justice and the distribution of welfare. Henry Sidgwick, for example, believed that although the total amount of societal happiness should be considered in the first place, it should be the situation with the most equitable distribution of happiness that must be selected from various situations with equal happiness (Sidgwick, 1877). Other utilitarians argue that the classical utilitarianism – with the emphasis on the greatest happiness for the greatest number of people – does not require such a clause, because it leads to a just and balanced distribution of welfare. The modern utilitarian Richard Hare mentions two reasons for this. First, a rich person experiences less added pleasure on average from an increase in income of 100 Euros than a poor person. This phenomenon is known in economics as decreasing **marginal utility** (the term marginal utility refers to the increase in utility with an increase in income for example). An improvement in income for poor people will sooner lead to maximization of happiness than an increase in income for people who already are rich. Second, inequality of income leads to jealousy and thus to pain and is thus to be avoided (Hare, 1982).

Marginal utility The additional utility that is generated by an increase in a good or service (income for example).

A third point of criticism is that utilitarianism ignores the personal relationships between people. In the hedonistic balance of Bentham each individual counts as an anonymous unit. Who receives the pleasure is irrelevant; it is only to total amount of pleasure that counts. In other words, the total happiness counts and not the individual happiness of specific persons. For this reason Mill called Bentham's followers reasoning machines. In daily life, some people's happiness has a greater impact on us than the happiness of others. If you were to be shipwrecked and had to make a choice between saving a friend or a famous surgeon, utilitarian theory dictates that saving the surgeon is the right thing to do, because he is more useful to society. This choice ignores the fact that it is *specific individuals* that want to be happy and that it really depends on *who* is made happier. The question, therefore, is whether we have special moral obligations to the people that we have a personal relationship with, and whom we want to make happy.

Finally, certain actions are morally acceptable even though they do not create pleasure and some actions that maximize pleasure are morally unacceptable. In the next section we will see that Kant always considers lying to be morally wrong, even if it results in more or maximal pleasure in certain situations. According to utilitarianism, even the most fundamental rules, such as the human rights formulated in the Universal Declaration of Human Rights (1948), can be broken if the positive consequences are greater than the negative ones: "the end justifies the means." On utilitarian grounds, an engineer could be asked to bend a fundamental rule of professional conduct because of the positive consequence it would have. Say, for example, that an engineer is asked to falsify the measurements he gave in a report by the party commissioning the work, because the correct measurement results would have major negative consequences, such as the payment of damages or bankruptcy. According to the traditional utilitarian view, this behavior would be justified in a certain situation. This traditional view is known as **act utilitarianism** because it judges the consequences of individual acts. A solution to this problem is proposed by one variant of

Act utilitarianism The traditional approach to utilitarianism in which the rightness of actions is judged by the (expected) consequences of those actions.

Rule utilitarianism A variant of utilitarianism that judges actions by judging the consequences of the rules on which these actions are based. These rules, rather than the actions themselves, should maximize utility.

utilitarianism: **rule utilitarianism**. Rule utilitarianism recognizes the existence of moral rules, if only because life would be very complicated without them. For each situation we would have to judge whether it was morally correct or not, because each situation is slightly different from another. Rule utilitarianism looks at the consequences of rules (in contrast with actions) to increase happiness. Though the falsifying of measurements may increase societal utility in a specific situation, a rule utilitarian will not allow it because the rule "measurement data should be presented correctly" generally promotes happiness within society. If such a rule withstands the test of promoting happiness then it is turned into a moral rule. Within rule utilitarianism there are a number of variants. There is a variant where the moral rules are viewed as conditional rules (they are more like rules of thumb), and a variant that views the rules as unconditional ones (they apply to all people in all circumstances without exception). Rule utilitarianism is close to duty ethics, which is the subject of the next section, although their conceptual foundations are very different.

3.7.4 Applying utilitarianism to the Ford Pinto case

In the Ford Pinto case the Ford company provided an act-utilitarian argument by making a cost-benefit analysis to justify that the defective vehicle model was not recalled and retrofitted by Ford. This cost-benefit analysis, according to Ford showed that the total social costs of retrofitting all the cars were higher than the social costs of the expected accidents. It is important to note that the cost-benefit analysis refers to social cost rather than to costs for Ford. For this reason, Ford's argument was utilitarian rather than egoistic.

The Ford Pinto case clearly illustrates some of the objections against utilitarianism. First, the amounts of money that Ford attached to different kinds of pain (dead, injuries) seem rather arbitrary, even if some of the amounts were based on government documents. Second, one might wonder how reliable the estimates of, for example, number of fatalities are. A change in these estimates may change the conclusion of the cost-benefit analysis. Apart from such more practical objections, the case also illustrates some of the more fundamental objections to utilitarianism. In making a decision solely based on considerations of overall welfare or happiness, Ford adopted a policy of allowing a certain number of people to die or be injured even though they could have prevented it. One could also argue that the Ford Pinto case reveals exploitation or abuse because the victims were sacrificed to optimize overall welfare (the ends justify the means). Moreover, the case shows how a utilitarian argument may lead to abandoning inherent principles, like "you cannot put a value on human life" or the freedom principle of Mill. According to the latter principle, Ford should have recalled and repaired the car.

Some of these objections might be overcome by applying rule utilitarianism to the case. Then, one should ask whether or not following rules like "companies must recall a car if it is unsafe" or "companies should produce safe cars" maximizes overall happiness. Since this seems to be the case, Ford was ethically obliged to recall the car, because this is required by rules from which everyone in the society would benefit most in the long run. So, in the case of rule utilitarianism, the fact that an action maximizes utility on a *particular* occasion does not show that it is right from an ethical point of view.

3.8 Kantian Theory

According to **duty ethics** (also known as deontological ethics), an action is morally right if it is in agreement with a moral rule (law, norm, or principle) that is applicable in itself, independent of the consequences of that action. There are two important points of difference between the various duty ethics theories. First, some theories rely on one main principle from which all moral norms can be derived (monistic duty ethics). Other theories, the pluralistic theories, are based on several principles that apply as norms for moral action. A second important difference concerns the foundation or origin of the moral rules. These rules can be given by God, such as in the Bible or the Koran, or they make an appeal to a social contract that the involved parties have implicitly agreed to (e.g., a company code), or they are based on reasonable arguments.

Duty ethics Also known as deontological ethics. The class of approaches in ethics in which an action is considered morally right if it is in agreement with a certain moral rule (law, norm, or principle).

The best-known system of duty ethics has been developed by Immanuel Kant. Since Aristotle, the basis for ethics had been sought in striving for happiness or welfare (e.g., Bentham and Mill). According to Kant, moral laws or normative ethics cannot be based on happiness. Happiness is an individual matter and changes for each person during his/her lifetime. Moreover, it is hard to determine what increases happiness, so striving for happiness can even lead to immorality. Thus, Kant argued that duty was a better guide for ethics.

A core notion in Kantian ethics is *autonomy*. In Kant's opinion man *himself* should be able to determine what is morally correct through reasoning. This should be possible independent of external norms, such as religious norms. The idea behind this is that we should place a moral norm upon ourselves and should obey it: it is our *duty*. We should obey this norm out of a *sense of duty* – out of respect for the moral norm. It is only then that we are acting with **good will**. According to Kant, we can speak of good will if our actions are led by the moral norm. Thus, the notion of good will is different from having good intentions.

Good will A central notion in Kantian ethics. According to Kant, we can speak of good will if our actions are led by the categorical imperative. Kant believes that the good will is the only thing that is unconditionally good.

Since a moral norm has validity independent of time and place, it means that a moral norm is unconditionally applicable (or categorically applicable) to everyone in all circumstances in Kant's view. Often a norm follows the form of "thou shalt …," such as "thou shalt not kill," or "thou shalt not lie." In contrast to a categorical norm, a hypothetical (conditional)

Hypothetical norm A condition norm, that is, a norm which only applies under certain circumstances, usually of the form "If you want X do Y."

norm only applies under certain circumstances. A **hypothetical norm** usually has the following shape: "if you wish to achieve this goal, then you will have to act in this way." An example of such a norm is "if you do not wish to betray your friend, then you may not lie," in which the rule of behavior ("you may not lie") is not unconditional but can only be applied under certain conditions ("you do not wish to betray your friend").

3.8.1 Categorical imperative

According to Kant there is one universal principle from which all moral norms can be derived, which makes his ethics a monistic duty ethics. This principle, which is the foundation of all moral judgments in Kant's view, is referred to as the **categorical imperative**. An imperative is a prescribed action or an obligatory rule. By arguing reasonably, any rational person should be capable of judging whether an optional action is morally right. The categorical imperative was formulated by Kant in different ways.

Categorical imperative A universal principle of the form "Do A" which is the foundation of all moral judgments in Kant's view.

Universality principle First formulation of the categorical imperative: Act only on that maxim which you can at the same time will that it should become a universal law.

The first formulation of the categorical imperative, the **universality principle**, is as follows:

"Act only on that maxim which you can at the same time will that it should become a universal law."

A maxim is a practical principle or proposition that prescribes some action. Kant states that the maxim should be unconditionally good, and should be able to serve as a general law for everyone without this giving rise to contradiction. We must oblige ourselves to follow generally applicable laws. Perhaps a woman decides to recycle her bottles and cans to help the environment. She should ask herself whether the maxim or rule behind her action – that one should recycle containers to help the environment – could be applied to all people. In this case, there is no apparent problem. She could consistently wish that everyone follow the rule or maxim behind her action. However, when you break a promise this is different. Sometimes people are in a situation where it would be more convenient to break a promise. Say that one wonders whether it is morally acceptable to break one's promise. The maxim of the action to be undertaken is "I may break my promises when doing so is convenient for me." The categorical imperative states that it is morally acceptable if I can wish everyone to break their promise without *contradiction*. Breaking a promise is only possible if people trust in the custom of making (and keeping) promises. If breaking a promise when convenient becomes a general law, no one would trust anybody to keep a promise. The contradiction now is that you cannot wish to break a promise and want the breaking of promises to become a general law. If the latter were to become true then promises would lose their meaning and it would be no use to make a promise.

According to Kant, the categorical imperative also implies a postulate of equal and universal human worth. His reflections on autonomy and self-legislation lead him to argue that the free will of all rational beings is the fundamental ground of human rights. The **equality postulate** is defined as the prescription to treat persons as equals, that is, with equal concern and respect (Dworkin, 1977, p. 370). To recognize that human beings are all equal does not mean having to treat them identically in any respects other than those in which they clearly have a moral claim to be treated alike. Opinions diverge concerning the question what these claims amount to and how they have to be balanced with competing claims (based on, for example, the principle of freedom). For example, how should goods be distributed if we set out to treat people as equals?

Equality postulate The prescription to treat persons as equals, that is, with equal concern and respect.

The second formulation of the categorical imperative is, according to Kant, equivalent to the first.

> The second formulation of the categorical imperative, the **reciprocity principle**, is as follows:
>
> "Act as to treat humanity, whether in your own person or in that of any other, in every case as an end, never as means only."

Reciprocity principle Second formulation of the categorical imperative: Act as to treat humanity, whether in your own person or in that of any other, in every case as an end, never as means only.

Humanity in this version of the imperative is presented as equivalent to "reason" or "rationality," for humans differ from things without reason (objects and animals) because humans can think. This imperative states that each human must have respect for the rationality of another and that we must not misguide the rationality of another. In other words, Kant here stresses the rational nature of humans as free, intelligent, self-directing beings. In saying they must never be treated as a means only, he means that we must not merely "use" them as means to our selfish ends. They are not objects or instruments to be used. To use people is to disrespect their humanity. Say I borrow money from someone and promise to pay him back although I know that I will not do so. In this case, I am using the person I made a promise to as a means and not as a goal. I am misleading him, or I am misleading his rationality. I have provided insufficient information about the fact that I will not keep my promise, so that he cannot make a rational choice. Probably he would not have lent me money if he had known that I did not intend to pay him back. I use his rationality as a means to achieve my own aim. The reciprocity principle is strongly anti-paternalistic by nature (on paternalism see Chapter 1), since, a person – as a rational being – should have the right to make up her or his own mind.

The reciprocity principle tells us that we should respect people *as* people, and not "use" them. However, we need to be careful in interpreting the idea of using people, or treating people merely as a means. The difference between treating someone as a means versus treating someone as a *mere* means is not always clear-cut. Suppose someone has religious objections to taking medication (a Christian Scientist, for example), and yet the doctor forces the person to be medicated for the person's own good. Now the doctor is treating the patient as a mere means to the patient's own welfare – paradoxical as it might sound – and that is unacceptable, according to the reciprocity principle. Note that to treat someone as an end does not simply mean doing what he or she wants. If a consumer argues about the purchase price of a car, and the salesman does not want to bargain about the price, this does not mean that the salesman treats the consumer not as an end. If the salesman informs the consumer about the price of the car and the condition of the car, the salesman treats the consumer as an end.

Immanuel Kant (1724–1804)

> Nothing can possibly be conceived in the world, or even out of it, which can be called good, without qualification, except a good will. Intelligence, wit, judgement, and the other talents of the mind, however they may be named, or courage, resolution, perseverance, as qualities of temperament, are undoubtedly good and desirable in many respects; but these gifts of nature may also become extremely bad and mischievous if the will which is to make use of them, and which, therefore, constitutes what is called character, is not good. (Kant, 2002 [1785])

Figure 3.5 Immanuel Kant. Photo: © INTERFOTO/Alamy.

Immanuel Kant, one of the most influential philosophers in history, was the fourth of nine children born to a poor saddle maker. He was born in 1724 in the university city of Koningsbergen in East Prussia, which was a rich trading place at the time. He was brought up in a tradition of devout Christianity that he strongly rejected in later life. After completing pre-university education he first studied theology and then philosophy, mathematics, and physics in Koningsbergen. After completing his studies in 1746, he became a teacher for various families. From 1755, when he attained the title of Magister, he became a private teacher at the University of Koningsbergen. At 46 he accepted a professorship in logic and metaphysics. He had great admiration for the enlightened king Frederick the Great of Prussia, but in 1794 he came into conflict with the King's successor due to his theological philosophy. He valiantly defended the right of scientists to think in freedom and to publish for fellow scholars. Kant died at the age of 80 (1804). His life was known to be highly disciplined – he had a great fervor for work and a strict daily routine. The inhabitants of Koningsbergen could set the clock by the time, when Kant passed by for his daily walk. The reason for this way of life was his poor physical health, which he tried to improve through his strict routine.

Kant's theory of mind represents a turning point in the history of philosophy, since it radically revised the way that we all think about human knowledge of the world. He built his systematic theoretical philosophy around the idea that the world as we experience it does not exist independently of us. Our own minds are responsible for its form and structure. This introduced the human mind as an active originator of experience rather than just a passive recipient of perception. As Kant puts it, it is the representation that makes the object possible rather than the object that makes the representation possible. This idea, in his words, effected a Copernican revolution. Before Nicolaus Copernicus (1473–1543) – the founder of modern astronomy – astronomical data were explained by assuming that the sun revolves around the earth. Reversing this, Copernicus explained the data by taking the earth to revolve around the sun. In moral philosophy, Kant proposed an equally revolutionary idea. In morality we are not required to obey laws imposed by God or eternal moral principles; instead we must understand morality as resting on a law that springs from our own practical rationality. Kant's ethics, which he expounded in the *Critique of Practical Reason* (1788) and the earlier *Groundwork of the Metaphysics of Morals* (1785), was based on the principle known as the "categorical imperative," an unconditional obligation derived from the concept of duty.

3.8.2 Criticism of Kantian theory

There are two primary criticisms of Kantian theory. According to Kant all moral laws can be derived from the categorical imperative. The question arises whether all these laws form an unambiguous and consistent system of norms. Often there are several contradictory norms, as we saw earlier in the case of the whistle-blower. Another example is the situation in which one can only save one's friend from an emergency situation by lying. It means breaking a norm: either you break the norm that you must always speak the truth or you break the norm about helping people when they need it. In Kant's theory there is no such thing as bending a rule. Kant does not allow for any exceptions in his theory.

To cope with this problem, William David Ross developed a pluralistic theory of moral obligations (Ross, 1930). Ross states that good is often situated on two levels: what seems to be good at first and that which is good once we take everything into consideration. The norms of the first level are called **prima facie norms** and those of the second level are called *self-evident norms* ("duties sans phrase"). Usually, the *prima facie* norms are our self-evident norms, but this is not necessarily the case. An example can illustrate this. Say you promise your students that you will check their work by the end of next week. Later on, a good friend of yours gets into trouble and needs aid. The fact that you have promised to check the work does

Prima facie norms Prima facie norms are the applicable norms, unless they are overruled by other more important norms that become evident when we take everything into consideration.

not disappear. Both norms are prima facie norms, but upon closer inspection only the norm "you must help your friend" is a self-evident norm while the one ("you must keep your promise") is not.

Note that here too we have to weigh the different norms: the norm to keep one's promise and the norm to help a friend. We are never certain that the norm we identify as the self-evident one truly is the self-evident norm. How we should weigh the norms remains unclear here too. Ross states that our choices are never more than considered judgments. Though this is perhaps not very satisfactory, it does pay respect to the complexity of our moral world. Examples of regular *prima facie* norms that are common in duty ethics include the following:

- Norms concerning faithfulness: freely given promises should be kept.
- Norms concerning reciprocity: this can refer to things like the Golden Rule in a positive or a negative sense (treat others as you would like to be treated/do not treat others as you would not like to be treated yourself).
- Norms of solidarity: help people in need regardless of their achievements or usefulness to society or to you as an individual.

second problem is that duty ethics, and thus Kantian theory, often elicits the objection that a rigid adherence to moral rules can make people blind to the potentially very negative consequences of their actions, as becomes clear in the child labor case.

Case Child Labor

The Socialist Party (SP) in the Netherlands started to boycott IKEA in 1998 and demanded that IKEA guarantee that children would never be involved with the production of IKEA products. According to some advocates for child workers, however, such as Theo Knippenberg from ChildRight Worldwide, such boycotts can actually harm the children in question, if their families have no other means of survival (Knippenberg, 1999). Knippenberg has found that many actions taken against child labor in the past have ended up doing more harm than good, because they take away a relatively good opportunity for children to provide themselves a living. As a result of losing a job, many of these children end up in slavery or prostitution. Moreover, trade and industry can contribute to the improvement of the working conditions of the children, such as working times, medical care, training, etc.

This case demonstrates, on the one hand, the value of adhering to a strict moral principle: that child labor should not be condoned. The SP believes in the moral force of this principle, regardless of the consequences, as is also witnessed in Kantian theory. On the other hand, utilitarians emphasize the negative consequences of such strict adherence to principle. According to Knippenberg, breaching the principle counts for nothing in the face of the consequences of abolishing child labor. As this example shows, both theories generally appeal to our moral intuitions, but they can become diametrically opposed concerning the moral correctness of an action. Ross' approach could offer a solution to the rigidity of Kantian theory. According to Ross the reason why a norm is a self-evident norm depends on the situation in which one finds oneself. We must do what is more of a duty in a given situation. The *prima facie norm* "child labor is not permitted" is not a self-evident norm in this situation, because the situation calls for us to provide a good future for the children and prevent them from becoming slaves or prostitutes. The norm "children should not be forced into slavery or prostitution" would be the self-evident norm instead of "child labor is not permitted."

3.8.3 Applying Kant's theory to the Ford Pinto case

To apply Kant's first categorical imperative, the universality principle, to the Ford Pinto case, we must examine whether the maxim of Ford: "Ford will market the Ford Pinto, knowing that the car is unsafe and without informing the consumers" can be universalized. To do this we have to explain whether this maxim can become a universal law, and can be willed without contradiction. The universal law would read as follows: "Marketing unsafe cars without informing the consumers is allowable." If this were to be a universal law marketing a car would become impossible because no rational person would buy a car anymore, because he or she could not trust that the car would be safe. It may be clear then that the maxim cannot be universalized and should, therefore, not be followed by Ford.

The second categorical imperative, the reciprocity principle, tells us that people should not be treated as mere means. As we have seen this principle implies respect for people's **moral autonomy** in making their own choices. From this, it follows that Ford should have informed its consumers because otherwise they cannot make an autonomous rational decision to buy the car or not. If consumers had known what Ford knew about the safety of the Ford Pinto, they would probably have thought twice before buying the car. By failing to inform them, the rational agency of the consumer was thus undermined, and they were used as *merely* a means (and thus not as an end) to achieve Ford's aim: increasing Ford's turnover. It is not just that Ford endangered people's lives; rather, it is that Ford did so without informing car drivers about the risks.

Moral autonomy The view that a person himself or herself should (be able to) determine what is morally right through reasoning.

3.9 Virtue Ethics

Utilitarianism and Kantian theory both are theories about criteria concerning action. Rather than taking action as point of departure for moral judgment, **virtue ethics** focuses on the nature of the acting person. This theory indicates which good or desirable characteristics people should have or develop and how people can achieve this. Virtue ethics is not exclusively aimed at reason, as the previous two theories were, but is more a mixture of ethics and psychology with an emphasis on developing character traits.

Virtue ethics An ethical theory that focuses on the nature of the acting person. This theory indicates which good or desirable characteristics people should have or develop to be moral.

Virtue ethics is based on a notion of humankind in which people's characters can be shaped by proper nurture and education, and by following good examples. The central theme is the development of persons into morally good and responsible individuals so that they can lead good lives. To this purpose, developing good character traits, both intellectual and personal character traits, is essential. These characteristics are called virtues. They not only indicate how to lead a good life but also what a good life is. Examples of virtues are reliability, honesty, responsibility, solidarity, courage, humor, and being just.

3.9.1 Aristotle

Virtue ethics stems from a long tradition and was already popular in ancient Greece with philosophers like Socrates, Plato, and Aristotle. Aristotle was the first to define virtue ethics as a field of inquiry in itself. According to Aristotle, the final goal of human action is to strive for the highest good: *eudaimonia*. This can be translated as "**the good life**" (or as "welfare" or "happiness"). This does not refer to a happy circumstance that brings pleasure (the goal of classical utilitarians), but a state of being a good person. It means leading a life as humans are meant to lead it; one should excel in the things that are part of being human. As only humans can reason, this is where happiness lies. If we wish to become happy as humans we must use our reasoning to its fullest extent. The good life is not only determined by activities related to reasoning, but is also realized by virtuous activities according to Aristotle. The good life therefore is an active life in agreement with the virtues necessary to realize one's uniquely human potential.

The good life The highest good or *eudaimonia*: a state of being in which one realizes one's uniquely human potential. According to Aristotle, the good life is the final goal of human action.

Aristotle (384–322 BC)

Virtue, then, is a state of character concerned with choice, lying in a mean, i.e. the mean relative to us, this being determined by a rational principle, and by that principle by which the man of practical wisdom would determine it. Now it is a mean between two vices, that which depends on excess and that which depends on defect; and again it is a mean because the vices respectively fall short of or exceed what is right in both passions and actions, while virtue both finds and chooses that which is intermediate. Hence in respect of its substance and the definition which states its essence virtue is a mean, with regard to what is best and right and extreme. (Aristotle, 1980 [350 BC])

Figure 3.6 Aristotle. Photo: © Argus/Fotolia.com.

Aristotle was born in Stageira, Macedonia in 384 BC. His father was the personal physician to the Macedonian King Amyntas II. As a result, he was sometimes referred to as the Stagerite. He came from a family of doctors, which probably explains his interest in physics and biology. In 367 BC Aristotle entered Plato's academy in Athens. He took lessons there for 20 years and taught there himself. Political circumstances made him leave Athens in 347 BC. He first moved to Assos (the north coast of Asia Minor) and then to Mitulene on the island of Lesbos. There he became fascinated with aquatic animals. Aristotle had a far greater interest in biological questions than his predecessors. He realized that the biology of humans could never be understood without studying the biology of lower animals. Up until the nineteenth century, Aristotle's research on water animals was unsurpassed in biological literature.

In 343 BC, Aristotle went to Pella in Macedonia to take up the duty of raising the 13-year-old Alexander the Great. Despite his election as head of the Academy in 339, he was only able to return to Athens in 334. Up to 323 BC he had his own philosophical school in the Lyceum, which was situated in the north-east of the city. It name, Peripatos, is taken from Aristotle's habit of teaching while he was walking, so that his pupils were often referred to as walkers (peripatol).

The news of the death of Alexander the Great in 323 BC resulted in a strong anti-Macedonian response in Athens, forcing Aristotle to flee "to stop the people of Athens committing a second atrocity against philosophy." Aristotle was referring to Socrates trial in 399 BC. A year later, in 322 BC, Aristotle died in Chalcis aged 63.

Aristotle is one of the most important founding figures in Western philosophy. He was the first to create a comprehensive system of Western philosophy, encompassing morality and aesthetics, logic and science, politics and metaphysics. For example, he is credited with the earliest study of formal logic, and his conception of it was the dominant form of Western logic until nineteenth-century advances in mathematical logic. His work *Ethica Nicomachea* is one of his most accessible texts. It is also the first systematic approach to ethics in Western philosophy. Though Christian Europe ignored him in favor of Plato until Thomas Aquinas reconciled Aristotle's work with Christian doctrine, this work was the origin of certain types of philosophical ethics: the so-called "happiness ethics," which was a dominant philosophy until the time of Immanuel Kant.

Each moral virtue (also referred to as a character virtue by Aristotle) holds a position of equilibrium according to Aristotle. A moral virtue is the middle course between two extremes of evil; courage is balanced between cowardice and recklessness for example, generosity between stinginess and being a spendthrift, and pride between subservience and arrogance. This is an expression of an old Greek notion: there is a certain ratio that is essential to humans that must be kept in balance and should not lean to the left or right if one wishes to achieve an optimal human state. A courageous person will not act as a coward in a dangerous situation, but he/she will also not be reckless and ignore the danger. According to Aristotle, moral virtues are not given to us at birth nor are they supernatural; they can be developed by deeds. In other words, they can be practiced just like all arts: "For the things we have to learn before we can do them, we learn by doing them, for example, men become builders by building and lyre players by playing the lyre; so too we become just by doing just acts, temperate by doing temperate acts, brave by doing brave acts" (Aristotle, 1980 [350 BC], 1130a).

People must seek a middle course, but this is not a simple matter. Aristotle believed that people know what they want instinctively, but not what they should do. Moreover, the middle course depends on the circumstances in a given situation. In other words, what is good in one case is not necessarily so in another. Unlike Plato (and, later, Kant), Aristotle argues that the good is sometimes ambiguous. However, people are not powerless in finding the middle course. The intellectual virtue sagacity or **practical wisdom** is aimed at making the right choices for action concerning what is good and useful for a successful life. According to Aristotle, a wise man can see what he has to do in the specific and often complex circumstances of life. Sagacity implies a capacity for moral judgment, which is the middle course. Moral virtues and the intellectual virtue go hand in hand.

Practical wisdom The intellectual virtue that enables one to make the right choice for action. It consists in the ability to choose the right mean between two vices.

The influence of Aristotle spread across Syria and through the Islamic world. From the thirteenth century on Aristotle's work started to influence Europe too, because the Christian philosopher Thomas Aquinas (1225–1274) reconciled the heathen virtue ethics with Christian doctrine. Thomas Aquinas distinguished seven virtues. These include the four cardinal virtues of prudence, temperance, justice, and fortitude. These virtues are natural and revealed in nature, and they are binding on everyone. There are also the three theological virtues of faith, hope, and charity. These are supernatural and are distinct from other virtues in their object, namely, God. From 1600 on virtue ethics was falling into oblivion because a new ethics was arising that was focused on rules and paid less attention to virtues. In recent years there is growing interest in the origins of virtue ethics; this is particularly due to the influence of the philosopher Alasdair MacIntyre.

3.9.2 Criticism of virtue ethics

William Frankena argues that virtue ethics is not essentially different from duty ethics (Frankena, 1973). According to him each virtue is accompanied by a moral rule for action and there is a virtue for each moral rule. However, it appears that not all obligations to act can be reduced to virtues and vice versa. Virtues characterize the person and provide insight into the background to action. A person's good character traits do raise expectations, but they do not provide a measure for judging an action. For example, the argument that the actions of an engineer are moral by definition because he is upstanding and reliable will not readily be accepted in a moral discussion. Moreover, it is hard to check whether the engineer acted with proper intentions. So, virtue ethics does not give concrete clues about how to act while solving a case, in contrast with utilitarianism and Kantian ethics. Opposite this we can argue that having the right virtues does facilitate responsible action, as will become evident in the LeMessurier case that is discussed later.

Finally, we can join Kant in wondering whether we can simply declare a moral virtue to be good in itself without any reservation. Kant's example for this is a cold psychopath whose virtues moderation of conscience and passion, self control and cool deliberation make him much more terrible than he would have been without those virtues.

3.9.3 Virtues for morally responsible engineers

Virtues as reliability, honesty, responsibility and solidarity, are quite general and most are virtues that morally responsible engineers need to possess too. If we look more specifically at the virtues engineers need, then we must focus on engineering practice. Michael Pritchard lists a number of virtues that are more specific than those mentioned above and that are required for morally responsible engineers (see box).

Virtues for Morally Responsible Engineers

- expertise/professionalism;
- clear and informative communication;
- cooperation;
- willingness to make compromises;
- objectivity;
- being open to criticism;
- stamina;
- creativity;
- striving for quality;
- having an eye for detail; and
- being in the habit of reporting on your work carefully. (Pritchard, 2001)

Stipulations in professional codes of conduct often refer to some of these virtues. The professional code of conduct of FEANI (Fédération Européenne d'Associations Nationales d'Ingénieurs or European Federation of National Engineering Associations) recognizes such virtues as integrity and impartiality. A list of virtues, however, does not say exactly how they are expressed in engineering practice, but the presence of certain virtues can have an important influence on the quality and ethical integrity of the work (see the LeMessurier case study).

Case LeMessurier

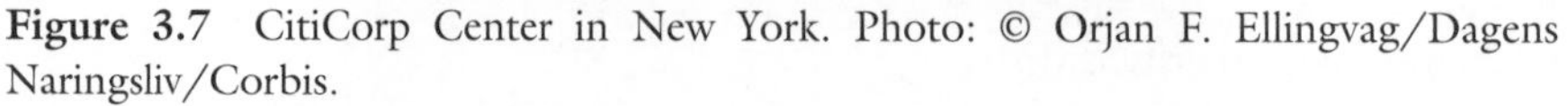

Figure 3.7 CitiCorp Center in New York. Photo: © Orjan F. Ellingvag/Dagens Naringsliv/Corbis.

William LeMessurier designed the Citicorp Center in Manhattan, which was built in 1977. It was an innovative design because the skyscraper had to be built on pillars nine floors high placed at the middle of the sides of the construction site. This unusual structure was designed to accommodate a church that was being resurrected under one corner of the Citicorp edifice.

During construction, the contractors decided to attach the supports with bolts rather than welding them due to the high costs of welding – this was done without LeMessurier being informed to avoid potential delays. The fact that this would result in much weaker connections was not viewed as a problem because the choice of using bolts was technically correct. In the original design welding was chosen because this would mean that there would be less movement in the skyscraper, which would improve the comfort of its future inhabitants. When LeMessurier heard about the alteration later, he did not worry about the safety risks because a connection using bolts met the safety requirements.

This changed when a month later, in 1978, he received a telephone call from a student whose professor had informed him that the construction was dangerous. The pillars should have been placed in the corners of the sites according to him. LeMessurier explained why the pillars were positioned in the middle of each side to allow the skyscraper to be better able to cope with storms than other standard constructions. Following this, he decided to deal with the technical aspects of his design and safety in his own lectures for building engineers.

Because LeMessurier took the remarks of the student and the professor seriously, he decided to carry out other wind-resistance tests beyond the standard ones. From this intellectual game, LeMessurier came to the conclusion that the building was not as safe as he had thought. A 16-year storm (one that passes every 16 years) could possibly rip loose one of the connections and the whole building could collapse. LeMessurier knew how to solve this problem however.

As soon as he made his finding, he informed the lawyers, insurance companies, the chief architect, the chief executive at Citicorp Center and the city hall. All parties (against all expectations) were highly cooperative and the corrections that LeMessurier advised were carried out. The building is now safer than the way it was originally planned when its construction began.

The media too had to be informed, because so much activity around a brand-new building could not go unnoticed. LeMessurier was highly dubious of being involved with the media, as the press could turn it into quite a story. After the news was published that the building was being altered to withstand more powerful storms, it received no further attention because the press happened to go on strike.

Source: Based on Morgenstern (1995).

For many engineers, LeMessurier's actions with regard to the Citicorp building exemplify the highest virtues of the engineering field. Nevertheless, many will wonder why LeMessurier deserves so much praise, since it was his professional duty to report mistakes to the authorities. Michael Pritchard, however, indicates that the way in which LeMessurier acted was exemplary – and therefore praiseworthy – for the following two reasons (Pritchard, 2001). First, much courage was needed to report the error, even though not reporting it would have been highly reprehensible. The report could have damaged his reputation considerably. Second, LeMessurier not only reported the problem, he also proposed a solution to it, which is characteristic of a virtuous engineer in Pritchard's opinion.

By taking seriously the objections of the student and his professor, LeMessurier also demonstrated another virtue: openness to criticism. Instead of ignoring the criticism because the construction met the safety requirements, LeMessurier decided to check everything and recalculate it. This demonstrates his dedication to safety of the general public. The case shows that virtues can direct the responsible actions engineers need to take in professional practice.

3.10 Care Ethics

Using utilitarianism and Kantian theory, we try to form a balanced moral judgment about the right way to act in a given situation. In these ethical theories an appeal is made to abstract and general principles, such as the utility principle and the reciprocity principle (or the universality principle). No attention is paid to the specific social context of the moral situation in question. These theories presuppose an independent and rational actor who makes decisions in a vacuum.

Care ethics An ethical theory that emphasizes the importance of relationships, and which holds that the development of morals does not come about by learning general moral principles.

Care ethics – initially inspired by the work of Carol Gilligan (Gilligan, 1982) – emphasizes that the development of morals does not come about by learning general moral principles. Its basis is that people learn norms and values within specific contexts and by encountering concrete people with emotions. By recognizing the vulnerability of the other and by placing yourself in his or her shoes to understand his or her emotions, you can learn what is good or bad at that particular time. Care ethics focuses attention on the living and experienced reality of people in which mutual relationships can be viewed from different perspectives, and where people's abilities and limitations impact moral decision-making.

3.10.1 The importance of relationships

Philosophers of care ethics argue that there is moral significance to the specific details of our lives, details that tend to be disregarded when we formulate "the good" in terms of general principles. We have seen that utilitarianism tends to ignore personal relationships, but that it does make a difference with whom we have a personal relationship. We have special relationships with our children, relatives, friends and colleagues. These relationships are coupled to special responsibilities and moral obligations. Moral problems are first and foremost understood in terms of responsibility of the individual with respect to the group. The solution of moral problems must always be focused on the maintenance of relationships the people have with each other. Besides people, companies can have special relationships too, such as the employees, suppliers, and people living close to a factory.

In care ethics the connectedness of people is key; the mutual responsibility and care for each other. People are connected to each other and through this connection there is attention for your fellow human being. People feel responsible for each other. Care arises from this involvement. Care encompasses all typically human activities that we carry out to maintain, continue, and repair our world, so that we can live in it as best as we can (cf. Tronto, 1993). Though care in this description is primarily described as an action, it is also important to look at care as a certain attitude or motivation.

Care ethics has some grounds in common with virtue ethics. Care ethics places the relationship central together with the acquired attitude of the person who can provide care. A proper attitude that one has acquired involves compassion, attention, and being caring. These virtues stimulate people to become emotionally involved and responsible. Moreover, both approaches are aimed at the good life; virtue ethics is based on good character traits and care ethics is based on care.

Care ethics specifically focuses the attention on the relationships that people have with each other. In relationships the recognition of vulnerability and dependence play an important role, especially if the relationships are asymmetrical, such as the relationship between parent and child, between employer and employee, or between doctor and patient. Thus, it is important to be conscious of the types of relationships we have and the role we have within them. Roles determine to what degree we can expect care from each other and they also determine whether we should take the other into account in our actions. Besides that, it is important to know how people respond to each other's vulnerabilities. The degree to which we respond appropriately and the way we shape our responsibility cannot be indicated in advance using rules, but has to be answered in the context in which the need for care arises.

3.10.2 Criticism of care ethics

A frequently-voiced criticism of care ethics is that it is philosophically vague. This is mainly due to the fact that it is unclear what "care" exactly entails. The term is used in numerous contexts and is also used to indicate more than one attitude or action. As a result it is not very normative. Care ethics assumes that caring is good in itself, thus it can tell us neither what makes a particular attitude or action right, nor what constitutes the right way to pursue them. Care ethics judges a situation by means of "good care" and not according to principles. But the question is what turns "care" into "good care"? Finally, care ethics like virtue ethics does not give concrete indications how one has to act in a particular situation, in contrast with utilitarianism or Kantian ethics. Care ethics is more focused on the attitude of the person who can provide care than on indications for ways to solve a concrete moral problem.

3.10.3 Care ethics in engineering

Although the attempt to develop a care ethics approach to engineering ethics is still in its infancy, care ethics' emphasis on care (e.g., for safety and sustainability) responsibility, and other concerns shared by engineers, suggests that is has a contribution to make to engineering practice. One of the possible applications is a social ethics of engineering (see box). The ideas of care ethics can also work for companies: they can contribute to the vision and mission of a company and can be a major influence on the practice of corporate social responsibility. One essential characteristic of a business situation is that one is working within an intersection of different relationships. A company has dealings with various parties and institutions, which have diverse and sometimes contradictory expectations. Employees have relationships with clients or contractors, with their employer, with consumers, with suppliers and sometimes even with the natural environment. The point is that an employer or employee has to ask himself/herself how he/she as part of the enterprise can best deal with the interests and rights of others. This has to be achieved through an attitude of compassion, attention and care.

Social ethics of engineering An approach to the ethics of engineering that focuses on the social arrangements in engineering rather than on individual decisions. If these social arrangements meet certain procedural norms the resulting decisions are considered acceptable.

Social Ethics of Engineering

Most of the approaches to ethics in engineering focus on the individual. Such approaches tend to neglect the relationships with others in which engineers enter in their work and that are morally relevant. A social ethics approach would pay more attention to such relationships and would inquire into the social arrangements in which engineering decisions are made. Relevant social arrangements include for example the way a design team or the engineering company is organized and the way that relations with stakeholders are structured. In the case of the Challenger disaster discussed in Chapter 1 it is, for example, striking that the crew was not involved in the discussion and was unaware of the launch debate the night before.

Richard Devon has proposed a number of norms of engagement for the participation of engineers in group processes, involving both engineers and non-engineers. These norms are:

- competency;
- cognizanze, requiring interdisciplinary skills and breadth built into the group;
- democratic information flows;
- democratic teams;
- service-orientation;
- diversity;
- cooperativeness;
- creativity; and
- project management skills

Some of these norms of engagement are rather similar to the virtues for morally responsible engineers mentioned by Michael Pritchard (see 3.9.3). The main difference is that whereas virtues are usually seen as individual character traits, these norms are understood at the level of group processes and social arrangements.

A social ethics approach emphasizes procedural criteria for dealing with moral problems in a group process rather than substantial moral norms that are to be applied by individuals. It leaves open the possibility "that the individual may be unhappy with the outcome but be able to accept it because the process was perceived as the most acceptable way for a group" (Devon, 1999, p. 91) to reach a decision.

Source: Based on Devon (1999), Devon (2004), and Devon and Van de Poel (2004).

In the case of child labor, in Section 3.7, the following reasoning would be applied from the perspective of care ethics. Since the children are involved in the production of IKEA articles, IKEA has a relationship with those children. The children are extremely vulnerable and dependent on IKEA: if IKEA stops with child labor then the consequences would be that the children would end up performing slave labor or becoming prostitutes. IKEA has the responsibility to care for these children. In this specific case this care could involve improving work conditions, providing medical care, and developing a schooling program. In this example of an application of care ethics, we see that the question whether child labor should be abolished fades into the background and that care ethics makes IKEA's involvement with the children the central issue. Another example is the care for employees by the employer in cases of mass unemployment, or mergers, takeovers, down-sizing or relocation of the enterprise. The employer has a relationship with the employees and from this relationship follow certain obligations of care between the employer and the employees. This care can consist of active involvement in transfer of employees within the company or finding places of work outside the company. From these two examples it becomes clear that a care ethics approach places high demands on an enterprise concerning responsible entrepreneurship.

In the Ford Pinto case, Ford had a relationship with the consumers. This relationship was asymmetrical since the consumers had in general no clear idea of all the relevant technical aspects of the Ford Pinto, and the consumers were dependent on the information Ford gave. Ford should recognize this vulnerability of the consumer, and therefore Ford had the responsibility to inform the consumer about the (un)safety of the car, or Ford should not have marketed the car.

3.11 Applied Ethics

Some philosophers believe that applied ethics is essentially the application of general moral principles or theories to particular situations (cf. Gert, 1984; Hare, 1988; and Smart, 1973). This view is, however, problematic for a number of reasons (cf. Beauchamp, 1984; and MacIntyre, 1984b). One is that no moral theory is generally accepted. Different theories might yield different judgments about a particular case. But even if there were one generally accepted theory, framework, or set of principles, it is doubtful whether it could be straightforwardly applied to particular cases. Take a principle such as distributive justice. In many concrete situations, it is not clear what distributive justice exactly amounts to. What does, for example, a just distribution of technological risks mean? Should everybody be equally safe?; should everybody have the same minimum level of safety?; or does someone's right to safety depend on the amount of taxes he or she pays? All these can be considered as an application of the principle of distributive justice to the distribution of risks, but clearly these answers reveal different moral outlooks. Without doubt, part of this confusion could be solved on the theoretical level, that is, by further elaborating the notion "distributive justice" and developing an ethical theory about it. It seems doubtful, however, whether this would solve all applications issues. This brings us to a third point. Theory development in ethics in general does not take place independent of particular cases. Rather, theory development is an attempt to systematize judgments over particular cases and to provide a rational justification for these judgments. So if we encounter a new case, we can of course try to apply the ethical theory we have developed until then to that case, but we should also be open to the possibility that the new case might sometimes reveal a flaw in the theory we have developed so far.

If ethical theories do not provide moral principles that can be straightforwardly applied to get the right answer, what then is their role, if any, in applied ethics? Their role is, first, instrumental in discovering the ethical aspects of a problem or situation. Different ethical theories stress different aspects of a situation; consequentialism for example draws attention to how consequences of actions may be morally relevant; deontological theories might draw attention to the moral importance of promises, rights and obligations. And virtue ethics may remind us that certain character traits can be morally relevant. Ethical theories also suggest certain arguments or reasons that can play a role in moral judgments. These arguments should be sound, since unsound arguments obstruct a rational discussion. Therefore, normative argumentation is the topic of the next chapter, which can help to distinguish good arguments from bad ones in ethical judgment.

3.12 Chapter Summary

While morality is the totality of opinions about what is good and right, ethics is the critical reflection on morality. Normative ethics not just describes what morality is but it judges morality and tries to formulate answers to questions like: what kind of person should I be?, and how should I act? Normative ethics, therefore, tries to come to certain normative judgments. However, it is not a manual or an unambiguous code in which you can look up the answer how to act in a difficult situation. Rather it is an area that is characterized by a variety of partly conflicting ethical theories about how to act. The three best known ethical theories in Western philosophy are consequentialism, deontology, and virtue ethics. Whereas virtue ethics focuses on the acting person and his/her character traits, deontology focuses on the actions themselves and consequentialism focuses on the consequences of actions.

Utilitarianism is a main variety of consequentialism. It measures consequences by their effect on one value: pleasure or human happiness. It is based on the so-called utility principle: the greatest happiness for the greatest number. Utilitarianism requires drawing up a moral balance sheet or a cost-benefit analysis to determine what the action with the best consequences is. Despite its intuitive attractiveness, utilitarianism has been heavily criticized. Typically, these criticisms have let to adaptations in the original theory. One criticism is that utilitarianism can lead to exploitation. To deal with this problem, John Stuart Mill has formulated the freedom principle: everyone is free to strive for his/her own pleasure, as long as they do not deny or hinder the pleasure of others. Another criticism is that actions are sometimes right or wrong independent of their consequences. Lying is an example. Rule utilitarianism is an attempt to deal with this criticism: it focuses on the utility of rules of action rather than on the utility of individual acts. Other criticisms of utilitarianism are that happiness is difficult to measure, that consequences are hard to predict, that it ignores the distribution of pleasures and pains and that it ignores personal relationships.

Immanuel Kant is the main representative of deontology. He formulated a principle for judging the rightness of actions that is independent of the actual consequences of those actions, the universality principle: Act only on that maxim which you can at the same time will that it should become a universal law. According to Kant this principle is basically the same as his reciprocity principle: Act as to treat humanity, whether in your own person or in that of any other, in every case as an end, never as means only. Two main criticism of Kant's theory are that it ignores conflicts between norms and that it is too rigid. A way of dealing with such criticisms may be to conceive of norms as prima facie norms rather than as universal norms that apply to each and every situation.

Virtue ethics focuses on the character of the acting person rather than his/her actions or the consequences of those actions. It goes back to Aristotle but is still relevant today for engineers. Relevant virtues for engineers include professionalism, objectivity, being open to criticism, stamina, creativity, and having an eye for detail. The main criticisms of virtue ethics include that it is does not tell you how to act and that virtues are not unconditionally good. To the first, virtue ethicists might reply that an engineer who possesses the right virtues acts differently from one who is lacking them. To the second, they would probably say that virtue ethics does not involve just isolated virtues but also practical wisdom: the ability to make ethical judgments in complex situations.

A fourth theory that was briefly discussed is care ethics: it does not focus on abstract principles but rather on the relations between people. This is relevant for engineers because they are often involved in complex projects with many stakeholders who partly rely on them. One way to apply care ethics to engineering is to look for norms that the social arrangements in engineering should meet to express due care to all relevant stakeholders.

As the diversity of ethical theories testifies there is not a single answer to the question what is right or wrong. However, one should not conclude from this that anything goes. Some things are morally good or morally bad according to all theories. Moreover, if theories disagree, they are still helpful in distinguishing the ethical questions in a concrete situation, in more precisely analyzing the situation and in suggesting possible reasons and arguments for acting in one way rather than the other. Even if using one or more of the theories is no guarantee for making the right decision, a moral decision that just ignores the ethical theories, and the underlying ethical concerns, is usually plainly unethical.

Study Questions

1 Mention a number of differences between values and norms, and between values and virtues.
2 Which of the following statements are descriptive, and which are normative?
   a. People should accept the risks of nuclear energy.
   b. The majority of your colleagues finds this proposal unacceptable.
   c. There is life on Mars
   d. Engineers who blow the whistle are usually in a weak position from a legal point of view.
3 Describe the main ideas of "normative relativism." What are the criticisms of normative relativism?
4 John Stuart Mill has argued that Kant's ethics is really a masked version of consequentialism because the consequences of actions do play an important role in his ethics – in spite of what Kant himself says. Why do you think that Mill is arguing this despite the fact that Kant himself denies that consequences are relevant to his theory? Do you agree with Mill? Explain why or why not.
5 Describe the main ideas of Bentham's utilitarianism. What criticism of Bentham's theory did Mill articulate?
6 What is rule utilitarianism? Describe how a rule utilitarian would go about determining whether I may "copy and paste" my essay from the Internet for a course. How would this differ from how an act utilitarian would reach a decision on this matter?
7 Describe the main idea of care ethics. What criticism do care ethicists have of utilitarianism and Kantian theory?
8 An engineer helped a colleague with her work; and this colleague happened to be rather pretty. The engineer thinks "Why not have an affair with her?" And since she is Kantian, the

engineer argues as follows: "If you really want to express respect for me, then you should join me for a drink in my room. Otherwise you would have treated me only as a means and – as a good Kantian – that is not something you can possibly want!"

a. What is it to treat someone only as a *means* according to Kant?
b. What is it to treat someone as an *end* according to Kant?
c. What is her answer if she is a good Kantian?

9 Suppose that it is possible to download copyrighted music through the Internet without paying for it. Suppose that the makers of such music (the artists) do not want their music freely copied in this way.

a. How would Kant address the question of whether it is morally permissible to download such music by such artists without paying for it? Explain in detail what a Kantian would say.
b. Would a utilitarian give a different answer? Why or why not?

10 James is an engineer working for the company AERO that produces aero-engines. The company is developing a new type of aero-engine called the FANX. James is responsible for the testing of the FANX. He is in the middle of conducting a range of crucial tests for the reliability of the new aero-engine. Yesterday, Bill – who is James' boss - has asked James to finish his test reports within a week because an important potential customer will visit AERO next week and wants to have a look at the first test reports. James first reaction is to refuse Bill's request: he is not able to finish the test report within a week; he first needs to do more tests. James considers these additional tests crucial for gaining good insight in the reliability of the FANX. Bill tells James to abandon the planned other tests and to start writing his report immediately. Later, there will be more time to do the other tests. Bill also tells James that if James refuses he will ask Eric to write the report. James says that he really needs more time. Moreover, he objects, Eric is not knowledgeable of the tests and will not be able to write a sound report. After the meeting, James contacts Eric who says that he agrees with Bill and that he will finish the test reports if Bill asks him to do so.

Suppose that James the next day decides to follow Bill's order and to finish the reports immediately, abandoning the other tests.

a. Can this choice of James be justified in utilitarian terms? Explain why or why not.
b. What should James do if he would try to apply Kant's categorical imperative to this situation? Argue your answer.
c. What virtues are relevant for an engineer doing tests like James? Mention at least four.
d. What action is supported by these virtues? Argue your answer.
e. Which normative theory is in your opinion best able to deal with this moral problem? Argue why.

Discussion Questions

1 Are there any absolute rules that should never be broken, whatever the circumstances? Defend your view.
2 Choose an event in your life where you believe you acted ethically. Discuss the event in terms of virtue ethics, Kantian ethics, and utilitarianism.
3 What makes a decision an ethical one?
4 How much should we take potential consequences into account when making an ethical choice? How much work should we put into making sure that the assessment of outcomes is correct?
5 Should we always do what is morally best? Is there a difference between morally decent and heroic behavior?

4

Normative Argumentation

Having read this chapter and completed its associated questions, readers should be able to:

- Describe sound arguments, and distinguish them from valid arguments;
- Describe the argumentation schemes and their associated critical questions: argumentation by analogy, means-end argumentation, causality argumentation, proof from the absurd, and characteristic-judgment argumentation;
- Identify the underlying argumentation schemes of the different ethical theories (utilitarianism, Kantian theory, and virtue ethics) for the assessment whether an action is acceptable;
- Evaluate the soundness of arguments (and identify fallacies) used in moral discussions on the acceptability of technological risks.

Contents

Ethics, Technology, and Engineering: An Introduction, First Edition.
Ibo van de Poel and Lambèr Royakkers.

4.1 Introduction

Case The Pros and Cons of the Golden Gate Bridge Suicide Barrier

Figure 4.1 Golden Gate Bridge in San Francisco. Photo: © Mike Norton / Fotolia.com.

When the Golden Gate Bridge in San Francisco was finished in 1937 it was – with its span of 1280 m (4200 ft) – the world's longest suspension bridge. It is considered to be one of the best and most beautiful examples of bridge design. The American Society of Civil Engineers has included the Golden Gate Bridge in its enumeration of the seven wonders of the modern world. However, the bridge is also the US's most popular place to commit suicide. Since 1937, at least 1300 people killed themselves by jumping off the bridge; an average of 20 to 25 per year.[1] Suicide prevention has in fact been a concern ever since the bridge was designed. Joseph Strauss, the chief engineer is quoted as saying in 1936 (a year before the opening of the bridge): "The guard rails are five feet and six inches high [i.e. about 1.7 m] and are so constructed that any persons on the pedestrian walk could not get a handhold to climb over them. The intricate telephone and patrol systems will operate so efficiently that anyone acting suspiciously would be immediately surrounded. Suicide from the bridge is neither possible nor probable." However, for some reason, the height of the railing was reduced in the detail design to 1.2 meter (4 feet), so making it not too difficult to climb the railing and jump off the bridge.

Already in 1940, the Board of Directors discussed an "anti-suicide screen" but decided against it on the basis of aesthetic and financial considerations. Another concern was that the screen might create dangerous wind resistance and make the bridge structurally unstable. Since then, proposals for a suicide barrier has been made every decade but were unsuccessful until recently. Below, we will consider some of the main arguments of the opponents (those who are against the installation of the barrier) and the proponents (those who are in favor of installation of the barrier).

Aesthetic argument

Opponents: The Golden Gate Bridge has been praised for its transparency and openness. Any barrier design would destroy the view. Architect Jeffrey Heller, for example, stated: "When you look straight out, you'll see through all this mesh, which will be sad enough, but looking straight down the roadway, it will become a cage. ... That is far too high a price for our society to pay."

Proponents: Aesthetic considerations can be accounted for in the design of the barrier.

Effectiveness argument

Opponents: a suicide barrier will not be effective, since people who want to kill themselves will simply go somewhere else.

Proponents: most suicidal people act on an impulse and, when prevented from actually killing themselves, they often seek help and loose the desire to die. According to a study by Dr Richard Seiden of UC Berkeley, the hypothesis that Golden Gate Bridge attempters will "just go someplace else" is unsupported by the data. He studied 515 people who, from 1937 to 1971, were prevented from jumping from the bridge, and found that only 6 percent went on to kill themselves.

Economic argument

Opponents: the barrier is not worth the costs. Even if the barrier would save 20 to 25 lives a year, it is not worth 50 million building costs plus the annual operation and maintenance costs.

Proponents: A human life cannot be expressed in monetary terms. Moreover, a cost-benefit analysis wrongly anonymizes the victims: "You may not think that 24 lives a year is worth spending $50 million ... well that is until it is your daughter, son, sister, brother, friend, husband, wife etc ..."

Autonomy argument

Opponent: one should not interfere with people's freedom to commit suicide: "If I want to jump off the bridge, I don't think it is anyone's job to stop me."

Proponent: the decision to kill oneself is usually made in an impulse. As one commenter puts it: "I do believe that people have free will and that society should not interfere with their free will. But … these people are mentally ill, and their illness can be treated. But, they cannot be treated if we let them jump off the bridge."

Responsibility argument

Opponent: it is not the responsibility of society or the directors of the bridge to try to prevent suicides. Moreover, it is wrong to use architecture to solve a social problem. In the words of Jeffrey Heller: "You can't correct all of the sadness and evil in the world."

Proponents: it is the responsibility of society and the bridge directors to act. As Jerome Motto, a past president of American Association of Suicidology expressed it: "If an instrument that's being used to bring about tragic deaths is under your control, you are morally compelled to prevent its misuse. A suicide barrier is a moral imperative." According to Motto, "It's not about whether the suicide statistics would change, or the cost, or whether [it] … would be as beautiful … A barrier would say, 'Society is speaking, and we care about your life.'"

On October 10, 2008 the Board of Directors voted 14 to 1 to install a stainless steel net which would be placed 6 m (20 ft) below the deck, and would collapse around anyone who jumped into it. The estimated costs are 50 million dollars. It is expected that construction of the net will take several years and will not start before a number of additional studies has been carried out.

Source: Mainly based on www.sfgate.com/lethalbeauty/ (accessed June 29, 2009). The quotations are from www.sfgate.com/cgi-bin/article.cgi?f=/c/ a/2005/11/03/MNGF1FCP631.DTL and www.mentalhelp.net/poc/view_doc.php?type=weblog&id=456&wlid=6&cn=9 Unless indicated otherwise, the (professional) background of the commentators is not indicated on the website.

As we can see from these examples, the purpose of argumentation is first and foremost to justify or refute a statement. Argumentation is an activity which can be directed towards defending an opinion (i.e., an attempted justification) or attacking an opinion (i.e., an attempted refutation). In their own ways, both activities contribute to the critical function of argumentation. In this chapter we will discuss some aspects of argumentation theory. In particular we will examine how you can justify or refute an opinion.

Ethical theories are clearly argumentative in their configuration. It has often been said that we need these theories to *justify* our moral judgments. Although, the necessity of argumentation in ethics is not debated, most literature on ethics pays no attention to formal argumentation. Our main purpose in this chapter is to give a picture of how the determination whether acts are right or wrong takes place in ethical theories on the basis of the typology of sorts of argumentation. We will restrict ourselves to the three main theories: utilitarianism, the theory of Kant, and virtue ethics (see previous chapter). We start from the normative standpoint that "action A is morally acceptable." We shall see that different argumentation schemes found in the argumentation literature (Van Eemeren and Grootendorst, 1992; Kienpointer, 1992; Schellens, 1985) are used in the ethical theories. To check whether a particular piece of argumentation is sound, it is of great importance to recognize the argumentation scheme, because certain critical questions belong with certain schemes. By means of these questions, we can see whether the argumentation can bear the scrutiny of criticism.

In Section 4.2 we will clarify some basic terms in **argumentation theory**, and go into the question when an argument is valid. The distinction between deductive and non-deductive arguments will be discussed in Section 4.3. In Section 4.4, we consider the assessment of arguments that are used in the three ethical theories: Utilitarianism; Kant's ethics; and virtue ethics; and we shall give more details on a frequently occurring argumentation scheme: the argumentation by analogy. In the last section, we discuss some fallacies, and especially fallacies with respect to the acceptability of technological risks.

Argumentation theory An interdisciplinary study of analyzing and evaluating arguments.

4.2 Valid Arguments

Before we go into the question when an argument is valid, it will be helpful to clarify some of the terms with respect to argumentation theory, that is, the study of how humans can, do, and should reach conclusions based on premises. The first term we define is *argument*, a fundamental logical concept.

> An **argument** is a set of statements, of which one (the conclusion) is claimed to follow from the others (the premises).

Argument A set of statements, of which one (the conclusion) is claimed to follow from the others (the premises).

So, the **conclusion of an argument** is the statement that is affirmed on the basis of the other statements of the argument, and these other statements, which are affirmed (or assumed) as providing support or reasons for accepting the conclusion, are the **premises** of that argument. A statement is a descriptive or normative judgment (see Chapter 3). Questions (e.g., What time is it?), orders (e.g., Stay!), and exclamations (e.g., GOAL!) are non-statements.

Conclusion of an argument The statement that is affirmed on the basis of the premises of the argument.

Premises The statements, which are affirmed (or assumed) as providing support or reasons for accepting the conclusion.

In general an argument can formally be expressed as follows:

$P_1, P_2, \ldots, P_n$, so C

where $P_1, P_2, \ldots, P_n$ are the premises and C is the conclusion. The number of premises varies, but in practice the number will be small.

Arguments can be judged on their effectiveness: To what extent is an argument useful to reach the intended result? In deciding this we can distinguish two extreme analyses:

1 the *rhetorical* analysis: is the argument persuasive for the audience?, to be proved right is more important than to be right.
2 the *logical* analysis: is the argument valid?

From a viewpoint of rational thinking, the primary purpose of an argument is to provide evidence for a conclusion, and that is why we leave out the rhetorical analysis.

Logic, as a discipline, is concerned with argumentation. Its concern is to distinguish a good argument from bad one, or better from worse. With that, the aim is to investigate, develop, and systematize principles and methods that can be used to distinguish valid and invalid arguments. In principle this distinction is not difficult. With **valid arguments**, the conclusion of the argument follows with necessity from its premises. And with invalid arguments, the premises do not entail its conclusion.

Valid argument An argument whose conclusion follows with necessity from its premises: if the premises are true, the conclusion must be true.

P_1: If it rains, the streets become wet.
P_2: It rains
C: The streets become wet.

A statement composed of two constituent statements and the connective "if .., then …" is called a *conditional*. The component statement that precedes "then" is called the antecedent, and the component following "then" is called the consequent. Premise P_1 is a conditional, where "it rains" is the antecedent, and "the streets become wet" the consequent.

For this argument it is inconceivable that the premises are true and the conclusion is false, because if the conclusion was false, then it does not rain (or premise P_1 is false). The argument forms an instance of the valid argument form **modus ponens**:

Modus ponens Form of a valid argument in which the conclusion "q" follows from the premises "p" and "if p then q."

If p, then q
p
So, q

p and q stand for any statement. An argument is valid when its premises, if true, do provide conclusive grounds for the truth of its conclusion. So, the premises and conclusion are so related that it is impossible for the premises to be true unless its conclusion is true also. Indeed, the definition of a valid argument is defined in terms of truth and falsehood, but this does not say anything about the actual truth of the premises. The premises are hypothetical: *Suppose* the premises are true, then the conclusion must be true in a valid argument.

The following argument, often confused with an instance of the *modus ponens*, is the *fallacy of affirming the consequent*:

P'_1: If it rains, I will stay at home
P'_2: I will stay at home
C': It rains

An error or deficiency in an argument is referred to as a **fallacy** (or specious argument). The argument above is invalid. That the premises are true does not guarantee the truth of the conclusion. For example, I will stay at home since I want to watch my favorite television program. In this case, both the premises are still true, but the conclusion need not be true, which is in conflict with our definition of a valid argument. The counter-example shows that the truth of the premises of this argument does not guarantee the truth of the conclusion. Using counter-examples, as done here, is a powerful method to show the invalidity of arguments.

Fallacy An error or deficiency in an argument.

Another well known valid argument form is **modus tollens**. An instance of this form is:

Modus tollens Form of a valid argument in which the conclusion "not-p" follows from the premises "if p then q" and "not-q."

P''_1: If it rains, I will stay at home
P''_2: I will not stay at home
C'': It is not raining.

The first premises of *modus tollens* is a conditional ("if it rains, then I will stay at home"), while the second ("I will not stay at home") denies the consequent of the first premise. The conclusion ("it is not raining") denies the antecedent, that is, the "if" clause of the conditional premise P''_1: "it rains". A *fallacy of denying the antecedent* occurs when one should conclude "I will not stay at home" from the premises "If it rains, I will stay at home" and "It is not raining."

As is usual in argumentation a conclusion can be challenged in two ways. We can show that a premise is false, or we can show that the argument is invalid and that the conclusion is premature. With respect to the last method we have seen an example: the fallacy of affirming the consequent. The conclusion "It rains" can be false, if the premises "If it rains, I will stay at home" and "I will stay at home" both are true. The conclusion of the first argument (an instance of the *modus ponens*) can only be challenged by showing that a premise is not true. To test the truth or falsehood of premises is the task of common sense or science in general, since premises may deal with any subject matter at all. The logician is not so interested in the truth or falsehood of statements or the content of the statements, but is interested in the determination of the validity of arguments (even of arguments whose premises might be false). The possibilities to challenge a conclusion are shown in Table 4.1.

For example, in the Golden Gate Bridge suicide barrier case the following argument could be produced from the autonomy argument:

P_1: Someone wants to commit suicide
P_2: If someone wants to commit suicide, then nobody has the right to stop him/her
C: Nobody has the right to stop a person (who wants to commit suicide).

Table 4.1 The possibilities of challenging a conclusion

	All premises are true	*Some premises are false*
The argument is valid	Conclusion cannot be challenged	Conclusion can be challenged
The argument is invalid	Conclusion can be challenged	Conclusion can be challenged

This argument is valid. So if we want to challenge the conclusion, we can only do this by showing that a premise in not true. In this example, the proponents dispute the second premise by stating that suicidal people are mentally ill, and that these people should be treated. And these people cannot be treated if we (the society) let them commit suicide.

In actual practice, one might see that some premises or the conclusion has not been explicitly stated. Because there is some common knowledge or background agreement between the presenter and the receiver of the argument, which must be assumed as part of the context of the argument. Consider the argument: "I'm not going to classes today, because I'm ill." Formally, we can express this as follows:

P_1 I'm ill
C I'm not going to classes today.

The premise P_1 does not supply adequate reasons for accepting the conclusion. We need to add a premise to guarantee the truth of the conclusion. We can reasonably supply for the *missing* or *unstated premise* (by relevant common knowledge) the statement: "If I'm ill, then I'm not going to classes", which makes the argument valid. So, if we reconstruct an argument for determining whether the argument is valid, we have to add all the missing premises or conclusion. It can happen, however, that a receiver supplies another premise for the unstated premise than the presenter, which can lead to misunderstandings.

4.3 Deductive and Non-Deductive Arguments

Valid arguments are of a *deductive* nature, that is, the conclusion is enclosed in the premises: the result, the conclusion, says no more and is not logically stronger than the totality of premises that the argument is based on. Valid arguments therefore are *monotonic*, that is, adding new premises cannot change anything in the logical validity of a conclusion that is drawn.

Many arguments from daily practice are not constructed deductively at all, since we often change our conclusions when new information is added. The conclusion that John inherits the money of his wife, from the premises "if John's wife dies, John will inherit her money (and nothing else is known)" and "John's wife dies" will change if we add the information that John has killed his wife. In these ***non-deductive arguments*** (also known as *non-monotonic arguments*), the conclusion is logically stronger than the premises. In other words, the premises – if true – give a limited amount of support to the conclusion. In such situations it is always the case

Deductive argument An argument which has a conclusion that is enclosed in (implied by) the premises.

that the truth of the premises does not guarantee the truth of the conclusion. Adding new premises can strengthen or weaken the conclusion. Accepting the conclusion is based on considerations that make use of the **plausibility principle** for non-deductive argumentation. For example, we must assume that John ate the cake between 8.00 and 9.00 a.m. because the cake disappeared in the mean time and he was the only person to have been in the room, animals had no access to the cake tray, and all other explanations for the cake's disappearance can be excluded. The sum of alternative explanations to be excluded must be continued until they lead to a plausible assumption, that is, that John must indeed have eaten the cake. In this non-deductive argumentation an attempt is made to be convincing through enumeration and supplementary argumentation, where all imaginable doubts have no base and the conclusion must be accepted, so that in this case John must have eaten the cake. A small degree of uncertainty remains in this argumentation due to its indirect nature (nobody actually saw John eating the cake).

Plausibility principle The principle that enumeration and supplementary argumentation in a non-deductive argumentation can make the conclusion plausible (acceptable).

A frequently occurring form of non-deductive argumentation in science is called **inductive argumentation**: argumentation from the particular to the general. Attempts are made to support the truth of an empirical law by means of measurement results. A standard example is Boyle's law of gas: pressure times volume is constant ($p \times V = c$). On the basis of a *limited* number of experiments, Boyle drew the conclusion that for *all* values of pressure and volume the product had to be constant. For this case we can check the degree of plausibility of this conclusion using the following **critical questions**:

Inductive argumentation A type of non-deductive argumentation. Argumentation from the particular to the general.

Critical questions Questions belonging to a certain type of non-deductive argumentation to check the degree of plausibility of a conclusion.

1 Were the experiments carried out relevant for the conclusion?
2 Were sufficient experiments carried out to support the conclusion?
3 Are there no counterexamples?

If all these questions can be answered positively, we can speak of a **sound argumentation** and not of a valid argumentation. It is only in the case of deductive argumentation that we can speak of valid argumentation. Due to the indirect nature of non-deductive argumentation, there always is a small degree of uncertainty: it cannot be excluded that the next measurement of a gas will not meet Boyle's law. Deductive argumentation completely excludes any possible doubt. They are direct (without enumerations and supplementary argumentation); if the premises are true, it is impossible for the conclusion to be untrue.

Sound argumentation An argumentation for which the corresponding critical questions can be answered positively and which therefore makes the conclusion plausible if the premises are true.

A fallacy related to inductive argumentation is the *rash generalization*. This occurs if the generalization is not sound, because too few observations were used as a basis for the general statement for example. Using particular argumentation schemes in the ethical arguments in the next section, we shall come across several fallacies.

4.4 Arguments in Ethical Theories

In this section, we consider the assessment of arguments that are used in three ethical theories: Utilitarianism; Kant's ethics; and virtue ethics, on the basis of the underlying argumentation schemes of the pragma-dialectical approach (Van Eemeren and Grootendorst, 1992; Kienpointer, 1992; Schellens, 1985). Before this we will discuss a frequently used argumentation scheme in ethical discourse, argumentation by analogy, which is often used to fill policy or moral vacuums, especially, surrounding modern technology.

4.4.1 Argumentation by analogy

Argumentation by analogy A type of non-deductive argumentation. An argumentation based on comparison with another situation in which the judgment is clear. The judgment is supposed also to apply to the analogous situation.

In an **argumentation by analogy** we try to reach a moral assessment of an unknown or new situation by basing that assessment on a situation in which the moral assessment is clear and the comparison is sufficient. Regarding moral questions related to modern technology, there often is a search for comparable traditional cases that have already been morally assessed. One example of this was the discussion in the early 1990s on hacking – the attempts to gain access to computer networks that belong to others. A number of hackers felt their behavior was morally acceptable, because their only purpose was to help system managers to trace errors in their systems. However, an opponent can make use of the following argumentation by analogy to negate the standpoint: "You do not go to a clothing store and set fire to the clothing there to see whether fire safety procedures are in place." This is based on the concealed premise that setting fire to clothing to test fire safety is morally unacceptable. Analogously, hacking is unacceptable too.

The argumentation by analogy has a non-deductive nature, because we draw a conclusion on the basis of comparison. We can never state with certainty that the same has to apply in a *comparable* situation (obviously something different from the *same* situation) as is the case in the selected example situation. Using critical questions, mentioned below, we can determine how plausible the assumption is and whether or not the argumentation is sound.

Argumentation by Analogy

If something is/was the case in an example case, then it also holds true for a comparable situation.

Formal description:

- Situation q is comparable with situation p (*the analogy premise*).
- If situation p occurs then r applies.
- So, if situation q occurs then r applies.

Critical questions:

1. Are the two situations comparable?
   - Are there important relevant similarities?
   - Are there no important relevant differences?
2. Is what is asserted about the example situation true? In other words, is it true that "if situation p occurs then r applies?"

If the argumentation by analogy is to be sound, all critical questions have to be answered positively. In the example given earlier, that is, hacking, the *proponent* (who is defending the standpoint) has to answer the second critical question positively, that is, the assertion that it is morally unacceptable to set fire to clothing in a clothing store to test fire safety must be accepted as true. In fact that there are important similarities is clear too: in both cases the aim is to test the safety of a certain system. However, the question whether there are no important and relevant differences is problematic. There is an important (significant) difference, because in the case of hacking – in contrast with cracking – no damage is caused and in the example situation there is damage. This difference is highly relevant, because causing damage to clothing is relevant for the moral acceptability while there is no material damage in hacking. The analogy therefore fails. We then speak of a fallacy: *false analogy.*

For a better analogy, the situation of hacking could be compared to unlawful entry: the deliberate entry of someone's home against his/her will. The important difference is now removed, because the causing of damage has been removed as an issue. People find it morally unacceptable for someone to enter their home without permission, irrespective of whether the intruder causes damage. We see this analogy being used in the Dutch Computer Criminality Act of 1993, in which hacking is viewed as a modern form of unlawful entry. This is also reflected in the term *unlawful computer entry* and its description: "the deliberate entering of a computer system or part of it against the will of the owner or administrator."

4.4.2 Arguments in a utilitarian plea

The assumption of utilitarianism to determine whether an action is acceptable, is as follows:

> An action is morally acceptable if and only if that action can be reasonably expected to produce the greatest happiness for the greatest number of people.

In a utilitarian plea, the **means-end argumentation** is at the forefront. From a given end the means are derived to realize that end. This argumentation is non-deductive,

Means-end argumentation A type of non-deductive argumentation. An argumentation in which from a given end the means are derived to realize that end.

as the conclusion is not embedded in the premises. There may be other means to realize the end, implying that the choice for means y need not be made (see box). So when we apply the means-end argumentation, we must make a reasonable case for the choice of means y and not for some other means that could also realize the end.

Means-End Argumentation

If you wish to achieve end x, then you must carry out action y.

Formal:

- x (the end)
- carrying out action y (the means) realizes the end x (*the means-end premise*)
- So: do y

Critical questions:

1 Does action y indeed realize end x?
2 Can action y be carried out?
3 Does execution of action y lead to unacceptable side effects?
4 Are there no other (better) actions to achieve x?
5 Is the end acceptable?

The end in a utilitarian argument is well-known: the greatest happiness for the greatest number; the means to this is an action. The utilitarian end is very abstract; depending on the consequences of the possible actions, the end is made more concrete. The action with the best consequences leads to the most happiness for the largest number of people. In utilitarianism, the end is formulated such that the fifth critical question is not relevant: the utilitarian end is worth striving for by definition, at least according to people adhering to utilitarianism. Depending on the context in which such an argumentation is used, the critical questions are of greater or lesser relevance. The most relevant critical question in a utilitarian argumentation is the first one: Does action y actually realize objective x? Two matters are of importance to answer this question. First, we must demonstrate that the action leads to the expected consequence, and second that the consequence is the best one, that is, "the greatest happiness for the greatest number." For the former, **causality argumentation** is of importance, because an expected consequence is derived from a certain action. In other words, there is a causal link between the action and the expected consequence. In the box, we give two formal descriptions of the causality argumentation. The first type is the non-deductive variant of this argumentation: the consequence (the conclusion) is not embedded within the premise (the action). This

Causality argumentation A type of non-deductive argumentation. An argumentation in which an expected consequence is derived from certain actions.

contrasts with the second type: the deductive type of causality argumentation. "If p is true and if it is true that p has q as a consequence, then q is true too." This argumentation shows some similarity to the *modus ponens.* In many cases, it is possible to convert a non-deductive argumentation into a deductive one. However, the problem of plausibility occurs in both. In the first description of the causality argumentation, the plausibility principle is clear from the fact that we can never be certain that q is the expected consequence of action p. In the second type, it is hidden within the causality premise.

We shall illustrate the causality argumentation in utilitarian reasoning by means of the Golden Gate Bridge suicide barrier case from the introduction. There the action "not to install a suicide barrier" is considered morally acceptable by the opponents since the barrier is not worth the costs. The causality argumentation contains the following causality premise: "the action in question leads to the following expected consequence: 20 to 25 people per year would not be saved, 50 million dollar building costs, and the annual operation and maintenance costs." The critical questions belonging to this causality argumentation are highly relevant here. In this case they are formulated as follows: does not installing the barrier indeed lead to the expected 20 to 25 people to die each year in comparison with installing the barrier? Do the building costs really amount to 50 million dollars? What was the basis for the estimation of the number of 20 to 25 people? And so on.

Causality Argumentation

In this argumentation, use is made of the fact that a certain expected consequence can be derived from a certain situation or action.

Formal description:

- p (action or situation)
- So: q (the expected consequence)

or

- p
- "p causes q" or "p has q as a consequence" (*the causality premise*)
- So: q

Critical questions:

1 Will the given situation or action indeed lead to the expected consequence?
2 Have no issues been forgotten, for example, with respect to the expected consequence?
3 How do you determine the expected consequence and can it be justified?

Next, we need to demonstrate that the consequence is indeed the best one. This is determined by means of a *comparative* assessment. This judgment can be made if the expected consequences of all possible actions have been determined (in which a causality argumentation is used each time), so that they can be compared. Using a kind of cost-benefit analysis, the best possible consequence is selected. Bentham felt money was a suitable means with which to express consequences for their comparison. By definition, the best consequence is the greatest happiness for the greatest number. Note that this comparative judgment is also underpinned by argumentation in which the critical question "Can we express all consequences in monetary terms?" is extremely relevant. This also played a role in the Golden Gate Bridge suicide barrier case and in the Ford Pinto case (see Section 3.1).

Because it is impossible to determine the expected consequences of all actions in many situations, some utilitarians make use of another end: the sum of the total usefulness (pleasure) and uselessness (pain) has to be positive. With this criterion, we can determine whether a certain action is suitable and morally acceptable without it being necessary to consider all other possible actions.

Frequently occurring fallacies in causality argumentations include the *post hoc ergo propter hoc* ("afterwards, so therefore") and the *slippery slope*. In the *post hoc ergo propter hoc*, a causal relationship is derived simply from the fact that two events occur after each other. For example, "I saw a black cat yesterday and then I saw a car accident. So black cats are bad luck." In the slippery slope wild argumentation is involved: far-reaching consequences are derived from a small cause. For example, "You should not gamble. If you start you will not be able to stop. Soon you will lose all your money and end up in the gutter." In means-end argumentation fallacies occur if the means does not lead to the end. This is comparable to the *post hoc ergo propter hoc*.

4.4.3 Argumentation in Kantian reasoning

The assumption in Kant's theory to determine whether an action is morally acceptable is as follows:

> An action is morally acceptable if and only if the action meets the first categorical imperative.

or

> An action is morally acceptable if and only if the action meets the second categorical imperative.

We shall start with the first categorical imperative and then deal with the second one in Kantian reasoning.

The first categorical imperative (the universality principle)

The first categorical imperative reads as follows: "Act only on that maxim which you can at the same time will that it should become a universal law." If we wish to defend that an action h is morally right, then we first take the negation of that action (not-h) and thus "not doing that action." Next, we show that the action not-h is morally unacceptable by showing that the maxim of that action (i.e., the principle that the action is either permitted or forbidden for you) leads to a contradiction as soon as you make a general law of it (cf. Korsgaard, 1996). Thus we can say that h is morally acceptable. Kant gives a number of examples on how to reach such a contradiction. His best-known example is the earlier mentioned example of false promises:

> Another finds himself forced by necessity to borrow money. He knows that he will not be able to repay it, but sees also that nothing will be lent to him unless he promises firmly

> to repay it in a definite time. He desires to make this promise, but he has still so much conscience as to ask himself: "Is it not unlawful and inconsistent with duty to get out of a difficulty in this way?" Suppose however that he resolves to do so: then the maxim of his action would be expressed thus: "When I think myself in want of money, I will borrow money and promise to repay it, although I know that I never can do so." Now this principle of self-love or of one's own advantage may perhaps be consistent with my whole future welfare; but the question now is, "Is it right?" I change then the suggestion of self-love into a universal law, and state the question thus: "How would it be if my maxim were a universal law?" Then I see at once that it could never hold as a universal law of nature, but would necessarily contradict itself. For supposing it to be a universal law that everyone in difficulty should be able to promise whatever he pleases, with the purpose of not keeping his promise, the promise itself would become impossible, as well as what it is supposed to accomplish, since no one would consider that anything was promised to him, but would ridicule all such statements as vain pretenses. (Kant, 2002 [1785])

In his example, Kant shows that the action "I will not keep my promise" is morally unacceptable if you are in need of money for example. He does this by showing that the maxim "if I am in need of money, I may break my promise" leads to a contradiction as soon as a general law is made of it: "anybody may break his/her promises if he/she is in need of money," which we abbreviate by "A." From this general law we can derive that *it makes sense to break my promise* (abbreviated by "p"), because then I will get out of my money problems. On the other hand, we can derive *it makes no sense to break my promise* ("not-p," the negation of "p"), because nobody will value his/her promises any more. Promises no longer make sense, because everybody is allowed to break their promises. From the contradiction that it both makes sense and no sense to break a promise, we can deduce "not-A": you cannot make a general law of "if I am in need of money, I may break my promise." Thus, one must keep one's promises.

This indirect method of proof is also referred to as **proof from the absurd** or *reductio ad absurdum* (Latin for "reduction to the absurd"). Here, it is assumed that the proposition to be demonstrated is not true and from this a contradiction is derived.[2] One of the oldest examples of this argumentation technique is the proof that $\sqrt{2}$ is not a rational number as explained in the box.

Proof from the absurd A deductive argumentation in which a certain proposition is proved by showing that the negation of the proposition leads to a contradiction.

The Square Root of 2

Assume that $\sqrt{2}$ is a rational number. This assumption implies that there exist integers m and n with $n \neq 0$ such that $m/n = \sqrt{2}$. Then $\sqrt{2}$ can also be written as an irreducible fraction m/n (the fraction is shortened as much as possible). This means that m and n have no common factor greater than 1. From $m/n = \sqrt{2}$ it follows that $m = n\sqrt{2}$, and so $m^2 = (n\sqrt{2})^2 = 2n^2$. So m^2 is an even number, because it is equal to $2n^2$, which is even. It follows that m itself is even. Because m is even, there exists an integer k satisfying $m = 2k$. We may therefore substitute $2k$ for m in $m^2 = 2n^2$, thereby obtaining the equation $(2k)^2 = 2n^2$, which is equivalent to $4k^2 = 2n^2$ and may be simplified to $2k^2 = n^2$. Because $2k^2$ is even, it now follows that n^2 is also even, which means that n is even (recall that only even numbers have even squares).

Then, m and n are both even, which contradicts the property that m/n is irreducible. Since we have found a contradiction, the initial assumption that $\sqrt{2}$ is a rational number is false; that is to say, $\sqrt{2}$ is irrational.

The indirect proof is a deductive form of argumentation. The conclusion "not-A" is completely supported by the premises "Assuming A leads to p" and "assuming A leads to not-p." For formal languages (such as mathematics), proof from the absurd is not problematic. However, when we convert languages of content (such as the language we use daily) into a formal language, the negation and particularly the double negation can be especially difficult. Say for example a man says to his wife "It is not true that I do not trust you." From a logical perspective, this should be equal to "I trust you," because not not-A is logically equivalent to A. Probably the man does not mean it this way though. The term "trust" implicitly refers to a value scale running from "very trustworthy" to "very untrustworthy." Searching for a contradiction in the application of the first categorical imperative, we come across this problem too: can we deduce that an action is morally acceptable from the negation of a morally unacceptable action? Saying that lying is viewed as a morally unacceptable action, does that mean that by definition a person has carried out a morally acceptable action by not lying? We would probably have to deny this, because the term "morally acceptable" refers to a value scale (see, e.g., Williams, 1973, esp. ch. 11). From the above it follows that as far as our daily use of language is concerned, the proof from the absurd cannot be a deductive argumentation by definition, because the conclusion is not always embedded in the premises upon closer inspection. By answering the critical questions, we can determine to what extent the argumentation is sound.

Proof from the Absurd

The proposition is proven by showing that the negation of the proposition leads to a contradiction (an inconsistent set of statements).

Formal:

- Assuming A (logically) leads to an inconsistent set of statements
- So: not-A

Critical questions:

1 Does assuming A indeed lead to an inconsistent set of statements?
2 Is not-A (or not-p) indeed the negation of A (or p)? In other words, can A be concluded from "not not-A?"

A nice example in literature of the proof from the absurd to show that this method is not only used in a scientific way stems from the novel *The Name of the Rose* by Umberto Eco, in which the Franciscan friar William of Baskerville hears from the abbot, that monk Adelmo, who died some days earlier – possibly by a fall out the window – is buried at the churchyard. William draws from this information the conclusion, that the window concerned has to be closed (see box).

The Name of the Rose

If the window had been open, you would immediately have thought he had thrown himself out of it. From what I could tell from the outside, they are large windows of opaque glass, and windows of that sort are not usually placed, in buildings of this size, at a man's height. So even if a window had been open, it would have been impossible for the unfortunate man to lean out and lose his balance; thus suicide would have been the only conceivable explanation. In which case you would not have allowed him to be buried in consecrated ground. But since you gave him Christian burial, the windows must have been closed. (Eco, 1983)

In this example, William assumes that the window was open. This assumption leads to a contradiction. On the one hand, if the window were open, monk Adelmo would not have been allowed to be buried in consecrated ground, and, on the other hand, he was allowed to be buried in consecrated ground (a Christian burial), since the abbot actually gave him a Christian burial. So, the initial assumption (the window was open) is false, which means that the window was closed.

Applying the theory of argumentation, and especially the proof from the absurd, to Kant's ethical reasoning pressed us to find out what the exact form is of the contradiction when we examine, for example, the maxim of Kant's man who deceitfully promises. According to Kant, the man's maxim is incapable of being conceived and willed without contradiction, however immoral actions do not seem very easily pronounced "contradictory" as Kant suggested.

Second categorical imperative (the reciprocity principle)

The second categorical imperative (the reciprocity principle) reads as follows: "Always act as to treat humanity, whether in your own person or in that of any other, in every case as an end, never as means only." This imperative means we should always respect our personal freedom and that of others to make well-considered choices.

> Secondly, as regards necessary duties, or those of strict obligation, towards others: He who is thinking of making a lying promise to others will see at once that he would be using another man merely as a mean, without the latter containing at the same time the end in himself. For he whom I propose by such a promise to use for my own purposes cannot possibly assent to my mode of acting towards him and, therefore, cannot himself contain the end of this action. (Kant, 2002 [1785])

In this argumentation, Kant uses an ends-means form of argumentation: if it is my aim to obtain money from someone without giving it back, I can realize this by means of a false promise or by misleading someone (the means). Next, we must ask the question whether the person I made the promise to would agree to the means and the end it is supposed to meet. If the answer is positive then the action is morally acceptable, but if the answer is negative (as in this case) the action is morally unacceptable.

Note that the concept "end" in the ends-means argumentation is used differently than in the second categorical imperative. In this imperative, the end is the respect for the freedom to make well-considered choices: you have to inform the person that you are using him as a means to achieve your end, such that the person in question can make a well-considered choice whether or not to cooperate.

4.4.4 Argumentation in virtue-ethical reasoning

In virtue ethics, the following assumption is made to determine whether an action is morally acceptable:

> An action is morally acceptable if and only if that action is what a virtuous agent would do in the circumstances. (Hursthouse, 1991)

The question is, of course, how we define a virtuous person. The definition is as follows: a virtuous person is one who acts virtuously, that is, one who has and exercises the virtues (Hursthouse, 1991). To determine whether an action is morally acceptable, not all virtues are equally relevant because this is determined by the position someone has and what is expected of him/her in that position. Thus, a virtuous salesperson must possess the virtues of friendliness and politeness. A virtuous designer needs to possess other virtues like creativity. The virtues friendliness and politeness are less relevant for a designer.

According to Dennis Moberg, the virtues responsibility, loyalty, and trust are relevant for a virtuous employee in an organization (Moberg, 1997). To show whether the action of an employee is morally right, we have to determine whether he/she would act that way if he/she was a virtuous employee. But how do you know how a virtuous employee would act? For this we need examples of virtuous employees. But for that you first need to decide whether the chosen example really is an example of a virtuous employee. To this purpose, we make use of a **characteristic-judgment argumentation**.

Characteristic-judgment argumentation
A type of non-deductive argumentation. An argument based on the assumption that a certain judgment about a thing or person can be derived from certain characteristics of that thing or person.

Characteristic-Judgment Argumentation

If someone or something X displays certain characteristics $s_1, s_2, \ldots, s_n$, then judgment A is justified for that person or thing.

Formal:

- X has the characteristics $s_1,s_2,\ldots,s_n$
- characteristics $s_1,s_2,\ldots,s_n$ are typical of A (*the characteristic-judgment premise*)
- So: A applies to X

Critical questions:

1 Do the characteristics mentioned justify judgment A?
2 Are the characteristics mentioned all typical of A?
3 Are there any other characteristics necessary for A?
4 Does X possess characteristics that justify the judgment not A?
5 Does X possess the characteristics mentioned?

To show that an employee is a virtuous employee (i.e., A applies to employee X so he/she must be virtuous), we need to demonstrate that the employee possesses the virtues responsibility, loyalty, and trust (X has characteristics $s_1,s_2,\ldots,s_n$). These virtues are characteristic of a virtuous employee (the characteristic-judgment premise). From the virtue-ethical assumption it follows that the actions carried out by this employee are morally acceptable.

The characteristics-judgment argumentation is of a non-deductive nature, because the conclusion does not necessarily follow from the premises. Thus, an employee may possess the characteristics typical of a virtuous employee, but may not have to be a virtuous employee. The first four critical questions are related to this. Whether the above argumentation is sound depends on how the critical questions are answered. Besides that, the soundness of the argumentation is linked to whether the employee really does possess the virtues – the final critical question is related to this.

4.5 Fallacies

In this book we define a fallacy as an error or deficiency in an argument. A fallacy seems to be correct but proves, on examination, not to be so. It is important to recognize fallacies, because they can mislead us. A person who recognizes fallacies is in a better position to argue successfully, and a person who cannot may fall prey to the illusions of verbal tricksters.

In this section we will discuss some common fallacies which are often used in an ethical debate and we will discuss some fallacies that can commonly be found in public debates on the acceptability of technological risks.

4.5.1 Some common fallacies in ethical discussions

There is no universally accepted classification of fallacies, but a common distinction is between *formal* and *informal fallacies.* Formal fallacies are determined solely by the form (hence the name) or structure of an argument. Any invalid argument (see

Section 4.2) is a formal fallacy. In this section we will focus on the informal fallacies. Informal fallacies are based on considerations of the context and content of the arguments. Informal fallacies which are often be used in ethical discussions are: attack on the person; confusion of law and ethics; straw person; wishful thinking; naturalistic fallacy; privacy fallacy; and ambiguity.

Attack on the person (Ad Hominem)
This fallacy is committed when an attempt is made to discredit an argument by bringing into question in some negative way the presenter of the argument instead of attacking the argument itself. This line of "reasoning" is fallacious because one must keep one's focus on the argument and ignore the qualities of the person who happens to make it. In some cases, an individual's characteristics can have a bearing on the question of the credibility of his/her claims. For example, if someone has been shown to be a pathological liar, then what he/she says can be considered to be unreliable. Such attacks, however, are weak, since even pathological liars might speak the truth on occasion.

Confusion of law and ethics
Confusion of law and ethics can simply be put as "If it isn't illegal, it is ethical." Ethics is, however, more encompassing than law. Law cannot forbid all inappropriate or harmful behavior, which is often not even desirable. Having an extramarital affair is not forbidden, but that is not to say that this is morally acceptable. The same holds for lying, betrayal, and nepotism.

Straw person
The straw person fallacy is committed when an attempt is made to miss-state a person's actual position and conclude that the original argument is a bad argument. Often this miss-stated argument is a distorted, exaggerated, or foolish version of the original one. Suppose that someone is critical of the use of nuclear energy, and his opponent react as follows: "Personally, I think the energy power supply is of the highest importance." With this the opponent suggests that the first one does not adhere to good energy power supply, but this cannot be derived from his criticism on nuclear energy.

Wishful thinking
Wishful thinking (or fallacy of desire) occurs when a person interprets facts, reports, events, perceptions, and so on, according to what he/she would like to be the case rather than according to the actual or rational acceptable evidence. So, the only evidence the presenter gives is his/her desire for several things to be the case. Examples can be often found in the attitudes, positions, and comments of people who are deeply committed to some cause. For example, "Surely God exists, because I have complete belief that He does."

Naturalistic fallacy
The naturalistic fallacy takes the form of deducing normative statements based only on descriptive statements. We can simply put it as: deriving *ought* from *is*. The (false) reason behind this fallacy is that we must always accept things as they are. For example, "Stealing bikes is morally acceptable in the Netherlands, because more bikes are stolen than bought in the Netherlands." So, a *sound* normative conclusion always needs some normative premises.

The privacy fallacy: "If you have done nothing wrong, you have nothing to worry about"

This argument suggests that privacy only protects people who have something to hide. Unfortunately, that argument does not hold for several reasons. First, erroneous information can dramatically affect your life. For example, individuals have been dismissed, or hampered in their careers, because of errors in the Criminal Records Bureau database. They have been stigmatized as criminals, because of database failings. This kind of errors by wrongly labeling people as criminals in a database is becoming increasingly common (Whitehead, 2008).

Second, a person can risk discrimination, if that person's information is publicly known. For example, if a person is tested HIV+ and this is publicly known, then an employer might be reluctant to hire him and an insurance company might be reluctant to insure him.

Ambiguity

Fallacies of ambiguity play with the meaning of words or phrases and therefore often are humorous. The next anecdote illustrates this:

> A captain got very annoyed at the excessive drinking of the first mate, and one day he wrote in the journal: "The first mate was drunk today." The first mate, seeking revenge on the captain, wrote in his journal, "The Captain was sober today." He suggests, by including the word "today" that the Captain is usually drunk. (Copi and Burgess-Jackson, 1990, p. 117)

A person commits this fallacy when he/she uses a word or phrase unclearly. That is if that word or phrase has more than one distinct meaning (ambiguous) or has no distinct meaning (vague). One must be careful if words or phrases can have different interpretations, because it can make a significant difference to the meaning of what is said and can therefore generate fallacious inferences.

4.5.2 Fallacies of risk[3]

In public debates on the acceptability of technological risks some specific fallacies can be identified. In this section we discuss some of these fallacies, where X and Y stand for an activity, product or technology. In Chapter 8 on the acceptability of technological risks, we will come upon the most of these fallacies.

The sheer size fallacy

X is accepted
Y is a smaller risk than X
So, Y should be accepted

Example: "You must accept nuclear energy, because the risks are smaller than that of driving of a car." This fallacy is also based on a false analogy. The analogy can only be made in a right way, if X and Y are alternatives in the same decision, but we do usually not choose between nuclear energy and driving a car.

The fallacy of naturalness

X is unnatural
So, X should not be accepted

This fallacy corresponds with the naturalistic fallacy. A normative statement is derived from only a descriptive statement. The idea behind this fallacy is that whatever is unnatural is wrong. The problem with this is that is not clear what is meant by "natural." The term "natural" is used in many different contexts. Is scalded milk natural? Is boiling contaminated water before drinking it natural? Is nuclear power natural? An extreme definition of natural could be that all products of our civilization are also products of nature, since we are also biological creatures. Or another extreme, only the products that were available to pre-civilization humans are natural.

The ostrich's fallacy

X does not give rise to any detectable risk or there is no scientific proof that X is dangerous
So, X does not give rise to any unacceptable risk

A counter-example of this fallacy is asbestos, an isolation product. During the course of time asbestos proved to have some extremely harmful side effects, such as cancers of the lung and stomach lining. The gist of the fallacy is that as long as a risk does not reveal, the risk does not exist.

The delay fallacy

If we wait we will know more about X
So, no decision about X should be made now

This is one of the most dangerous fallacies concerning the acceptability of technological risks, since the premise "If we wait we will know more about X" is almost always true. As a consequence, the decision can always be postponed to avoid risk-reducing actions. It may very well be better to make an early decision on fairly incomplete information than to make a more well-informed decision at a later stage, since the problem may get worse.

The technocratic fallacy

It is an engineering issue how dangerous X is
So, engineers should decide whether or not X is acceptable

Indeed, engineers have competence in determining the nature and the magnitude of technological risks, but this competence is not the same as competence in deciding whether or not a technological risk is morally acceptable. The reflection on the acceptability of technological risks requires not only technological knowledge, but also ethical competence. See also subsection 1.5.2.

The fallacy of pricing

We have to weight the risks of X against its benefits
So, we must put a price on the risks of X

The Ford Pinto case in the previous chapter is an illustration of this fallacy. There are many things we cannot easily value in terms of money, including those that involve the loss of human lives. Furthermore, there is no sensible price that can be meaningfully assigned to technological risks, since it also depends on the circumstances. In

Chapter 6 we will discuss some alternatives to weigh the pros and cons of technological risks without expressing these in money.

4.6 Chapter Summary

In an ethical debate we are regularly confronted with arguments put forward by ourselves and other people. To assess these arguments is often the best way to justify our own opinion and, if necessary, to refute the arguments of others. In this chapter, we have become acquainted with some basic principles of argumentation theory which concern is to distinguish good arguments from bad ones.

We have distinguished two main branches in argumentation theory: deductive and non-deductive argumentation. In deductive argumentation we can divide arguments in valid and invalid arguments. An argument is valid if and only if the conclusion of the argument follows with necessity from its premises. However, many arguments used in ethical debates are non-deductive: the premises (if true) give a limited amount of support to the conclusion. We call an argument sound if the conclusion is plausible given the premises. To assess non-deductive arguments we have used the argumentation schemes of the pragma-dialectical approach. In this chapter we have seen several argumentation schemes often used in ethical debates:

- Inductive argumentation
- Argumentation by analogy
- Means-end argumentation
- Causality argumentation
- Proof from the absurd
- Characteristic-judgment argumentation

With the help of the argumentation schemes we have constructed how the ethical theories (utilitarianism, Kant's ethics, and virtue ethics) come to conclusions about whether an action is morally acceptable or not.

Each argumentation scheme has its own critical questions. These questions can be used to check whether a particular part of argumentation is sound and therefore can bear the scrutiny of criticism. If an argument is not sound, then an error or deficiency occurs in it. This we have called a fallacy. We have mentioned several fallacies in relation to the acceptability of technological risk, such as the fallacy of naturalness and the technocratic fallacy.

Argumentative skills are valuable because they contribute to fruitful moral deliberation. Here the recognition of bad arguments is important, since we must try to avoid persuading others on insufficient grounds. But it is also important to be able to persuade others with good arguments – which can be defended – to agree on what is the best morally acceptable action or decision.

Study Questions

1 What is the difference between a valid and sound argumentation? Give an example of a valid argumentation and an example of a sound argumentation.
2 Describe the naturalistic fallacy, and give an example.

3 Give a context in which the use of the naturalistic fallacy can be justified.
4 Explain why the phrase "you're either with us, or against us" is a fallacy.
5 On July 25, 2000 the very first fatal accident involving a Concorde occurred when a Concorde bound from Paris to New York crashed during take-off and all 100 passengers and nine crew and four people on the ground were killed. After the accident, the Concorde was taken out of service for some time. Suppose, someone argues that: "It makes no sense to take the Concorde out of service. The safety record of the Concorde is rather good; it was the first accident in 31 years. Flying the Concorde is per kilometer travelled much safer than riding in a car. Car accidents are generally accepted, so there is no reason to suppose that the risks of the Concorde are unacceptable."
   What do we call this fallacy? Argue, why this argumentation is not sound.
6 In a utilitarian plea, the mean-ends argumentation is at the forefront. What is the means, and what is the end in such a plea?
7 What assumption is made to determine whether an action is morally acceptable in virtue ethics?
8 The following appeared as part of an annual report sent to stockholders by Olympic Foods, a processor of frozen foods: "Over time, the costs of processing go down because as organizations learn how to do things better, they become more efficient. In color film processing, for example, the cost of a 3-by-5-inch print fell from 50 cents for five-day service in 1970 to 20 cents for one-day service in 1984. The same principle applies to the processing of food. And since Olympic Foods will soon celebrate its 25th birthday, we can expect that our long experience will enable us to minimize costs and thus maximize profits."
   a. What do we call this argumentation scheme?
   b. What are the critical questions to decide whether this argumentation is sound?
   c. Is this argumentation sound?
9 In the case of The Golden Gate Bridge someone argues: "A suicide barrier will not be effective, since people who want to kill themselves will simply go somewhere else."
   a. What do we call this argumentation scheme?
   b. What are the relevant critical questions to decide whether this argumentation is sound?
   c. Is this argumentation sound?
10 Explain that the responsibility argument of the proponents in the case of The Golden Gate Bridge can be founded on virtue ethics. What are the critical questions you could ask to attack this argument?

Discussion Questions

1 Debate the strengths of the arguments used by the opponents and the proponents for the Golden Gate Bridge suicide barrier.
2 One claim is that pollution is often the cost of development: In Europe and the United States industrial waste was largely unregulated during times of significant economic expansion. Do you think that there are sound arguments to justify restraint on the part of developing countries today? Defend these arguments.
3 Can all moral disagreements be solved by arguments? If not, how should they then be solved?

Notes

1 Some people believe the actual number is higher because many cases are not reported.
2 In intuitionist logic the proof from the absurd can not be derived, since intuitionist logic does not accept the Aristotelian law of excluded middle ('A or not-A' is always true).
3 This section is based on Hansson (2004a).

5
The Ethical Cycle

Having read this chapter and completed its associated questions, readers should be able to:

- Understand that moral problems are ill-structured;
- Explain the analysis steps of the ethical cycle;
- Explain the role of the ethical cycle in moral decision-making;
- Apply the ethical cycle to concrete moral problems in engineering;
- Analyze and evaluate the complex consequences and motives that typically attend moral issues in engineering practice;
- Describe wide reflective equilibrium and its relation with the analysis step reflection;
- Deliberate and discuss moral issues with other people.

Contents

Ethics, Technology, and Engineering: An Introduction, First Edition.
Ibo van de Poel and Lambèr Royakkers.

5.1 Introduction

Case Gilbane Gold

Figure 5.1 Sewage treatment plant. Photo: © nonameman/Fotolia.com.

The city of Gilbane has been processing its waste water into manure for agriculture for 75 years. This yields a tax benefit of $300 per year per household. Given this tax advantage, the manure produced is called "Gilbane Gold." For the past 15 years, the city also has a company producing computer parts: Z-Corp. The city attracted the company to come by offering tax benefits. The company is important for the city because it creates opportunities for employment. However, the production process produces lead and arsenic, which are emitted through the waste water of the factory.

Lead and arsenic are heavy metals that built up in organisms and may cause negative health effects. If concentrations of arsenic and lead would cumulate in Gilbane Gold, this might have negative long-term effects. Therefore, the restrictions that the city has set on the concentration of arsenic and lead in the waste water are about ten times as strict as Federal Regulations.

Independent environmental consultant Tom Richards, who has been hired by Z-Corp, has found out that the conventional method for measuring arsenic and lead in the waste water used by Z-Corp measures lower concentrations than a new, more reliable, method. However, the old method is the one prescribed by City Regulations and city officials, after being informed on the issue, have not objected to the continued use of that method. Moreover, Z-Corp can easily stay within the limits of the

City Regulations even with the new measurement method by diluting the waste water because the regulations only refer to concentrations and not to absolute amounts. Some consider this, however, "a major loophole in the law." When Richards further pursues the matter, Z-Corp decides not to renew his contract.

A young engineer, David Jackson, now becomes responsible for Z-Corp's emissions of arsenic and lead into the waste water. In the meantime, Z-Corp signs a contract with a Japanese firm which will result in a 500 percent increase in production. Jackson, who is really concerned now, raises the issue with management but is told that there is no money available to solve the problem: the factory is hardly profitable. Moreover, manager Diane Collins argues, as long as Z-Corp meets the law, it has no broader responsibility. Jackson nevertheless fears that the waste water treatment plant may not be able to deal with the larger amounts of arsenic and lead. Since he has a duty as professional engineer "to hold paramount the safety, health and welfare of the public," it might be appropriate to speak out in public. Indeed Jackson is approached by Channel 13 a local television station about the issue.

Source: This is a fictional case based on a video produced by the National Society of Professional Engineers and the National Institute for Engineering Ethics.

Gilbane Gold is a fictional case in which a young engineer has to decide how to act in a difficult situation. It is the kind of situation you may also find yourself in after starting working as an engineer. Such situations call for moral judgment, using the tools we have introduced in the preceding chapters. However, moral judgment is not a straightforward or linear process in which you simply apply ethical theories to find out what to do. Instead it is a process in which the formulation of the moral problem, the formulation of possible "solutions," and the ethical judging of these solutions go hand in hand. This messy character of moral problems, however, does not rule out a systematic approach. In this chapter we describe a systematic approach to problem-solving that does justice to the complex nature of moral problems and moral judgment: the ethical cycle. Our goal is to provide a structured method of addressing moral problems which helps to guide a sound analysis of these problems. In Section 5.3, we will describe the ethical cycle. We will illustrate the usefulness of this cycle with an example in Section 5.4. In Section 5.5, we will discuss how the ethical cycle, which is mainly part of individual moral judgment, can be integrated into collective deliberations on moral issues. But, first we will pay attention to the fact that moral problems are ill-structured, which explains their messy and complex character.

5.2 Ill-Structured Problems

Moral problem-solving is a messy and complex process, like a design process. The design analogy has been introduced by engineering ethicist Caroline Whitbeck.[1] Mainstream ethics, Whitbeck argues, has been dominated by rational foundationalist approaches (Whitbeck, 1998b). As a result, ethics has focused primarily on the analysis of moral issues and on a quest for the ultimate rational foundations of morality. This means that the field of ethics as we know it now is typically searching for one, or a limited number of, basic moral principle(s), and tends to build on unrealistic decision-making problems. The rational foundationalist approach, according to Whitbeck, is unnecessarily reductive and therefore misleading. She holds that moral philosophy should be tolerant towards different approaches, and should overcome the idea that dealing with moral problems is only about analyzing preset moral problems, and selecting the one best option through justified principles.

One need not agree completely with Whitbeck's criticism on moral philosophy in general to appreciate the alternative she seeks to offer with her design analogy. This analogy can best be understood by considering the central notion of ill-structured problems. Whereas well-structured problems (such as basic arithmetical calculations), usually have clear goals, fixed alternatives to choose from, usually maximally one correct answer and rules or methods that will generate more or less straightforward answers, **ill-structured problems** have no definitive formulation of the problem, may embody an inconsistent problem formulation, and can only be defined during the process of solving the problem. In cases of ill-structured design problems, thinking about possible solutions will further clarify the problem and possibly lead to reformulation of the problem (Cross, 1989). Moreover, ill-structured problems may have several alternative (good, satisfying) solutions, which are not easily compared with each other (cf. Cross, 1989; Rittel and Webber, 1984; and Van de Poel, 2001). This is due to the fact that for ill-structured problems, no single criterion exists to order uniformly the possible solutions from best to worst (Simon, 1973). Another characteristic is that it is usually not possible to make a definitive list of all possible alternative options for action (Simon, 1973). This means that solutions are in some sense always provisional.

Ill-structured problem A problem that has no definitive formulation of the problem, may embody inconsistent problem formulations, and can only be defined during the process of solving the problem.

For Whitbeck, the fundamental mistake rational foundationalists make is that they fail to see that moral problems are *ill-structured*. By framing moral problems as "multiple-choice" problems (where we have a fixed number of possible alternatives to choose from, of which only one is right), moral philosophers implicitly suggest that moral problems are well-structured. As an alternative, Whitbeck proposes to take the ill-structured nature of moral problems as a starting point for considering moral problem-solving. Given the fact that designers have to deal with ill-structured problems all the time, Whitbeck holds we can learn a lot from designers and engineers when dealing with moral problems in domains that are not traditionally associated with "design."

The most important lesson to be learned from designing is that practical problem-solving is not only about analyzing the problem and choosing and defending a certain solution, but also about finding (new) solutions. Whitbeck calls this "synthetic reasoning." Designers engage in a design *process*, during which new information may arise, uncertainties and unknowns are taken to be defining characteristics of the problem situation, and several possible solutions are pursued simultaneously. Another lesson from designing is that designers seem well able to satisfy apparently conflicting demands at once. Whitbeck maintains that even though some moral problems may be irresolvable, it is misleading to present moral problems as such from the start, "because it defeats any attempt to do what design engineers often do so well, namely, to satisfy potentially conflicting considerations simultaneously" (Whitbeck, 1998a, p. 56).

Apart from these characteristics, which moral problems share with design problems (and other ill-structured problems), moral problems have their own peculiarities which make them even more messy and complex. One of them is that in identifying a moral problem one needs a conception of what morality and ethics are. Such a conception is partly theory-dependent as different ethical theories emphasize different parts of reality as morally relevant. Nevertheless, despite such differences, there is much common ground in ethical theories on what are moral concerns or problems. As a first approximation it will often be possible to define a problem based on common sense and one's own theoretical commitments. This formulation may later be refined during the process of moral problem-solving.

A second peculiarity of moral problems is related to the first one. The different ethical theories are not only relevant in identifying and formulating moral problems but also in judging them. The diversity of theories also reveals a diversity of reasonable moral opinions among different people on moral issues. This does, however, not mean that any solution to a moral issue will do. Solutions are better if they are based on systematic reasoning about the moral problem, on the taking into account of different viewpoints and theories, and on the exercise of a critical and reflective attitude.

5.3 The Ethical Cycle

Moral problem-solving is thus a messy and complex process. This does, however, not preclude the possibility of a systematic approach to the identification, analysis and solution of moral problems. A systematic approach might even be required to avoid the reduction of moral judgment to mere gut-feeling without any attempt to understand the moral problem or to justify one's actions. The approach we propose, the ethical cycle, aims at an improvement of moral decision-making or at least it tries to avoid certain shortcuts. Such shortcuts for example consist in neglecting certain relevant features of the problem or in just stating an opinion without any justification.

The **ethical cycle** is a helpful tool in structuring and improving moral decisions. The cycle helps you to make a systematic and thorough analysis of the moral problem and to justify your final decisions in moral terms. Ultimately, moral problem-solving is directed at finding the morally best, or at least a morally acceptable, action in a given situation in which a moral problem arises. It is, however, hard to guarantee that the ethical cycle indeed delivers such a solution, albeit because people may reasonably disagree about what is the morally best, or a morally acceptable, solution. We will discuss this further in Section 5.5.

Ethical cycle A tool in structuring and improving moral decisions by making a systematic and thorough analysis of the moral problem, which helps to come to a moral judgment and to justify the final decision in moral terms.

The ethical cycle consists of a number of "steps" (Figure 5.2). It is important to stress that by distinguishing these steps we do not want to suggest that moral problem-solving is a linear process. Rather, it is an iterative process, as the feedback loops in Figure 5.2 already suggest. The cycle, for example, starts with formulating a moral problem. In many actual cases, the moral problem only becomes clear after further delving into the facts of the situation, by distinguishing stakeholders, looking at ethical theories, et cetera. In other words, formulating a good problem statement is an iterative process that continues during the other steps. Nevertheless, it is important to start with formulating a moral problem to get the process going.

AGORA

AGORA is a web-based tool for education in ethics and technology (see www.ethicsandtechnology.com). AGORA has been developed as part of an ICT innovation project, which has been undertaken by a consortium of the three Dutch Universities of Technology (Delft University of Technology, Eindhoven University of Technology and University of Twente) and has been financially supported by the Dutch SURF Foundation. The program enables students to pass through the ethical cycle and to exercise their moral understanding and skills extensively. The main part of the program is dedicated to exercises in which the analysis of cases according to the ethical cycle is central. The steps in the ethical cycle are represented in AGORA as All Possible Steps (see Figure 5.3). This can be seen as a container full of steps from which teachers can choose some building blocks for the models of analysis as they think best in their didactic situation or for the purposes they want to achieve. So, instead of all the steps being carried out by the students, a teacher can choose, for example, only for a utilitarian analysis.[2]

5.3.1 Moral problem statement

Moral problem Problem in which two or more positive moral values or norms cannot be fully realized at the same time.

Moral dilemmas A moral problem with the crucial feature that the agent has only two (or a limited number of) options for action and that whatever he chooses he will commit a moral wrong.

The start of the ethical cycle is the formulation of a **moral problem**. A characteristic of a moral problem is that there are two or more positive moral values or norms that cannot be fully realized at the same time. Ethicists often call situations like these **moral dilemmas** instead of moral problems. Originally "dilemma" means "double proposition" implying that there are only two options for action. The crucial feature of a moral dilemma is, however, not the number of actions that is available but the fact that all possible actions are morally unsatisfactory. The agent seems condemned to moral failure; no matter what she does, she will do something wrong (or fail to do something that she ought to do). A well-known

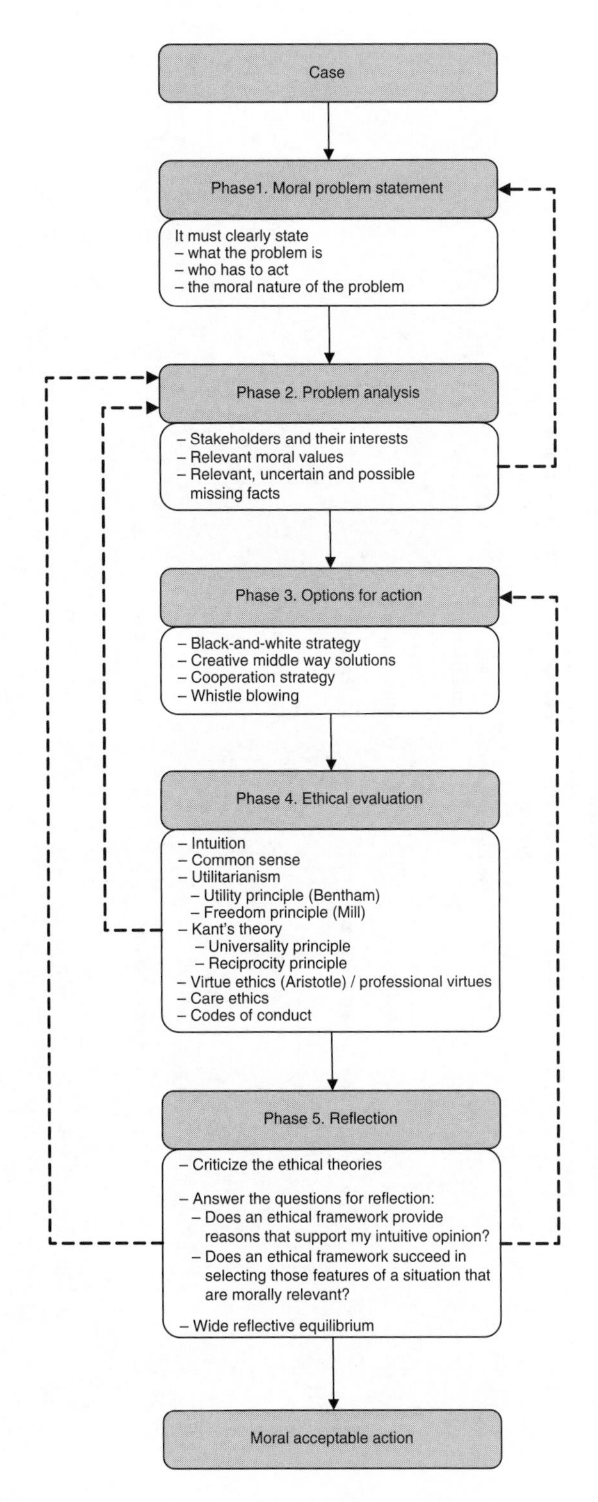

Figure 5.2 The ethical cycle.

Case description	Problem statement	Problem analysis	Options for action	Ethical evaluation	Reflection	Discussion
Case description	Problem statement	Stakeholders	Options for action	Common sense	Compare to intuition	
		Interests	Redefine problem	Ethical codes	Reflection and concl	
		Facts	Intuitive answer	Act-utilitarianism	Further questions	
		Lacking information		Rule-utilitarianism		
		Uncertain facts		Pareto		
		Values		Kant's analysis		
		Conflicting values		Respect-test		
		Responsibility		The freedom-test		
				The virtues		
				Virtue analysis		

Figure 5.3 An overview of all analysis steps in Agora corresponding to the steps of the ethical cycle.

example of a moral dilemma is taken from William Styron's *Sophie's Choice* (Styron, 1979). Sophie and her two children are at a Nazi concentration camp. On arrival, she is "honored" for not being a Jew by being allowed a choice: One of her children will be allowed to live and one will be killed. But it is Sophie who must decide which child will be killed. Sophie can prevent the death of either of her children, but only by condemning the other to be killed. The guard makes the situation even more excruciating by informing Sophie that if she chooses neither, then both will be killed.

Although some moral problems are real dilemmas, many moral problems are not. Often the problem is not an impossible choice between two or more evils. Therefore, we will use the term "moral problem" instead of moral dilemma. In order to apply the ethical cycle successfully, it is important that the moral problem is stated as precisely and clearly as possible. This can best be done by formulating a moral question. A good moral question meets three conditions: 1) it must clearly state what the problem is; 2) it must state who has to act; and 3) the moral nature of the problem needs to be articulated. Sometimes, the second condition is not relevant; for example when we ask a general question about the moral acceptability of a particular course of action or a technology. An example of such a question is: Is cloning morally acceptable? or, more precisely: Under what conditions – if any – is cloning morally acceptable?

Gilbane Gold: Moral Problem Formulation

One possible problem formulation is:

> Will the lead and arsenic in Z-Corp's waste water cause negative health effects?

Although it is essential to ask and answer questions like this one in dealing with the moral problem, so that one knows what one is talking about, the question is not a good moral problem formulation because it is a factual question rather than a moral question (condition 3 in the text).

Another possible formulation is:

> How can the city of Gilbane secure both Z-Corp and Gilbane Gold as a source of welfare?

Again this is not a sound moral problem formulation, because this is a practical question about how to achieve a given goal (securing both sources of welfare) rather than a moral question about what to do given a range of (potentially conflicting) moral considerations.

One possible problem formulation that meets all three criteria in this case would be:

> Should David inform the public about the potentially excessive levels of arsenic and lead in Z-Corp's waste water even if Z-Corp management does not consider it a serious problem?

Often it will not be possible to formulate a definitive formulation of the moral problem at this stage. The reason is that at later stages analyses will be made, like the identification of relevant values, which are crucial for a good problem formulation. Nevertheless, one can start with a somewhat vaguer notion of the moral problem and try to make the formulation of the moral problem clearer and more precise once some of the other steps have been carried out. We will illustrate this below in the boxes.

5.3.2 Problem analysis

During the problem analysis step, the relevant elements of the moral problem are described. Three important elements can be distinguished: the stakeholders and their interests, the moral values that are relevant in the situation, and the relevant facts. These elements are to be described during this step because they give a good impression of the current situation with respect to the moral problem; moreover, they are indispensable for the carrying out of the later steps of the ethical cycle.

Gilbane Gold: Relevant Values

- Public health
- Environmental care
- Public welfare
- Honesty (speaking the truth)
- Loyalty to the company
- Integrity (i.e., living by one's own moral standards and commitments)

These values can also be used to reformulate the moral problem a bit, for example:

> What should David Jackson do given on the one hand moral considerations of public health, environmental care, honesty, and integrity and on the other hand his loyalty to the company and the importance of Z-Corp for public welfare in Gilbane?

This problem formulation places more emphasis on the relevant moral values than the one before and it does not directly focus on one possible solution, so leaving more room for creatively looking for solutions that meet the various moral concerns.

Stakeholders Actors that have an interest ("a stake") in the development of a technology.

Stakeholders are both the people who can influence the options for action being chosen and the eventual consequences of this action as well for the people suffering or profiting from those consequences. Stakeholders can be individuals, like colleagues, groups, like the design team, organizations, like a company or society, as far as it concerns the common interest. For each of the stakeholders, it is to be indicated what interests he or she has.

Gilbane Gold: Main Stakeholders and their Interests

- *Management of Z-Corp*: increasing production in a profitable way, meeting legal requirements, good reputation
- *City officials and council*: protecting the safety of the inhabitants, maintaining income from Gilbane Gold, maintaining employment opportunities, protecting the environment
- *Farmers*: safe and reliable fertilizer
- *Inhabitants of city*: health, employment, low taxes, environmental protection
- *David Jackson:* being a reliable and honest engineer, keeping his job, meeting the law

Stakeholders may disagree about the facts. Usually, not all facts are undisputed in a moral problem situation. Facts can also be uncertain or unknown. Disputed, uncertain, or unknown facts are certainly not irrelevant for the analysis of the moral problem. In later steps, they can make a distinct difference. One way to deal with such facts is to make explicit assumptions about them. Naturally, different people will often make different assumptions. Since the final option chosen at the end of the ethical cycle can depend on the assumptions made with respect to facts, it is advisable to formulate the moral standpoint sometimes in a hypothetical form: "If x is the case, than option for action A is morally acceptable; but if it turns out that y is the case then option B is morally acceptable."

Gilbane Gold: Some Unknown or Disputed Facts

- The city waste water treatment plant will not be able to handle the increased levels of arsenic and lead in Z-Corp's waste water and this will cause environmental and health risks (contaminated vegetables).
- Expanding the waste water treatment of Z-Corp is too expensive. It would threaten the profits and might mean job losses or even bankruptcy.
- Jackson will lose his job if he goes public.

5.3.3 Options for actions

After the analytic step in which the moral problem is formulated, a synthetic step follows in which possible solutions for action are generated in the light of the formulated problem analysis. Often a moral problem is formulated in terms whether it is acceptable to engage in a certain action or not. In this **black-and-white-strategy**

Black-and-white-strategy A strategy for action in which only two options for actions are considered: doing the action or not.

only two options for actions are considered, doing the action or not, other actions are simply not considered. While this strategy may be helpful in better understanding and formulating the moral problem, in many more complex situations it is too simplistic. In real life, options are usually not given but have to be thought out or "invented" by the agent. In fact, by thinking out new options for action a seemingly irresolvable moral dilemma can sometimes be resolved or made less dramatic. During this step creativity is therefore of major importance. It can invite us to find options for actions that bridge seemingly conflicting moral values playing a role in the moral problem.

Strategy of cooperation The action strategy that is directed at finding alternatives that can help to solve a moral problem by consulting other stakeholders.

Also, the **strategy of cooperation** can be helpful in thinking out possible options for action. This strategy is directed at finding alternatives that can help to solve the moral problem by consulting other stakeholders. Sometimes, such cooperation and consultation can lead to win-win situations – solutions which make nobody worse off. Often such win-win situations are not self-evident and one should creatively look for new options for action.

Whistle-blowing (speaking to the media or the public on an undesirable situation against the desire of the employer, see Section 1.5.3), is a last resort strategy because it usually brings large costs both to the individual employee and to the organization. Nevertheless, some situations may require whistle-blowing, for example, if human safety or health is at stake and there are no other options of actions available.

Gilbane Gold: Options for Action

Our original problem formulation was:

> Should David inform the public about the potentially excessive levels of arsenic and lead in Z-Corp's waste water even if Z-Corp management does not consider it a serious problem?

This formulation suggests a black-and-white-strategy: either Jackson should inform the public or he should not. In this black-and-white strategy one of the options is whistle-blowing because it is clear that Z-Corp is against making the information public.

Now consider the reformulated problem:

> What should David Jackson do given on the one hand moral considerations of public health, environmental care, honesty, and integrity and on the other hand his loyalty to the company and the importance of Z-Corp for public welfare in Gilbane?

This formulation suggests a range of other options, including:

1 Develop a better but inexpensive treatment method;
2 Contact the engineering society for advice and help;
3 Contact city council to inform them about the new problem and ask them to take action; or
4 Contact people from the city's waste water treatment plant to see how serious the problem is.

Most of these additional options employ the strategy of cooperation and pay more attention to the relationship of David, and his company Z-Corp, with relevant stakeholders. The fourth option is especially relevant in the light of the disputed or unknown facts we have identified.

5.3.4 Ethical evaluation

In this step, the moral acceptability of the various options for action is evaluated. This can be done on the basis of both formal and informal moral frameworks. Formal moral frameworks are based on professional ethics discussed in Chapter 2: the codes of conduct, and the main ethical theoretical backgrounds discussed in Chapter 3: utilitarianism, Kant's theory, and virtue ethics.

Ethical evaluation also can be based on more informal ethical frameworks. We distinguish two such frameworks here: intuitions and common sense. The **intuitivist framework** is rather straightforward: indicate which option for action in your view is intuitively most acceptable and formulate arguments for this statement. The **common sense method** asks to weigh the available options for actions in the light of the relevant values. In a specific case, it might, for example, be possible to argue that although making a profit is important, the value that is really at stake (or dominant) is public safety. In determining which value is dominant, certain guidelines can be followed, such as, "dominant values are usually intrinsic values and not merely instrumental values," and "if more people find a value important, it is more likely that it is a dominant value." Once the dominant value has been selected, the option can be chosen that best meets that dominant value (Brady, 1990).

Intuitivist framework The ethical framework in which options for action are evaluated on basis of one's view about what is intuitively most acceptable and that formulates arguments for this statement.

Common sense method The method that weighs the available options for actions in the light of the relevant values.

The fourth step results in moral evaluations about the various options for action. These judgments need not be the same because different frameworks can result in different preferred options for action in a given situation.

An Illustration of Conflicting Ethical Frameworks

To see how different ethical frameworks may lead to conflicting recommendations, consider the following case[3]:

> Jasmine is director of a building department in a big city. Due to budget constraints, the city has been unable to hire a sufficient number of qualified individuals to perform building inspections. This makes it difficult for the inspectors to do a good and thorough job. At the same time, a new and tougher building code was adopted by the city. While this code promotes greater public safety than the last one, it also contributes to the difficulty inspectors have in doing a good and thorough job.
>
> Jasmine sets up an appointment with the chairman of the city to discuss her concerns. The chairman agrees to hire additional code officials for the building department on the condition that Jasmine agrees to permit certain specified buildings under construction to be inspected under the older, less rigid enforcement requirements. Should Jasmine agree to concur with the chairman's proposal?

Applying Kant's universalization test to this case would yield an argument like this: If Jasmine would comply with the older, less rigid requirements, she acts from the maxim "apply less rigid requirements when you are pressed to do so." This maxim cannot be universalized. That is, if everybody would act like that, rules would become meaningless. Apart from that, the profession of building inspector would be rendered meaningless. If rules are altered when one is put under pressure, what would a building inspector inspect?

Applying an act utilitarian framework, the main question is how Jasmine can achieve the best results. In this case, it seems obvious that the best results are achieved if she agrees with the chairman's proposal. Otherwise she will have too few inspectors and will not even be able to inspect all building according to the old less strict code.

If is worth noting that according to the Kantian framework, the actual consequences of her action are irrelevant, while in the utilitarian framework these consequences are crucial for the ethical judgment.

5.3.5 Reflection

Since the different ethical frameworks, including the informal frameworks, do not necessarily lead to the same conclusion, a further reflection on the outcomes of the previous step is usually required. The goal of this reflection is to come to a well-argued choice among the various options for actions, using the outcomes of the earlier steps.

> **Wide reflective equilibrium** Approach that aims at making coherent three types of moral beliefs: 1) considered moral judgments; 2) moral principles; and 3) background theories. Also the resulting coherent set of moral beliefs is often called a wide reflective equilibrium.

The approach to reflection we want to advocate here is known as the method of **wide reflective equilibrium** (Daniels, 1979, 1996). This approach aims at making coherent three types of moral beliefs:

1) considered moral judgments; 2) moral principles; and 3) background theories. The background theories include ethical theories, but also other relevant theories such as psychological and sociological theories about the person. The inclusion of theories is important because they block the possibility of simply choosing those principles that fit our considered judgments. Achieving wide equilibrium forces us to bring our judgments not only into coherence with principles but also with background theories. Because such theories also apply to other cases, our various considered moral judgments become connected, so that we are forced to examine critically our various considered judgments and eventually have to achieve coherence between the different layers of our moral beliefs.

The basic idea is that in a process of reflection different ethical judgments on a case are weighed against each other and brought into equilibrium. As we see it, this process is not so much about achieving equilibrium as such, but about arguing for and against different frameworks and so achieving a conclusion that might not be covered by one of the frameworks in isolation.

Central to the reflection step is thus argumentation (see Chapter 4). Arguments for or against ethical frameworks can be positioned at two levels. One level is the general criticism of the ethical frameworks. Utilitarianism can, for example, be criticized for neglecting duties or moral rights, while deontological theories might be criticized for not taking into account the consequences of actions. Such criticisms are well-known in moral philosophy and might be helpful for the reflection in this step. The second level of criticism is the concrete situation in which a certain option for action has to be chosen. It might for example be the case that a certain general objection to an ethical theory is not so relevant in a particular case. For example, a general objection against the utility principle ("the greatest happiness for the greatest number") of classical utilitarianism is that it neglects distributional issues (see Section 3.7.3), but it might be that in the particular situation different options for actions hardly have distributional effects, so that in that situation this objection is not so relevant. In general, we suggest two types of questions for reflection on this second level:

- Does an ethical framework provide reasons that support my intuitive opinion? If not, do I have other reasons that support my intuitive opinion? If I have other reasons are they strong enough to override the reasons within the ethical framework? If not, do I have to revise my intuitive opinion and in what way?
- Does an ethical framework succeed in selecting those features of a situation that are morally relevant? Are there any other moral relevant features that are not covered? Why are these relevant and how could they be accounted for?

The result of the fifth step is the choice for one of the options of action; a choice that can be argued in relation to the different ethical frameworks.

5.4 An Example

Above we already applied parts of the ethical cycle to the Gilbane Gold case. We will now apply the whole ethical cycle to a more extensive example. As an example, we adopt a case presented by Harris, Pritchard and Rabins in their book *Engineering Ethics: Concepts and Cases* (see box).

Case Highway Safety

David Weber, age 23, is a civil engineer in charge of safety improvements for District 7 (an eight-county area within a US Midwestern state). Near the end of the fiscal year, the district engineer informs David that delivery of a new snow plow has been delayed, and as a consequence the district has $50 000 in uncommitted funds. He asks David to suggest a safety project (or projects) that can be put under contract within the current fiscal year.

After a careful consideration of potential projects, David narrows his choice to two possible safety improvements. Site A is the intersection of Main and Oak Streets in the major city within the district. Site B is the intersection of Grape and Fir Roads in a rural area.

Pertinent data for the two intersections are shown in Table 5.1.

Table 5.1

	Site A	*Site B*
Main road traffic (vehicles/day)	20 000	5 000
Minor road traffic (vehicles/day)	4 000	1 000
Fatalities per year (3 year average)	2	1
Injuries per year (3 year average)	6	2
PD* (3 year average)	40	12
Proposed Improvement	New signals	New signals
Improvement Cost	$50 000	$50 000

* PD refers to property damage only accidents.

A highway engineering textbook includes a table of average reductions in accidents resulting from the installation of the types of signal improvements David proposes. The tables are based on studies of intersections in urban and rural areas throughout the United States, over the past 20 years (see Table 5.2).

Table 5.2

	Site A	*Site B*
% reduction in fatalities	50	50
% reduction in injuries	50	60
% reduction in PD	25	–25*

* Property damage only accidents are expected to increase because of the increase in rear-ends collisions due to the stopping of high-speed traffic in rural areas.

David recognizes that these reduction factors represent averages from intersections with a wide range of physical characteristics (number of approach lanes, angle of intersection, etc.); in all climates; with various mixes of trucks and passenger vehicles; various approach speeds; various driving habits; and so on. However, he has no special data about Sites A and B that suggest relying on these tables is likely to misrepresent the circumstances at these sites.

Finally, here is some additional information that David knows about.

1 In 1975, the National Safety Council (NSC) and the National Highway Traffic Safety Administration (NHTSA) both published dollar scales for comparing accident outcomes, as shown in Table 5.3.

Table 5.3

	NSC ($)	*NHSTA ($)*
Fatality	52 000	235 000
Injury	3 000	11 200
PD	440	500

A neighboring state uses the following weighting scheme:

Fatality 9.5 PD
Injury 3.5 PD

2 Individuals within the two groups pay roughly the same transportation taxes (licenses, gasoline taxes, etc.).

Which of the two site improvements do you think David should recommend? What is your rationale for this recommendation?

Source: Harris, Pritchard and Rabins, 2005, pp. 325–326. (Reprinted with permission.)

This case description is, we must admit, rather stylized. We have chosen, however, to leave out certain complexities and uncertainties as to be able to show more clearly and straightforwardly how the ethical cycle would proceed in a case like this. In particular, we show that the ethical cycle by including a reflection step moves beyond the simple opposition between a consequentialist and a deontological ethical approach for which this case description was originally devised.

5.4.1 Moral problem statement

In the original case, the moral problem statement is already given: "Which of the two improvements do you think David should recommend?" This is not the only possible moral problem statement in this case. One might for example wonder whether making this decision is actually David's responsibility. The case concerns spending of public funds and it might be argued that such a decision is to be made by the relevant city council or state council. One might formulate as problem statement: "Is it David's (moral) responsibility to make this decision?" We will, however, restrict ourselves here to the problem statement formulated by Harris, Pritchard, and Rabins. This problem statement meets two of the three earlier mentioned conditions for a good problem statement: it is clear what the problem is (which option to choose) and it is clear who has to act (David). It is not clear from the statement itself, however, why it is a *moral* problem. Maybe this is simply a practical decision about what to do or an economic decision about how to spend public fund most efficiently. In fact, many people respond to this problem by stating that it is obvious that option A should be chosen because it results in the highest level of reductions in fatalities and injuries. However, from a deontological point of view it might be argued that site B is the best option because it is fairer to reduce the risk for people who are now subject to the highest risk factor (see Section 5.4.4 for more details). So, considering the case from a deontological point of view helps to realize that there is a potential moral problem here. Ethical theories thus help in recognizing the moral relevant characteristics of a situation and in formulating the moral problem. This also underlines the iterative character of the ethical cycle: it might well be that someone only recognizes the fairness considerations in step 4 of the ethical cycle. He or she might then go back to step 1 and reformulate the moral problem and redo steps 2 and 3.

5.4.2 Problem analysis

Now, we have to state the relevant facts, stakeholders, and interests and values. The main facts are already listed in the detailed case description. Some facts are uncertain. It is, for example, not known whether the general reduction factors for municipal and rural intersections apply to the specific case. There are no indications to the contrary, but this does not guarantee that these factors do apply. Such uncertainties could make a difference for the final judgment on the case.

Apart from David, drivers and their passengers, tax payers, and the relevant city or state council could be distinguished as relevant stakeholders. As a first approximation, one could say that the main interests of these stakeholders are safety (drivers and their passengers), minimal costs (tax payers) and highest safety for lowest cost (city or state council). On closer examination, these stakeholders are not really uniform. Some drivers will use only the city intersection, some – but probably less – only the rural, some will use both, some will use neither; which might result in different preferences about where to place traffic lights. Moreover, some drivers will prefer speeding above safety and will maybe prefer that no traffic lights are placed at all! Most drivers will, as tax payers, prefer minimal costs, which may conflict in this case with increasing safety. To determine which option of action is the "best," it is necessary to make compromises concerning the various interests: you trade off a certain level of safety for a certain level of costs.

Although it is difficult to draw up a definitive list of stakeholders and interests, the above analysis is helpful in distinguishing relevant values. In the formulation of the problem, we have already distinguished two relevant values: safety and fairness. We might now add a third one which is related to the interest of keeping costs low. Low cost is, however, hardly a moral value as such. The moral value at stake here seems to be something as "public utility," which in this particular case implies that higher costs, ultimately resulting in higher taxes, may pay themselves back in higher public utility through higher safety.

5.4.3 Options for actions

In this case, the options for action are already given in the problem formulation. One might, however, wonder whether these two options are really the only ones. Whitbeck, for example, comments on this case:

> Notice ... that the problem is presented as a forced choice between spending all the remaining resources on one intersection and spending it all on the other. In fact, there would likely be many other choices. For example, putting up traffic signs at both intersections may be an alternative to installing traffic lights at either one. (Whitbeck, 1998a, p. 65)

So, it might be useful to think of other options in the light of the relevant values. In Section 5.3.3, we suggested a number of strategies that could be helpful in devising options. The black-and-white-strategy has been chosen in the original formulation of the problem. This has probably been done for didactical considerations, that is, illustrating the difference between a consequentialistic – in particular a utilitarian ethical – framework and a deontological framework. While this may be illuminating, it might also give the wrong impression that the actual problem is best solved by a black-and-white strategy, which is usually not the case. Another strategy, for example, could be the cooperation strategy, which is directed at finding alternatives that can help solve the moral problem by consulting other stakeholders. In this case, it might be useful, for example, to consult drivers and people who live in the neighborhood of the intersections because they may have more specific knowledge about why and what accidents occur at the intersections, or may have creative solutions. Whistle-blowing is not really relevant here because there is not a hidden abuse that needs to be uncovered.

5.4.4 Ethical evaluation

Common sense

According to this approach we first look at the relevant values. In this case the values at stake are safety, fairness, and public utility. You might argue that the dominant value in this case is safety and, consequently, you could argue that the best option for action is the action that reduces the most fatalities and injuries in *absolute* numbers. In this interpretation of safety – using the data that are given in the case description – you should recommend site A. However, other interpretations are possible as well, which can lead to other recommendations. For example, the best option is the action that reduces the most fatalities and injuries in *relative* numbers: a reduction of 0.5 fatalities per 6000 vehicles/day for site B (which corresponds with a reduction of 2 fatalities per 24 000 vehicles/day) is "better" than a reduction of 1 fatality per 24 000 vehicles/day for location A. In this case, the recommendation will be site B. The common sense approach gives no clear-cut answer, but it stresses the importance of the interpretation of safety (assuming that this is the dominant value). So, you have to look for arguments concerning relative versus absolute numbers to motivate and justify your choice.

Table 5.4 Gross benefit per year of placing traffic lights at the two sites using different "pricing schemes"[4]

	Site A	*Site B*
NSC	$ 65 400	$ 58 400
NHTSTA	$ 273 600	$ 129 440
Neighboring state	30 PD	5.95 PD

PD refers to property damage only. The numbers in the table indicate the expected reduction expressed in the unity "property damage only" according to the pricing scheme of the neighbouring state mentioned in the case description.

Utilitarianism

The utilitarian framework selects the option that brings "the greatest happiness for the greatest number." The expected social utility can be calculated with a cost-benefit analysis using the different "pricing schemes" suggested in the case description where money is used to express quantities of pleasure (benefits) or pain (costs). For the sake of simplicity, we leave out the effect of uncertainty in making these calculations, but it is important to recognize that such uncertainties might affect your final judgment.

As Table 5.4 shows, the available data suggest that site A in the city area is to be chosen. In all calculations site A has the largest gross benefit, and also the largest net benefit, since the costs of $ 50 000 is the same in all calculations. The data in the calculations according to the pricing schemes of NSC and NHTSTA, moreover, suggest that the costs of $ 50 000 are recovered within one year for both choices.

Kantian theory

The application of Kantian theory in this case is based on fairness considerations. Kant's first categorical imperative "Act only on that maxim which you can at the same time will that it should become a universal law" implies the equality postulate, that is, the duty to treat persons as equals, that is, with equal concern and respect (see Section 3.8). One could argue that, as a consequence of this postulate, everybody has a right to the same level of protection, so that the same maximum risk factor applies to everyone. In this case, individuals approaching intersection B face a higher risk than individuals approaching intersection A (see Table 5.5). A choice for site B would therefore be fairer, since this decreases the current inequality in risk factors.

Kant's second categorical imperative (the reciprocity principle): "Act as to treat humanity, whether in your own person or in that of any other, in every case as an end, never merely as a means" is difficult to apply to this case. This imperative states that each human must have respect for the rationality of others and that we must not misguide the rationality of others, but in this case the rationality of others is not an issue.

Virtue ethics

From a virtue ethics point of view, one might try to formulate a list of virtues that are relevant for engineers (see Section 3.9.3). One may then ask how a virtuous engineer, employing the relevant virtues, would act in this situation. For example, how can a

Table 5.5 Current risk of fatality and of injury for individuals approaching the intersection per year (under the assumption that there is one person in each vehicle).[5]

	Site A	*Site B*
Fatalities	2.3 E-07	4.6 E-07
Injuries	6.9 E-07	9.1 E-07

virtuous engineer make the decision *objectively*? This might reveal new relevant moral considerations, or might even lead to a reformulation of the moral problem (step 1 of the ethical cycle). One might, for example, begin to wonder whether it is desirable that David makes this choice himself or whether he should merely inform the public authorities who then make the decision.

Professional ethics
If we look at the code of conduct of the *National Society of Professional Engineers*, the following article is relevant to David:

2. Engineers shall perform services only in the areas of their competence.

David has competence in determining the nature and the magnitude of the safety improvements, but this competence is not the same as competence in deciding which location is the better choice. This requires not only engineering knowledge, but also moral competence. You could argue in this case that David should not make the choice himself, but that he should inform completely the public authority so that they can make a conscious choice.

5.4.5 Reflection

Since the applied ethical frameworks provide different outcomes, further reflection is required. First of all, in this case, one could reflect internally on the frameworks. For convenience, we will leave aside common sense, the virtual ethical framework and the professional ethics, and focus on the utilitarian and Kantian framework. With respect to the utilitarian framework, one could for example question whether the provided data on the monetary value of a human life, injuries, and property-damage only accidents are adequate. Nevertheless, the various monetary schemes and the weighing scheme of the neighboring state all suggest the choice of site A over site B. In fact, it is not possible to devise a monetary scheme in which site B would score better unless one's weights human lives negatively and/or injuries and property damage positively. So the outcome that the utilitarian test selects site A is rather robust.

This is less so for Kant's approach or the fairness test. The rural intersection is more dangerous in terms of the probability of a fatality or injury per vehicle approaching the intersection. However, we do not know the average number of people in a car and whether this number is the same for the rural and urban intersection. The data therefore do not rule out that the individual risk of a car driver or passenger in expected fatalities *per year* is actually higher on site A than on site B, contrary to what Table 5.5 suggests.

There are also other reasons to doubt whether fairness considerations necessarily suggest the choice for section B. If fairness is understood in terms of a right to protection, this is perhaps best understood in terms of an equal level of minimal safety for everyone. It might well be that that level is already met at both intersections. Alternatively, one could understand fairness in terms of equal absolute safety. This would mean that everybody has a right to the same absolute level of risk. This would have rather absurd consequences, however. It would, for example, imply that if someone would be very safe off, for example due to chance, everybody would have the right to that level of safety, even if that would be very hard, if not impossible, to realize. It would even imply that it would be desirable to make the safest person less safe, even if that would increase the safety of nobody else, because in this way a more equal distribution of risks is achieved.

The last remarks already make clear that applying only the Kantian framework without considerations of overall safety or public utility does not make much sense in this case. Conversely, one might argue that public utility or overall safety considerations alone are also not enough, which would mean that the utilitarian framework alone is too narrow to judge this case. What seems required then is a certain balancing of the various moral frameworks or considerations, including possibly also one's intuitive opinion and common-sense considerations

The approach that we advocate here is that of wide reflective equilibrium. Suppose that someone has the considered judgment that location A is best (belief a). He or she might defend this choice by referring to the principle "the greatest happiness for the greatest number" (belief b). This principle, in turn, might be justified on basis of the ethical theory of utilitarianism (belief c). Utilitarianism is not only a theory about where to place traffic lights but a much broader theory that is related to a whole range of moral judgments, including the judgment that – for the sake of comparison – we can express human lives in a common value like money (belief d). The same person judging that location A is best (belief a) might reject the moral judgment that we can express the value of human live one way or the other in money (belief d). In this case, the set of beliefs a, b, c and not-d is incoherent.

There are several ways you can solve the incoherence between a, b, c and not-d. We mention some:

- You could give up the belief not-d. After all, you might come to the conclusion that human life is not priceless, even if you intuitively thought so. So you might choose to adopt the belief d.
- You could also look for another ethical theory (c) or another ethical theory with moral principles (b and c), which would still justify a, but would not imply other moral judgments, like d, that you consider dubious.
- You might also try to look for a theory that better fits your judgments about valuing human life. You might, for example, have the considered moral judgment that since we cannot put a price on human lives, you should treat humans equally and

respect their freedom. On the basis of such a belief you might embrace – at least for the moment – a deontological ethical theory and some principle of fairness. On that basis, you might revise your initial belief a about the case, and now choose site B.

This list does not exhaust the possibilities. One could also try to combine utility and fairness considerations in several ways. One could for example argue that fairness considerations imply that all drivers and passengers have a right to the same minimal level of safety. One might then argue that this level is actually met at both intersections, so that one can choose without scruples the option with the highest public utility – location A.

The important point about this example, however, is not how you solve the incoherence between your different beliefs. The important thing is that by trying to achieve a wide reflective equilibrium you are forced to engage in a broader and more systematic theoretical consideration of the case, including a range of arguments and reasons.

5.5 Collective Moral Deliberation and Social Arrangements

The emphasis in the ethical cycle is on individual judgment. However, in many, if not most, situations in real life, other people will be involved in and affected by your choices. You might doubt whether in such situations, individually achieving a conclusion on how to act is justified. In particular, you might wonder why others, especially people affected by your actions, should accept your individual conclusion on how to act. Of course, if you have used the ethical cycle, you will be able to argue your choice, but given the nature of moral reflection and the diversity of ethical frameworks that might give conflicting advices, it seems doubtful that every person using the ethical cycle would come to the same conclusion as you did. The natural inclination of many ethicists would be to look for a better, overarching moral framework. Even if one would believe that such an endeavor is worthwhile, it certainly does not solve the problem if you have to act here and now. We therefore propose a more practical solution – engaging in a moral deliberation with other people involved and possibly affected.

Engaging in deliberation is also useful for other reasons. If you are confronted with moral problems you often have to act in a situation in which you depend on others to achieve certain options for action. A certain support from others is therefore required to be able to act in a morally effective way. This is certainly true if you are working in a company. Therefore, deliberation and discussion with others are important additions to the ethical cycle.

The final step in the ethical cycle is reflection, leading to a well-argued choice for an option of action. This choice, however, needs not be your final choice; it can also be seen as a provisional choice that you can revise in discussion with others. The objective of such deliberations is to make public your reasons for a certain choice and to expose them to criticism by others. Such discussion and criticism can result in a revision of your choice, for example because your arguments turn out not to be adequate after all, or because certain arguments have been overlooked. So conceived, deliberation is mainly a tool to improve one's moral judgment.

Moral deliberation An extensive and careful consideration or discussion of moral arguments and reasons for and against certain actions.

However, as already suggested above, one could also argue that **moral deliberation** is essential for more fundamental reasons. Philosopher Jürgen Habermas has argued that moral judgments are legitimized by them being the result of a moral deliberation that meets certain standards (Habermas, 1981). This includes the standard that the discussion should not be decided on the basis of authority or power, but on the basis of arguments. Other requirements for rational discussion or deliberation are that people should be honest and sincere, and should argue their point of view. The idea is that if deliberation meets such requirements, we have good reason to believe that the outcomes are sound.

A somewhat comparable idea has been formulated by the political philosopher John Rawls. Rawls (1971) embraces the wide reflective equilibrium approach that we earlier described in this chapter (Sections 5.3.5 and 5.4.5). However, he realizes that it is very well possible that people come to different reflective equilibriums (especially in his later work, see Rawls, 1993, 2001). He nevertheless believes that people might often agree on moral issues even if they disagree on how their moral judgments are exactly to be justified. He calls this situation an **overlapping consensus**. Here we shall understand an overlapping consensus as agreement on the level of moral judgments, while there may be disagreement on the level of moral principles and background theories. In the case of safety improvements described in Section 5.4, two people might for example agree that a choice should be made for site A but one person might justify this choice in terms of utilitarianism while the other adopts a notion of fairness that requires a minimal level of safety that is in his view already met at both intersections so that a choice can be made for the one that is best in absolute numbers.

Overlapping consensus An agreement on the level of moral judgments, while there may be disagreement on the level of moral principles and background theories. Each of the participants should be able to justify the overlapping consensus in terms of his or her own wide reflective equilibrium.

An overlapping consensus is different from a compromise because it requires that each of the discussants can justify the overlapping consensus in terms of his or her own reflective equilibrium. In case of a compromise, you sometimes accept an outcome because you think it is the best you can get given the preferences of the others involved. But how can we achieve an overlapping consensus? Rawls himself believes that the achievement of an overlapping consensus is easier if all parties involved accept a reasonable degree of pluralism among moral opinions. When we do, Rawls believes, we are also able to distinguish between private reasons and public reasons in moral discussions. Private reasons are reasons that are important for how I want to live my life and make private moral decisions that do not directly affect others. I may, however, recognize that others do not, and should not, share my private reasons. Public reasons are reasons that we think apply to everyone. Rawls believes that when our focus in moral discussions is, as much as possible, on public reasons we are more likely to achieve an overlapping consensus.

In addition to Rawls, it might be argued that the achievement of an overlapping consensus on certain moral issues is more likely if the social arrangements in which we, for example, develop technologies, meet two procedural criteria (Van de Poel and Zwart, 2010).

1 *Learning.* A distinction can be made between first-order learning, in which the people involved learn better to achieve given goals, and second-order learning, in which learning takes place with respect to what goals to strive for and what moral values to take into account.
2 *Inclusiveness and openness.* Inclusiveness means that all relevant perspectives are included in the debate, for example by engaging a diversity of relevant stakeholders. Since inclusiveness is usually relative – what is relevant may change in the course of time or maybe subject to disagreement – also openness is important. Openness means that new considerations and parties can enter the debate.

Learning is important because it makes it more likely that actors change their opinion (reflective equilibrium), so that an overlapping consensus may become achievable where it previously was not. Inclusiveness and openness are important to avoid the circumstance where a consensus is achieved by leaving out certain relevant considerations.

Both the perspectives of Habermas and Rawls stress the importance of procedural criteria for arriving at a moral judgment, and both require social arrangements that meet certain norms. In this respect, both fit well with approaches such as Constructive Technology Assessment (CTA; Section 1.6, see also Section 7.5) and the social ethics approach to engineering (Section 3.10.3) that were briefly discussed before. This emphasis is different from the ethical cycle in which the various substantive ethical frameworks play a much more important part. We think that this need not be seen as incompatible, however. To engage in a moral deliberation, it is desirable that you have a well-argued moral opinion. Of course, you should be willing to revise your opinion, but in order to have a debate at all, you should first have your own well-argued opinion. For this purpose, the ethical cycle, including the use of substantial ethical frameworks to arrive at a moral opinion, is very useful.

5.6 Chapter Summary

Moral problem-solving is a difficult and complex process because moral problems are usually ill-structured. They do not have a clear-cut problem formulation, need to satisfy different, often conflicting, moral constraints and have not one best solution. In these respects moral problems are like design problems. Solving moral problems therefore does not only require analysis but also synthetic reasoning (devising new options) and creativity.

The complex nature of moral problem-solving does not preclude a systematic approach. The approach that we have introduced in this chapter is called the ethical cycle. It consists of five basic steps:

1 Formulating the moral problem;
2 Analyzing the problem in terms of stakeholders and their interests, values and facts;
3 Identifying and devising options for action with the help of strategies such as the black-and-white strategy and the cooperation strategy;
4 Ethical evaluation of the various options for action with the help of various ethical frameworks;
5 Reflection on the outcomes of the evaluation phase, resulting eventually in a well-argued choice for one of the options for action.

With respect to the reflection step, we have proposed the wide reflective equilibrium approach that aims at coherence between moral beliefs at three levels: 1) considered moral judgments; 2) moral principles; and 3) background theories. The important thing is that by trying to achieve a wide reflective equilibrium you are forced to engage in a broader and more systematic theoretical consideration of the case, including a range of arguments and reasons. It is precisely because this reflection involves *theories*, that such reflection becomes broader and more encompassing. This suggests that theories have an important role to play in moral judgment. However, this role is more complex than simply applying the theory to the case at hand.

In addition to coming to an individual moral judgment by using the ethical cycle, discussion with others is important. The goal of such discussion is to make public the reasons you have for a certain judgment and possibly to revise those reasons and your conclusion in debate with others. Such discussion is also important because other stakeholders in technological development might not agree with your conclusions, and including their point of view may improve the decision made. Deliberating with them makes the final result more legitimate, because you are then also respecting the moral autonomy of other stakeholders, and are expressing due care to them. Moreover, it is likely to make you as an engineer more effective because you will often need the cooperation of others to live by your moral judgments.

Study Questions

1 Why are moral problems ill-structured problems?
2 Why are most moral problems not real dilemmas?
3 XYZ orders 5000 custom made parts from ABC for one of its products. When the order is originally made ABC indicates it will charge $75 per part. This cost is based in part on the cost of materials. After the agreement is completed, but before production of the part begins, ABC engineer Christine Carsten determines that a much less expensive metal alloy can be used while only slightly compromising the integrity of the part. Using the less expensive alloy would cut ABC's costs by $18 a part. Christine brings this to the attention of ABC's Vernon Waller, who authorized the sales agreement with XYZ. Vernon asks, "How would anyone know the difference?" Christine replies, "Probably no one would

unless they were looking for a difference and did a fair amount of testing. In most cases the performance will be virtually the same – although some parts might not last quite as long." Vernon says, "Great, Christine, you've just made a bundle for ABC." Puzzled, Christine replies, "But shouldn't you tell XYZ about the change?" "Why?" Vernon asks, "The basic idea is to satisfy the customer with good quality parts, and you've just said we will. So what's the problem?"

a. What exactly is Christine's problem here? Explain why it is a moral problem.
b. Which moral values or principles are at stake here?
c. Mention three things Christine can do to deal with his problem.
d. What would an analysis of this problem according to (Bentham's) classical utilitarianism look like? What would the utilitarian advice to Christine be?
e. Would John Stuart Mill's modified form of utilitarianism lead to a different advice? Why or why not?
f. What would a Kantian ethicist recommend to Christine? Motivate your answer.
g. What do you think is the right thing to do for Christine? Motivate your answer and also explain why you do not accept (some of) the advice from d, e, and f.

4 Apply the ethical cycle to the Challenger case from Section 1.1. Explain how each step would be applied in this case.
5 Apply the ethical cycle to the BART case from Section 2.1. Explain how each step would be applied in this case.

Discussion Questions

1 What do you consider appropriate grounds for overriding someone's personal decisions? Would you, for instance, prevent the sale of home body piercing kits or child pornography, and if so, on what grounds?
2 Motivate what you would do if you were David Jackson (see Section 5.1).
3 According to the wide reflective equilibrium approach, people should aim at coherence between the different levels of their (moral) beliefs. Is coherence indeed as important as this approach presupposes? Are coherent beliefs never wrong? Can a belief be right and nevertheless be incoherent? Can you think of an approach to ethical judgment in which coherence is not important at all?
4 Do discussions with others (moral deliberation) lead to better moral judgments? Why or why not?

Notes

This chapter is based on Van de Poel and Royakkers (2007)

1 Whitbeck provides us with a compelling sketch of what a designer-perspective on moral problems could offer, but the analogy was not fully developed. In Dorst and Royakkers (2006) this analogy is constructed more carefully and completely.
2 For the main considerations in the development of Agora and the features of this program, we refer to Van der Burg and Van de Poel (2005).
3 This is case 98–5 of the Board of Ethical Review (BER) of the National Society of Professional Engineers (NSPE). Available at http://www.niee.org/cases/ (accessed September 29,

2009). The case description here is adapted a bit and can also be found in Agora (www.ethicsandtechnology.com).

4 As an example we will figure out the gross benefit of site A according to the pricing scheme of NSC:

$$((50\% \text{ of } 2) \times \$52\,000) + ((50\% \text{ of } 6) \times \$3\,000) + ((25\% \text{ of } 40) \times \$440) = \$65\,400.$$

5 As an example we will figure out the current risk of fatality of site A:

$$2/((20\,000 + 4000) \times 365) = 2.3 \text{ E-07}.$$

6

Ethical Questions in the Design of Technology

Having read this chapter and completed its associated questions, readers should be able to:

- Identify ethical issues at the different stages of the design process;
- Understand how conflicts between design requirements in design may amount to value conflicts;
- Describe the various methods for dealing with value trade-offs in design and their pros and cons;
- Apply these methods to engineering design problems;
- Describe the difference between normal and radical design and the moral relevance of this distinction;
- Discuss to what extent engineers can rely on existing regulative frameworks in making ethical decisions in engineering design.

Contents

Ethics, Technology, and Engineering: An Introduction, First Edition.
Ibo van de Poel and Lambèr Royakkers.

6.1 Introduction

Case High Speed Train Disaster in Germany

Figure 6.1 Intercity Express train crash. Photo: © DPA/Press Association.

On June 3, 1998, the German high speed Intercity Express (ICE) train 884 "Wilhelm Conrad Röntgen" derailed at a speed of about 200 kilometers an hour and ran into a bridge that fell down on the train. The subsequent cars of the train jackknifed into the crashed cars in a zig-zag pattern. Overall, 101 people were killed and 88 severely injured. Investigations after the accident showed that the disaster was to a large extent due to a change in the wheel design of the train. Originally the train had been equipped with monobloc wheels, which are single casted. Such wheels decrease the rolling friction between the wheels and the rails, so increasing energy efficiency and lowering the costs of electricity consumption. However, such wheels may also decrease the comfort and may result in material stresses in the wheel. In some high speed trains, the comfort problem is solved by using a bogie (the framework carrying the wheels) with air suspension. The German ICE however was based on conventional bogies with steel springs. Once in service, it turned out that the wheel system caused severe resonance and vibration. This was considered a problem, especially in the dinner car where dinnerware "walked" over the tables. Engineers of the German railways then proposed to solve the problem with a new wheel design that had a rubber ring between the steel tire and the wheel body. This wheel design had been used in trams, but at significantly lower speeds. The new wheel design was not tested at high speed before it came into service, but was based on existing experience and materials theory. Nevertheless, after introduction the wheel solved the vibration problems and did not show any major problems until the fatal accident.

The new wheel design consisted of a steel wheel body surrounded by a 20 mm thick rubber damper and then a relatively thin steel tire. It was this tire that eventually failed and let to the disaster. This could have been caused by two mechanisms. One is metal fatigue due to the fact that the metal tire around the wheel is deformed a bit every rotation. The other mechanism is that the rubber damper extends due to heating up. Since rubber extends more than steel due to heat and since the rubber tire is locked in between the wheel body and the metal tire, this will result in a high pressure on the steel tire and, as a result, cracks may form from the inside in the metal tire. The first of these mechanisms was known at the moment that it was decided to change the wheel design, the second one not.

The wheels of the train were routinely checked every day. First, it was measured whether the wheels were still round enough or had become ellipsoids. The wheel out-of-roundness should not be larger than 0.6 mm. In practice, higher values were measured and allowed, probably because an out-of-round wheel was not seen as a safety issue but rather as an issue of comfort and wear. This might have been true for monobloc wheels, but it was certainly not true for the new wheel design. Second, the total wheel diameter should be at least 854 mm. A new wheel had a diameter of 920 mm. At the last check before the accident the wheels had a diameter of 862 mm. Later investigations of the Fraunhofer Institute in Darmstadt suggested that a norm of 890 mm would have been more appropriate given the new wheel design. Again, the norm was probably based on monobloc wheels, neglecting the peculiarities of the new wheel design. Third, the wheels were also inspected for cracks. Initially, advanced testing machines were used to inspect the wheels. These produced, however, a large number of false positives, that is, indications of a defect in the wheel while there actually was no problem. As result of these false positives, the use of the advanced inspection apparatus was discontinued and the wheels were only inspected visually with a flash light and audibly by sledging a hammer against the wheels. The latter method, which is often used for monobloc wheels, was probably inadequate for the new wheel design because the rubber damper will absorb the hammer stroke.

Three engineers, two from the German railways (Deutsche Bahn) and one from a supplier, who all had been involved in the certification of the wheels were tried and charged with manslaughter. After 53 days, the judge concluded that the three had not been grossly negligent. Three reasons were given for this. First, among experts there was not a principled objection against the new type of wheels. Second, the metal cracks that could cause an accident like this can develop rather quickly, so that adequate inspection procedures might not have prevented an accident like this. Third, it was not considered proven that the engineers had made gross mistakes in calculating the load on the wheels. Therefore the case was dismissed on the condition that each defendant paid a fee of 10,000 Euros (a possibility that is specific to German criminal law).

Source: Case description is based on Brumsen (2006) and http://en.wikipedia.org/wiki/Eschede_train_disaster (accessed August 18, 2009).

The disaster with the German ICE train illustrates the importance of design to engineering. In this case, as in other cases, the design phase was crucial for the proper working of a technology and possible risks and other side effects. The example also shows the importance of activities that are closely related to design like testing, certification, and inspection. Second, the example shows that most design involves trade-offs. In wheel design for trains a trade-off is faced between energy (and cost) efficiency and comfort. Monobloc wheels are rather efficient because they cause little rolling resistance between the wheel and the track but they may negatively influence comfort (especially at high speeds). The compound wheels provided more comfort against a somewhat higher rolling resistance and, although unforeseen, lower safety. Finally, the example shows the importance and risks of innovative or radical design. The new wheel design was innovative in the sense that it had not been used for high speed trains. Although such innovative design may potentially solve important design problems, it may also create new risks as the example clearly illustrates.

All three issues will be covered in this chapter. We start with a description of what designing is and what type of ethical issues may arise in the various stages of the design process (Section 6.2). Next, we shall take a closer look at one aspect of the design: the choice between different conceptual designs in the light of design requirements and trade-offs (Section 6.3). Finally, we look at the difference between normal and radical design and the moral relevance of this difference (Section 6.4).

6.2 Ethical Issues During the Design Process

The design process is a central area where ethical considerations concerning technology arise. The reason for this is that crucial decisions regarding technology are made in the design process. To an important extent the design of a technology determines how it will be produced and be used, what maintenance will be required, and how the product is to be scrapped. Obviously, later choices, by for example users, are also of importance, but choices in the design process greatly influence the social consequences of a product. Therefore, nearly all ethical questions related to technology development that engineers are confronted with are reflected in the design process in some way or another.

Designing can be described as an activity in which engineers translate certain functions or aims into a working product or system. A ferry can be conceived of as the translation of the function "transporting people from one side of the river to the other." In most cases a function or social goal can be translated into a technical solution in several ways. If you want to achieve transport between two riverbanks, you can choose among a series of possible technical solutions, such as a bridge, a tunnel, a ferry, or a cable-lift. The solution chosen not only depends on the function to be realized, but on a series of additional design requirements, such as speed of transport, costs, building time, sustainability, and safety.

Engineering design The activity in which certain functions are translated into a blueprint for an artifact, system, or service that can fulfill these functions with the help of engineering knowledge.

Design process An iterative process in which certain functions are translated into a blueprint for an artifact, system, or service. Often the following six stages are distinguished: problem analysis and formulation; conceptual design; simulation; decision; detail design; and prototype development and testing.

Engineering design is thus the process in which certain functions are translated into a blueprint for an artifact, system, or service that can fulfill these functions. Engineering design is usually a systematic process in which use is made of technical and scientific knowledge. The **design process** is an iterative process that can be divided in different stages, like:

- *Problem analysis and formulation*, including the formulation of design requirements;
- *Conceptual design*, including the creation of alternative conceptual solutions to the design problem and possible reformulation of the problem;
- *Simulation* of one or more concept design to test how well they meet the design requirements;
- *Decision*: choice of one conceptual solution from a set of possible solutions;
- *Detail design:* the design is further detailed;
- *Prototype development and testing*, in which a prototype is developed and tested. This testing may lead to adaptations in the design.

Below we go through the various stages in the design process and indicate at every stage what the most important ethical issues are that you can encounter as a designer. In addition, we will briefly pay attention to ethical issues in manufacture and construction.

6.2.1 Problem analysis and formulation

Problem analysis stage The stage of the design process in which the designer or the design team analyses and formulates the design problem, including the design requirements.

Design requirements Requirements that a good or acceptable design has to meet.

During the **problem analysis stage**, the designer or the design team conceptualizes the design problem. This stage results in a certain formulation of the design problem and of certain **design requirements** that a good or acceptable solution has to meet. Findings in later stages can sometimes result in the revision of the problem formulation or the design requirements.

Problem formulation and perspective

The formulation of the design problem is of great importance, because it determines the framework in which the problem will be approached during the rest of the design process.

Often the party commissioning the design only has a vague notion of what the problem really is. The designers then play a crucial role in explicitly defining the problem.

In formulating the design problem, a certain perspective may be implicitly or explicitly chosen. An example is the design of a search engine for the Internet. In this case, you can take the perspective of the company that provides the search engine. For them, the search engine will not only have to operate properly and be easy to use, but it will have to store information on the search behavior of visitors so that the search engine provider can sell banner adverts to potentially interested parties. From the perspective of the user of the search engine having to deal with banners is probably undesirable. This is not only because the users do not wish to be bothered by banners, but also because they may consider the storage of search data a violation of their privacy. The choice of perspective for the design problem in the example mentioned has ethical relevance too. The question is which interests, from which of the parties involved, must be taken into account from a moral viewpoint, and which moral norms and values such as the respect for privacy are a stake.

Formulation of design requirements

Next to the design problem, the design requirements are formulated. On the basis of professional and corporate codes, a number of ethical considerations can be mentioned that should be taken into account in formulating the design requirements. This concerns matters like safety, health, the environment, sustainability, and the social consequences of technologies. Design requirements derived from ethical considerations are also sometimes laid down in legislation. For example, there is much legislation in the fields of safety, health, and the environment that enforces requirements on the design process. Next to general legislation, there often are **technical codes and standards** that are relevant for the design process. Technical codes are legal requirements that are enforced by a governmental body to protect safety, health, and other relevant values. Technical standards are usually recommendations rather than legal requirements that are written by engineering experts in standardization committees. Codes and standards have two main functions (Hunter, 1997). The first is standardization and the promotion of compatibility. This results in, for example, the design drawings being understandable and clear for others and spare parts being compatible. A second aim of codes and standards is guaranteeing a certain quality or protecting public values. Though ethical considerations usually are not explicitly stated in codes and standards, ethical considerations concerning matters like safety, health, and the environment often are the foundation for the content of codes and standards. In some engineering areas, designs have to be certified before they may enter production and use. **Certification** is the process in which it is judged whether a certain technology meets the applicable technical norms and standards and is, for example, safe enough. In the case of the new wheel design for the ICE

Technical codes and standards Technical codes are legal requirements that are enforced by a governmental body to protect safety, health, and other relevant values. Technical standards are usually recommendations rather than legal requirements that are written by engineering experts in standardization committees.

Certification The process in which it is judged whether a certain technology meets the applicable technical codes and standards.

train discussed at the beginning of this chapter, the question was raised whether these wheels should have passed certification. Some of the engineers involved in certification were prosecuted, although they were, eventually, not convicted.

Specifying design requirements

Another point at which ethical questions can arise is the specification of the design requirements. A requirement like safety can be specified in various ways. In the design of a chemical plant, one can look at the safety of employees and that of people living close to the plant for example. Ethically, it would not be acceptable to limit safety to only the employees. Obviously, people in the direct environs of such a plant experience the consequences of the design choices made without having had the opportunity to agree or disagree or to benefit directly from the plant.

6.2.2 Conceptual design

Conceptual design stage The stage in which the designer or the design team generates concept designs. The focus is on an integral approach to the design problem.

Creativity The virtue of being able to think out or invent new, often unexpected, options or ideas. Creativity is an important professional virtue for designers.

In the **conceptual design stage** the aim is to generate concept designs. The focus is on an integral approach to the design problem. The designer does not try to realize each design requirement independently, but works on a combination of design requirements and searches for a total concept that can bring about this combination.

During the conceptual design stage the **creativity** of designers is of major importance. Creativity is a prime virtue for designers. Designers are better able to do their work the more creative they are. Next to creativity, virtues like competence, precision, honesty, accuracy, and reliability are important for designers (see Chapter 3). Creativity is a professional virtue, but it is not a moral one. This means that it is an important professional characteristic for a designer. However, creativity does not make you a morally better person. Nevertheless, creativity can also from a moral point of view be an important virtue for designers. It can help to bridge seemingly opposed moral values that play a role in the design process. A good example is the design of the storm surge barrier in the Eastern Scheldt, in the Netherlands, where the moral values of safety and ecological care were at stake.

Case The Design of the Storm Surge Barrier in the Eastern Scheldt

After a huge flood disaster in 1953, in which a large number of dikes in the province of Zeeland, the Netherlands, gave way and more than 1800 people were killed, the Delta Plan was drawn up. Part of this Delta Plan was to close off the Eastern Scheldt. From the end of the 1960s, however, there was growing societal opposition to closing off the Eastern Scheldt. Environmentalists, who feared the loss of an ecologically valuable area because of the desalination of the

Figure 6.2 Eastern Scheldt storm surge barrier. Photo: © Daniel Täger/Fotolia.com.

Eastern Scheldt and the lack of tides, started to resist its closure. Fishermen also were opposed to its closure because of the negative consequences for the fishing industry. As an alternative they suggested raising the dikes around the Eastern Scheldt to sufficiently guarantee the safety of the area.

In June 1972, a group of students launched an alternative plan for the closure of the Eastern Scheldt. It was a plan that had been worked out as a study assignment by students of the School of Civil Engineering and the School of Architecture of the Technical University of Delft and the School of Landscape Architecture of the Agricultural University of Wageningen. The aspects the students focused on were safety and ecological care. On the basis of these considerations, they proposed a storm surge barrier, that is, a barrier that would normally be open and allow water to pass through, but that could be closed if a flood threatened the hinterland. The flood barrier was a creative compromise to balance the two moral values, safety and ecological care, that were at stake.

At first the Rijkswaterstaat, the governmental body responsible for waterways in the Netherlands, discarded the idea because it was not considered feasible technically. However, pressure from political developments – parliament too started to resist closing off the Eastern Scheldt – made the Rijkswaterstaat take the option more seriously and after some time it was decided to build a storm surge barrier. Though the storm surge barrier turned out to be much more expensive than the original solution – and also exceeded the original budget – many still consider the design to be a creative and acceptable compromise between safety and ecological values.

Source: Van de Poel (1998).

6.2.3 Simulation

Simulation stage The stage of the design process in which the designer or the design team checks through calculations, tests, and simulations whether the concept designs meet the design requirements.

The concept designs are checked in the **simulation stage** to see whether they meet the design requirements. This takes place in a number of ways, for example, through calculations, carrying out computer simulations, and doing tests with prototypes. An important question in this stage is how reliable the predictions are that are made in the design process about the later behavior of the designed product. The reliability of predictions is mainly a methodological issue and, therefore, falls outside the scope of this book. However, moral considerations play a partial role in how much reliability in predictions is desirable or acceptable. The answer to this question depends in part on what is morally at stake. In the case of the design of a nuclear power plant, where an accident can result in thousands of deaths and an area can become uninhabitable for an extended period, the demands placed on the reliability of a statement concerning the probability that an accident will occur are considerably higher than say for a can opener.

Computer models

Computer models are often used in simulations. Next to the fact that such models sometimes make use of knowledge that is not completely reliable, computer simulations can be unreliable for a number of reasons (Petroski, 1982, ch. 15):

- Computer models can contain mistakes or errors that the users of the model are unaware of. One example is Ariane 5's failure to launch in June 1996 (see box).
- The assumptions made in drawing up a computer model can be wrong even if no explicit errors or mistakes are made. The problem is often compounded because users of the models are unaware of the assumptions made by the model-makers and the possible unreliability of these assumptions.
- The users of computer models are sometimes unaware of the limited domain of application for such models. There is a danger that the computer models used are extrapolated to fields of application in which their predictions are less reliable.

These problems often increase because computer models evoke an appearance of precision and certainty. As a result, engineers often place more trust in the reliability of predictions based on computer models than is justified. This misconception corresponds with the fallacy of wishful thinking (see Section 4.5.1).

Case The Explosion of the First Ariane 5 Rocket

On 4 June, 1996, an unmanned Ariane 5 rocket exploded just after its launch. The rocket and its load together had a value of 450 million Euros. An investigation showed that the software had instructed the rocket to self-destruct.

Figure 6.3 Ariane 5 shuttle, Photo: copyrightESA/CNES/ARIANESPACE-Activité Photo Optique Video CSG.

The problem could be traced to the fact that the software had not been properly adapted and still was based on the Ariane 4 rocket. The Ariane 5 had a more powerful engine than the Ariane 4, so that the software had to cope with far larger numbers. The error occurred when the horizontal speed was measured and stored as a 64-bit real number and translated into a 16-bit real number. However, the value was higher than 32768, which is the highest number that can be described in 16 bits, so that instruments were fed the wrong values.

6.2.4 Decision

Decision stage The stage of the design process in which various concept designs are compared with each other and a choice is made for a design that has to be detailed.

In the **decision stage**, various concept designs are compared with each other and a choice is made for a design that has to be detailed. The results from the simulation stage are used for this comparison. Evaluation of the different possible designs usually takes place in terms of the design requirements that resulted from the analysis stage. Following that various designs can be evaluated in terms of the positive and negative social impact they may have. In general, there will not be just one design that meets the various design requirements best. To have a precise understanding of why this is the case, it is useful to consider the different shapes that design requirements can take on. Some requirements are formulated in such a way that they are met completely or not at all. Some requirements are so formulated that products meet them to a greater or lesser extent. This is usually the case for requirements like safety, sustainability, costs, and ease of use. Characteristically, such requirements are formulated in terms like "as much as possible" (safety, sustainability, and ease of use) or "as low as possible" (costs). These kinds of design requirements are often referred to as **design criteria**. For most design processes, none of the possible designs scores best for all of the design criteria. The safest car design usually is not the cheapest. To determine which design is the "best," it is necessary to make compromises concerning the various design criteria. This is also referred to as a **trade-off** between design criteria. You trade off a certain level of safety for a certain level of costs. The crucial question is how we can determine the most desirable or acceptable trade-offs between the different design criteria. In many cases this question has an ethical side to it, because different design criteria like safety, sustainability, and ease of use have a moral motivation. In such cases trade-offs may amount to value conflicts, which we will discuss in more detail in Section 6.3.

Design criteria A kind of design requirements which are formulated in such a way that products meet them to a greater or lesser extent. Design criteria are often used to compare and choose between different concept designs.

Trade off Compromise between design criteria. For example, you trade off a certain level of safety for a certain level of sustainability.

Decision-making in design raises other ethical issues as well. One issue is who to include. Design potentially affects not only the lives of users but of a range of other stakeholders as well who might have quite different perspectives on the design problem and the desirability of potential solutions. These perspectives are not only relevant during the formulation of the design requirements but in the decision stage as well, because different stakeholders may have, for example, conflicting opinions about what are acceptable trade-offs in design. In the design literature, various approaches for including users and other stakeholders in design have been developed, such as participatory design (e.g., Schuler and Namioka, 1993). Also the approach of Constructive Technology Assessment (CTA) that was described in Section 1.6 is relevant here. The important norm here is that decision-making in design should be *inclusive* with respect to relevant stakeholders and moral considerations.

Another issue is that many choices are made in designing that are not explicit decisions but that are simply seen as the best way to deal with this problem or to proceed.[1] Although implicit choices are not necessarily bad choices, a range of incremental and implicit choices may result in a situation that nobody would have wanted had it been the result of an explicit one-shot choice. In fact, many moral problems in design seem to stem not so much from a deliberate immoral decision, but from a range of decisions that in themselves are at worst morally dubious (Lloyd and Busby, 2003).

A good illustration is the interpretation of the Challenger case by the sociologist Diane Vaughan (Vaughan, 1996). In Section 1.1, we described the Challenger case primarily as a conflict between managers and engineers. According to Vaughan, however, the decision to launch the Challenger should be seen in the light of a pattern of earlier, partly implicit, decisions in which the interpretation of what was technically occurring when O-rings eroded on earlier flights and what was still an "acceptable risk" shifted. The decision to launch the Challenger at the night before the fatal disaster fitted in this pattern, although it once again was a reinterpretation, and broadening, of what was still an acceptable risk.

Vaughan describes the Challenger launch decision as a case of "**organizational deviance**": norms that are seen as deviant or unethical outside the organization are seen within the organization as normal and legitimate. Most people believe, with hindsight, that the Challenger never should have been launched while the decision to do so fitted many of the implicit norms and rules that had evolved within the organization. The reason for this organizational deviance was not that NASA did not care about safety; in fact safety concerns are important within the organization and NASA has tried to create a range of organizational procedures to safeguard safety. The deviance rather came about as the result of a pattern of incremental and partly implicit decisions with respect to the O-rings.

Organizational deviance Norms that are seen as deviant or unethical outside the organization are seen within the organization as normal and legitimate.

Vaughan's analysis of the Challenger disaster illustrates a more general point: decisions – also incremental and implicit ones – tend to commit us to certain courses of actions and frame subsequent decisions (Darley, 1996). We find it very hard to revise a decision even if we know it has been a wrong decision. While this is true for individual decisions, it is usually even more difficult to revise collective decisions. One important lesson is that adequately organizing decision-making during the design process is essential to good design. David Collingridge has suggested four criteria for such decision-making:

1 corrigibility of decisions;
2 choose systems that are easy to control;
3 flexibility of the decision; and
4 insensitivity of the decision to error. (Collingridge, 1980, pp. 32–42)

6.2.5 Detail design

Once the choice has been made for a particular design, it has to be elaborated on and detailed. Also in the **detail design stage** ethical questions can arise. The importance of this design stage is underlined by the case of Golden Gate Bridge (Section 4.1) in which during detail design it was decided to decrease the height of the railings from 1.7 m to 1.4 m, so making the bridge much more suitable for committing suicide. In this stage, for example, choices have to be made about which materials to use. Materials differ in terms of risks, health effects, and environmental impact. One example is the use of impoverished uranium as a stabilizer in airplanes. Functionally this is a suitable material, but it certainly is accompanied by certain health risks. To what extent these are acceptable is an ethical issue.

Detail design stage The stage in which a chosen design is elaborated on and detailed.

Another example is the use of tropical hard-wood for dams in rivers or canals. Again, the material itself has desirable properties but from the perspective of sustainability – the preservation of the rain forest – the use of such a material can be undesirable.

6.2.6 Prototype development and testing

Test The execution of a technology in circumstances set and controlled by the experimenter, and in which data are gathered systematically about how the technology functions in practice.

After the design is detailed, often a prototype of the design is constructed and tested. A **test** is the execution of a technology in circumstances set and controlled by the experimenter, and in which data are systematically gathered about how the technology functions in practice. The importance of testing is underscored by the disaster with the German high speed ICE train. Lack of proper testing was an important contributing factor to the disaster. However, tests are fallible too. One problem is that tests are not always representative of the circumstances in which the product eventually has to function. You need to know what circumstances are relevant in actual practice and which are irrelevant for performing a good test. An example in which certain aspects of actual use were erroneously ignored as irrelevant concerns the herbicide 2,4,5-T (see box). Toxicologists who asserted that use of the product was acceptable assumed that the product was produced and used as prescribed. Under those specific conditions the product was safe. In the circumstances in which it was actually used, however, it was unsafe.

Case The Herbicide 2,4,5-T

In the 1970s there was some unrest among agricultural workers in England concerning the safety of using herbicide 2,4,5-T on a large scale. The scientific Pesticides' Advisory Committee maintained that 2,4,5-T was not a health risk. They based themselves on laboratory studies on the toxicity of the substance. Complaints issued by the National Union of Agricultural and Allied Workers were set aside as imaginings. In its assessment, the scientific committee assumed that the laboratory situation was representative of actual practice: "pure 2,4,5-T offers neither hazards to users nor the general public ... provided that the product is used as directed" (cited in Wynne, 1989, p. 36).

More specifically, the following assumptions were the basis for believing that 2,4,5-T was harmless:

1 The production process is such that dioxin and other toxic substances never contaminate the main product.
2 The containers of 2,4,5-T always reach their destination with full and clear instructions for users (farmers and agricultural workers).
3 Farmers and agricultural workers always use the right solvents, pressure valves, spray nozzles and safety clothing despite the inconvenience.

Figure 6.4 Pesticide spraying. Photo: © Tyler Olson/Fotolia.com.

All three of these assumptions were problematic, but the third one in particular proved unfeasible. A scientist who investigated the working conditions of the farmers in question, described the scientists from the pesticide committee as "living in cloud-cuckoo land behind the laboratory bench" (cited in Wynne, 1989, p. 37). In contrast to what the pesticide committee stated, there were good reasons to believe that 2,4,5-T was a hazard to the health of agricultural workers – at least under the conditions in which they normally had to work.

Source: Wynne (1989).

6.2.7 Manufacture and construction

After a product or construction has been designed, it needs to be manufactured or constructed. Manufacture and construction may raise their own ethical issues, some of which can be anticipated and addressed in design. We mention a number of ethical issues that may arise during manufacture and construction.

One issue is labor conditions during manufacture and construction. Especially in competitive markets, there is often strong pressure to reduce the costs of production. This may result in worse labor conditions; raising the question what labor conditions are ethically desirable or acceptable. A related issue is whether it is desirable or acceptable to outsource production to low-wage countries, where labor conditions may be worse and where use may be made of child labor. Labor conditions here also relate to issues of safety and health protection of workers. The production of certain technologies may be dangerous or employees may be subject to certain hazardous or toxic emissions during production processes. Safety and health precautions may be more difficult to uphold in some developing countries. Here the global codes of conduct that were discussed in Section 2.4 may provide useful guidelines.

The production and construction phase may also raise ethical issues with respect to the environment and sustainability. Production processes may produce waste or pollute the environment. Such issues can often be dealt with by adequate planning and design of the production process in conjunction with the design of the product. Environmental life cycle analysis of products, including their production phase, may provide a useful tool; these methods are discussed in Section 10.6.2.

Construction work is known to be of a dangerous activity. The US construction worker, with a death rate of 39 per 100 000 employees is subject to a risk on the job that is five times larger than the average worker in the United States (MacCollum, 1995, p. 3). Construction safety can, to a large extent, be improved by better planning and design and should, therefore, be considered during the design phase (Gambatese, Behm, and Rajendran, 2008). Design for construction may also be important for the safety of users and the public. If designed constructions are difficult to construct in practice, it may well happen that the actual constructions deviates from the designed construction, so introducing safety or other risks. An example is the collapse of two walkways of the Hyatt Regency Hotel in Kansas City.

Case Hyatt Regency Hotel Walkway Collapse

Figure 6.5 Hyatt Regency walkway collapse. Photo: Dr. Lee Lowery, Jr.

On July 17, 1981, two walkways in the Hyatt Regency Hotel in Kansas City in the USA collapsed causing 114 fatalities. One reason for the collapse was that, as turned out during the investigation following the disaster, the construction did not meet the requirements of the Kansas City Building Code. As a result, two engineers lost their engineering registration.

However, the investigation showed also another major cause of the disaster that was due to lack of communication between designers and constructors. Figure 6.6 shows the original design of one of the walkways and the design as it was eventually implemented. In the implemented design the load on the bolts of the upper walkway are about twice as high as in the original design. This means that the walkway would probably not have collapsed had the design not been changed. The building contractors changed the design because the original design was difficult to build. They changed the drawings, but these changes were not noticed by the structural engineers who approved unwittingly the changes.

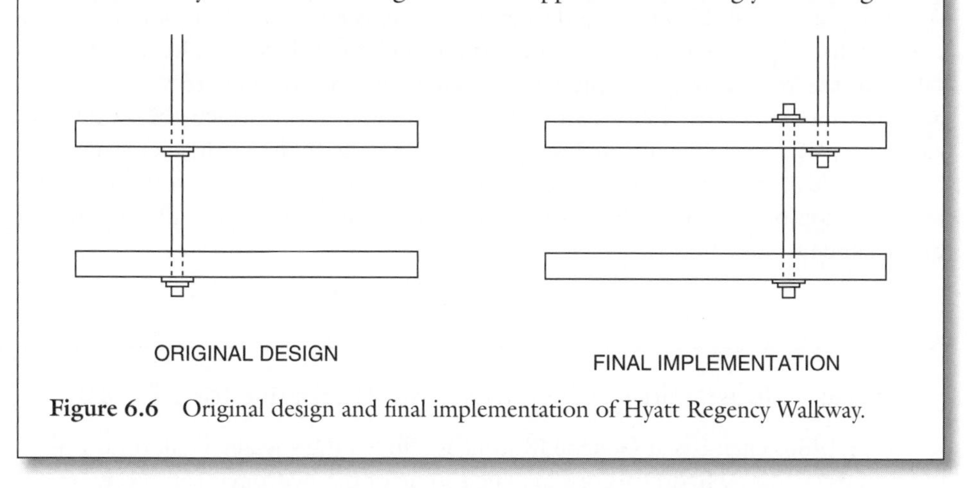

Figure 6.6 Original design and final implementation of Hyatt Regency Walkway.

6.3 Trade-offs and Value Conflicts

As we have seen, in the decision stage, the design is assessed in terms of the design requirements. A choice has to be made between different designs. This choice is often far from simple, because the different designs usually score well on different criteria. Trade-off decisions are difficult decisions anyway but they may become particularly problematic if the different design criteria that conflict correspond with different moral values. In such cases making trade-offs in design may amount to what we will call a **value conflict**. More precisely, we will define a value conflict as the situation in which all of the following conditions apply:

Value conflict A value conflict arises if (1) a choice has to be made between at least two options for which at least two values are relevant as choice criteria, (2) at least two different values select at least two different options as best, and (3) the values do not trump each other.

1 A choice has to be made between at least two options for which at least two values are relevant as choice criteria.
2 At least two different values select at least two different options as best. The reason for this condition is that if all values select the same option as the best one, we do not really face a value conflict.

Trumping (of values) If one value trumps another any (small) amount of the first value is worth more than any (large) amount of the second value.

3 The values do not trump each other. If one value **trumps** another any (small) amount of the first value is worth more than any (large) amount of the second value. If values trump each other, we can simply order the options with respect to the most important value and if two options score the same on this value we will examine the scores with respect to the second, less important, value. So if values trump each other, there is not a real value conflict.

Value conflicts can be morally problematic because they may well result in the situation that the designers cannot do justice to all relevant moral values simultaneously. In such cases, a value conflict amounts to a moral problem (see Section 5.3). An example of a value conflict in design is given in the case study about alternative coolants for CFC 12 (see box). A crucial question here is: how should environmental concerns regarding the design of new coolants for refrigerators be weighed against safety concerns? Below, we shall discuss five ways in which this evaluation can take place: cost-benefit analysis; multiple criteria analysis; the determination of thresholds for design criteria; reasoning about values; and the search for new technical solutions. For each method we shall present the main advantages and disadvantages.

Case Household Refrigerators – An Alternative for CFC 12

In the 1930s chemists at General Motors developed the so-called chlorofluorocarbons (CFCs) – hydrocarbons in which some of the hydrogen (H) atoms are replaced by chlorine (Cl) or fluorine (Fl) atoms. Due to their thermodynamic properties CFC turned out to be excellent coolants. Moreover, they are non-toxic and non-flammable. For household refrigerators, the most commonly used CFC coolant became CFC 12. In the 1980s it was discovered that CFCs are main contributors to the hole in the ozone layer. In 1987, the Montreal Protocol called for a world-wide reduction in the production and use of CFCs. Subsequently, in the 1990s, CFCs were forbidden in many countries.

As a consequence of the ban on CFCs in the 1990s, an alternative had to be found to replace CFC 12 as a refrigerant in household refrigerators. Three moral values played an explicit role in the formulation of design requirements for alternative coolants: safety, health, and environmental sustainability. In the design process, safety was mainly interpreted as non-flammability, and health as non-toxicity. Environmental sustainability was equated with low ODP (Ozone Depletion Potential) and a low GWP (Global Warming Potential). Both ODP and GWP mainly depend on the atmospheric lifetime of refrigerants. In the design process, a conflict arose between those three values. This value conflict can be illustrated with the help of Figure 6.7.

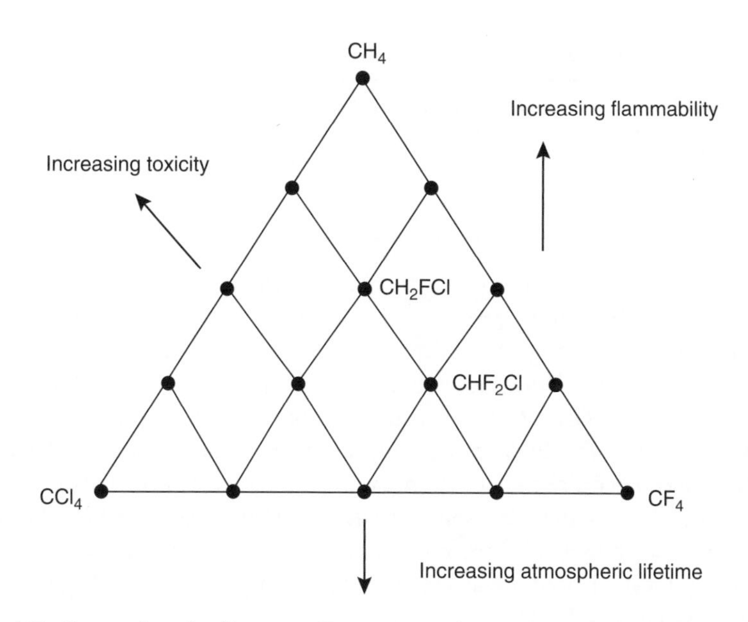

Figure 6.7 Properties of refrigerants. From McLinden, and Didion (1987). Copyright ASHRAE: ASHRAE, 1791 Tullie Circle, N.E., Atlanta, GA 30329, ©1987, ASHRAE (www.ashrae.org). Used with permission from ASHRAE.

Figure 6.7 is a graphic representation of CFCs based on a particular hydrocarbon. At the top, there is methane or ethane, or another hydrocarbon. If one moves to the bottom, the hydrogen atoms are replaced either by chlorine atoms (if one goes to the left) or fluorine atoms (if one goes to the right). In this way, all the CFCs based on a particular hydrocarbon are represented. The figure shows how the properties of flammability (safety), toxicity (health) and all environmental effects depend on the exact composition of a CFC. As can be seen, minimizing the atmospheric lifetime of refrigerants means maximizing the number of hydrogen atoms, all of which increases flammability. This means that there is a fundamental trade-off between flammability and environmental effects, or between the values of safety and sustainability.

Main alternatives to CFC 12 (CCl_2F_2) that were considered are HFC 134a ($C_2H_2F_4$) and hydrocarbons like isobutane (HC 600a or C_4H_{10}). Table 6.1 shows the ODP and GWP of these substances. Initially the industry preferred the alternative HFC 134a. Although this substance has a larger GWP, and thus contributes more to the greenhouse effect, than hydrocarbons like HC 600a, it is inflammable while hydrocarbons are flammable. HFC 134a was also attractive for the chemical industry because it could be patented, in contrast to the existing hydrocarbons, and therefore would be much more profitable to produce.

Table 6.1 ODP and GWP of some coolants

Coolant	*ODP (compared to CFC 12)*	*GWP (compared to CO2)*
CFC 12	1	8 500
HFC 134a	0	1 300
HC 600a (isobutane)	0	3

In parts of Europe, the tide has turned against HFC 134a since Greenpeace in the early 1990s found a refrigerator firm from former East Germany, Foron, willing to develop a refrigerator with hydrocarbons as coolant. When Greenpeace and Foron in August 1992 succeeded in collecting more than 50 000 orders for Foron's so-called *Greenfreeze*, within months the main German refrigerator firms switched to the hydrocarbon isobutane as coolant. In December 1992, the *Greenfreeze* acquired safety approval from the German certification authorities. Although there has been some discussion about the energy consumption of refrigerators with hydrocarbon as coolants, current studies seem to suggest that refrigerators with isobutane as coolant are at least as energy efficient as those using HFC 134a.

Source: Based on Van de Poel (2001).

6.3.1 Cost-benefit analysis

Cost-benefit analysis A method for comparing alternatives in which all the relevant advantages (benefits) and disadvantages (costs) of the options are expressed in monetary units and the overall monetary cost or benefit of each alternative is calculated.

Discount rate The rate that is used in cost-benefit analysis to discount future benefits (or costs). This is done because 1 dollar now is worth more than 1 dollar in 10 years time.

Cost-benefit analysis is a general method that is often used in engineering. What is typical of cost-benefit analysis is that all considerations that are relevant for the choice between different options are eventually expressed in one common unit, usually a monetary unit, like dollars or Euros. There are various types and variants of cost-benefit analysis (see, e.g., Mishan, 1975). If we consider the costs and benefits for society as a whole, this is usually referred to as a social cost-benefit analysis. Cost-benefit analysis can also be limited to the costs and benefits of a company that is developing a product and looking to market it.

Cost-benefit analysis may be an appropriate tool if one wants to optimize the expected economic value of a design. Still, even in such cases, some additional value-laden assumptions and choices need to be made. One issue is how to discount future benefits from current costs (or vice versa). One dollar now is worth more than one dollar in 20 years time, not only because of inflation but also because a dollar now could be invested and would then yield a certain interest rate. To correct this, a **discount rate** is chosen in cost-benefit analysis. The choice of discount rate may have a major impact on the outcome of the analysis. Another issue is that one might employ different choice criteria once the cost-benefit analysis has been carried out. Sometimes all of the options in which the benefits are larger than the costs are considered to be acceptable. However, one can also choose the option in which the net benefits are highest, or the option in which the net benefits are highest as a percentage of the total costs.

Cost-benefit analysis is more controversial if non-economic values are also relevant. Still, the use of monetary units does not mean that only economic values can be taken into account in cost-benefit analysis. In fact, approaches like **contingent validation** have been developed to express values like safety or sustainability in monetary units. Contingent validation proceeds by asking people how much they are willing to pay for a certain level of safety or, for example, the preservation of a piece of beautiful nature. In this way, a monetary price for certain safety levels or a piece of nature is determined. Approaches like contingent validation have serious limitations and are often criticized, because they are believed to commit the fallacy of pricing (see Section 4.5.2). However, it would be premature to conclude that cost-benefit analysis necessarily neglects non-monetary or non-economic values. When employing cost-benefit analysis, different ethical criteria might be used to choose between the options (Kneese, Ben-David, and Schulze, 1983; Shrader-Frechette, 1985). One might, for example, choose an option with which nobody is worse off. By selecting a specific choice criterion, ethical considerations beyond considering which options bring the largest net benefits might be taken into account.

Contingent validation An approach to express values like safety or sustainability in monetary units by asking people how much they are willing to pay for a certain level of safety or sustainability (for example, the preservation of a piece of beautiful nature).

In terms of values, cost-benefit analysis might be understood to be the maximization of one overarching or super value. Such a value could be an economic value like company profits, or the value of the product to users but it could also be a moral value like human happiness. If the latter is chosen, cost-benefit analysis is related to the ethical theory of utilitarianism. With Bentham's classical variant of utilitarianism (see Section 3.7.1), for example, the assumption is that all relevant moral values can eventually be expressed in terms of the moral value of human happiness. One might question this assumption, however. One issue is that it is often difficult to indicate to what extent values like safety, health, sustainability, and aesthetics contribute to the value of human happiness, and to furthermore express this in monetary terms. A second, more fundamental issue is that such an approach treats all these values as instrumental values, whose worth should ultimately be measured on the basis of their contribution to the intrinsic value of human welfare. One might wonder whether values like human health, sustainability, and aesthetics do indeed have only instrumental value or are intrinsically valuable.

It might, however, well be possible to employ cost-benefit analysis in a more instrumental way, that is, as a mere technical way to compare alternatives in the light of heterogeneous considerations or values. Although expressing everything in terms of money presupposes a common value, it could be maintained that this value is only a means of comparison, rather than a substantial value like human happiness. Nevertheless, this still presupposes that various criteria can be measured or expressed on a common scale (Hansson, 2007b). According to some ethicists, however, the existence of such a common scale is problematic because some values are incommensurable (see box) (see, for example, some contributions in Chang, 1997). Moreover, different people think differently about the relative importance of values like safety, welfare, and sustainability.

Value Incommensurability and Trade-Offs

Incommensurability Two (or more) values are incommensurable if they cannot be expressed or measured on a common scale or in terms of a common value measure.

Two or more values are **incommensurable** if they cannot be expressed or measured on a common scale or in terms of a common value measure. Incommensurable values cannot be traded off directly. It has been suggested that incommensurability and a resistance to certain trade-offs is constitutive of certain values or goods (Raz, 1986). Consider, for example, the following trade-off: for how much money are you willing to betray your friend? It may well be argued that accepting a trade-off between friendship and financial gain undermines the value of friendship. On this basis it is constitutive of the value of friendship to reject the trade-off between friendship and financial gain. It has also been suggested that values may resist trade-offs because they are "protected" or "sacred" (Baron and Spranca, 1997; Tetlock, 2003). This seems especially true of moral values and values that regulate the relations between, and the identities of, people. Trade-offs between protected values create an irreducible loss because a gain in one value may not always compensate or cancel out a loss in the other. The loss of a good friend cannot be compensated by having a better career or more money.

Some philosophers have denied the existence of value incommensurability. They believe that all values can ultimately be expressed in terms of one overarching or super value. Utilitarianism often attributes such a role to the value of human happiness, but a similar role may be played by the value of "good will" in Kantianism. The notion that there is ultimately only one value that is the source of all other values is known as value monism. Value monists do not necessarily deny the existence of more than one value but they believe that value conflicts can essentially be solved by having recourse to a super value.

6.3.2 Multiple criteria analysis

A second method to weigh different design criteria is **multiple criteria analysis**. Similar to cost-benefit analysis there are various types and variants. We shall restrict ourselves to the main outlines. Multiple criteria analysis is based on a comparison of different options with each other with respect to a number of criteria. Usually, the relative importance of the criteria is determined first, because usually not all criteria are equally important. Next, each option is weighed for all the criteria and a numeric value is awarded on a scale from 1 to 5 for example. Finally, the value for each option is calculated according to the following formula: $w_j = \Sigma g_i * v_{ij}$ over I, where w_j is the value of the jth option, g_i is the relative weight of the ith criterion, and v_{ij} is the score of the jth option on the ith criterion. The option with the highest value is then selected.

Multiple criteria analysis A method for comparing alternatives in which various decision criteria are distinguished on basis of which the alternatives are scored. On basis of the score of each of the alternatives on the individual criteria, usually a total score is calculated for each alternative.

Multiple criteria analysis does not demand that all criteria are translated into one overarching criterion or value, such as human happiness or welfare. However, multiple criteria analysis does demand that we determine the relative importance of the different design criteria in some way or another. Like cost-benefit analysis, multiple criteria analysis thus presupposes the commensurability of values. Compared to cost-benefit analysis, the comparison between options in multiple criteria analysis is vaguer because no explicit attempt is made to translate all criteria to a common unit (like money), which may result in flawed decision-making because the result depends on the scale chosen, as can be shown with the help of an example.

Let us have a look at the example of the coolants in refrigerators. Say that we assess the options HFC 134a and isobutane for the criteria of safety (flammability) and environmental impact (ODP and GWP). Moreover, we feel both criteria are equally important. One possible result from this multiple criteria analysis could be as in Table 6.2, in which higher numbers mean that the option scores better on that value, that is, it is safer or more environmentally friendly:

Here, the score for the criteria options ran from 1 to 3. Usually this scale is understood as an ordinal scale, in which only the order of the items is relevant (see also the box for explanation). So the only relevant information that Table 6.2 contains is that HFC134a is safer than isobutane (because 3 is larger than 2) and that isobutane is more environmentally friendly than HFC134a (because 3 is larger than 1). If we convert this to an assessment on an interval scale from 1 to 5. We could have Table 6.3 as a result. Note that this table contains the same information as Table 6.2: HFC 134a is safer than isobutane (5 is larger than 2) and isobutane is more environmentally friendly than HFC134a (4 is larger than 2). Now, however, the total score of HFC 134a is larger than that of isobutane.

This example shows that by changing solely the choice of scale, we can change the chosen option. So our choice depends on the chosen scale instead of on the inherent properties of the different options, which is undesirable. Avoiding this would require measuring both criteria (safety and the environment) on a ratio scale with the same

Table 6.2

	HFC134a	*Isobutane*
Safety	3	2
Environment	1	3
Total score	4	5

Table 6.3

	HFC134a	*Isobutane*
Safety	5	2
Environment	2	4
Total score	7	6

unit (see box). It seems, however, unlikely that we can measure safety and environmental effects on the same scale due to value incommensurability. The box explains the background of this flaw in more details.

Multiple Criteria Analysis and Measurement Scales

To understand why the results of multiple criteria analysis sometimes depend on the measurement scale chosen, we first need to distinguish different measurements scales:

Ordinal scale A measurement scale in which only the order of the items of the scale has meaning.

Interval scale A measurement scale in which in addition to the order of items also the distance between the items has meaning.

Ratio scale A measurement scale in which the ratio between items on a scale has meaning.

- An **ordinal scale** is a scale in which only the order of the items of the scale has meaning. An example is an ordering of the tastefulness of meals.
- An **interval scale** is a scale in which in addition to the order of items also the distance between the items has meaning. An example is the temperature scale Celsius (or Fahrenheit). It is meaningful to say that the difference between 10 °C and 20 °C is the same as between 20 °C and 30 °C.
- A **ratio scale** is a scale in which also the ratio between items on a scale has meaning. An example is distance measured in meters (or feet). It makes sense to say that 2 m is twice as long as 1 m, whereas it does not make sense to say that 20 °C is twice as hot as 10 °C. The reason for this is that the Celsius scale lacks an absolute point of zero. It would be different if we measure temperature in Kelvin because 0 Kelvin is defined as the lowest possible temperature. It is theoretically impossible to have a temperature below 0 Kelvin (as it is impossible to have a distance below 0 meter).

Each of these scales allows for different sets of mathematical operations:

- On an ordinal scale, arithmetical operations like addition, subtraction, multiplication and division are not allowed. Options can only be compared in terms of better and worse.
- On an interval scale, the arithmetical difference between two options has meaning, so that addition and subtraction are allowed, while multiplication and division are not.
- On a ratio scale, all arithmetical operations (addition, subtraction, multiplication and division) are allowed.

In the variety of multiple criteria analysis we consider here the overall worth of an option is calculated with the formula wj = Σgi * vij over I, where wj is the value of the jth option, gi is the relative weight of the ith criterion, and vij is the score of the jth option on the ith criterion. This formula is only meaningful if vij is measured on a ratio scale. Multiple criteria analysis therefore places great demands on how precisely we can measure the value of options on individual criteria. In many cases it is not that hard to determine which option scores better for which criterion, that is, to order the options on an ordinal scale. However, the method does require that we can express the value of the options for each criterion on a ratio scale. It should also be noted that this ratio scale should have the same unit for all criteria; otherwise we cannot meaningfully add up the scores on individual criteria. (One cannot add up meters and degrees Celsius for example.)

Take, for example, the design of an elevator. One relevant design criterion is the travelling time with which we can move from one floor to the next. It is not hard to rank various potential elevator designs according to this criterion. However, can we unequivocally translate this ranking into a relative valuation on a ratio scale? We can obviously measure travelling time in seconds which is a ratio scale. This is, however, not enough. Suppose that one of the other criteria is maintenance costs measured in Euros. Obviously, we cannot add up seconds and Euros; so we either need to convert seconds to Euros or the other way around or to convert both to a third scale like "goodness." But can we measure the travelling time of the elevator in terms of "goodness" on a ratio scale? For example, if travelling time is reduced from 30 seconds to 20 seconds, is that as good as saving 10 seconds by reducing travelling time from 20 seconds to 10 seconds, and is 10 seconds twice as good as 20 seconds? In most cases it is not.

6.3.3 Thresholds

Threshold The minimal level of a (design) criterion or value that an alternative has to meet in order to be acceptable with respect to that criterion or value.

A third way to cope with conflicting design criteria is to set a **threshold** for each criterion. For each separate criterion (e.g. safety, health, costs, and sustainability) a threshold is determined for what is acceptable. Setting thresholds not only occurs in the design process, but also in legislation

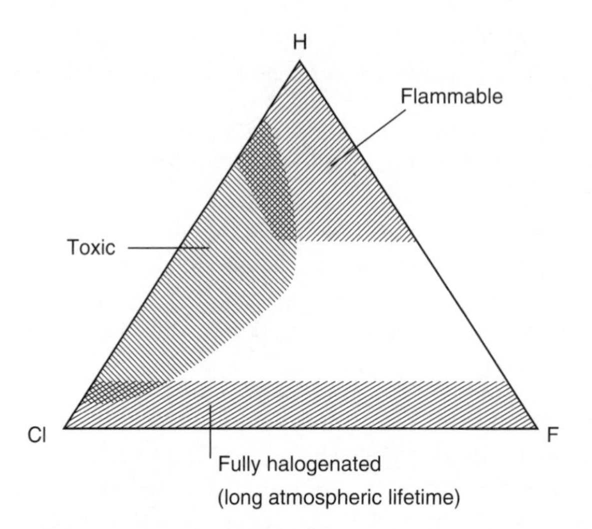

Figure 6.8 Properties of refrigerants. From McLinden and Didion (1987). Copyright ASHRAE: ASHRAE, 1791 Tullie Circle, N.E., Atlanta, GA 30329, ©1987, ASHRAE (www.ashrae.org). Used with permission from ASHRAE.

(standardization) and in technical codes and standards. A minimal level of safety is often defined this way for example.

An example of setting thresholds is to be found in the case of the design of new refrigerants. In this case, the engineers McLinden and Didion from the *National Bureau of Standards* in the USA drew Figure 6.8 with respect to the properties of CFCs. According to McLinden and Didion the blank area in the triangle contains refrigerants that are acceptable in terms of health (toxicity), safety (flammability) and environmental effects (atmospheric lifetime). Note that by drawing the blank area in the figure, McLinden and Didion – implicitly – establish threshold values for health (toxicity), safety (flammability) and the environment.

An advantage of setting thresholds is that the acceptable threshold is considered for each criterion without making direct trade-offs between different design requirements. This may, for example, be helpful to guarantee a minimal level of for example safety in the design process. However, the question is whether it is possible or desirable to determine thresholds in complete isolation from other concerns. If, for example, the government draws up safety standards, this usually takes place in the light of the costs involved to achieve that level of safety and what else we have to sacrifice in terms of other moral values such as welfare, a good life, or sustainability. Setting a threshold per criterion in the design process also occurs with reference to other criteria and what is technically feasible. Taking into account other criteria in setting thresholds may make sense, but there is a danger that thresholds are selected that make a design possible under all circumstances. The question is whether this is ethically acceptable. It would for example not always be desirable to adapt your assessment of the desired degree of sustainability to what is achievable.

Another possible disadvantage of setting thresholds is that you limit yourself as an engineer to realizing these values, while more can be achieved with a given design in terms of environmental impact or sustainability for example. This was also the case with the design of coolants for refrigerators. The alternative HFC 134a did meet the thresholds set by the engineers McLinden and Didion, but according to others the chemical's environmental impact was too large. What also played a role was the fact that in the existing standards, like the *ASHRAE Code for Mechanical Refrigeration* (ASHRAE Standard 15–1978), coolants in equipment for household applications were not allowed to be flammable. The acceptance of flammable coolants in this instance required the reformulation of the standard in question. Thus, the ASHRAE Code was reformulated in 1994 (ASHRAE Standard 15–1994), so that household equipment was allowed to contain a maximum of 3 kg of flammable coolant in certain cases (Van de Poel and Van Gorp, 2006). In this case, environmental concerns – partly based on moral considerations – led to the reformulation of existing standards.

6.3.4 Reasoning

The approaches to dealing with trade-offs that have already been discussed are all calculative approaches. They strive to operationalize and measure the value of a design in one way or another. Of these approaches, the setting of thresholds does not aim at calculating the overall value of an option, but it does presuppose that the value of an option can be measured for each of the individual design criteria. We will now look at an approach that does not share this calculative approach, but which emphasizes judgment and reasoning about values. This approach aims at clarifying the values that underlie the conflicting design requirements, and consist of three steps: 1) identifying relevant values; 2) specifying the values; and 3) looking for common ground among values. As illustration, we will look at the values involved in the design of automatic seatbelts: safety and freedom (see box).

Case Automatic Seatbelts

A car with automatic seatbelts will not start if the automatic seatbelts are not put on. This forces the user to wear the automatic seatbelt. One could say that the value of driver safety is built into the technology of automatic seatbelts. This comes at a cost, however: the user has less freedom. Interestingly, there are various seatbelt designs which exist that would imply that there are different trade-offs in terms of safety and user freedom. The traditional seatbelt, for example, does not enforce its use, but there are various systems that give a warning signal if the seatbelt is not being worn. This does not enforce seatbelt use, but it does encourage the driver to wear his seatbelt.

Identifying relevant values

The first thing to do when one wants to exercise judgment in cases of trade-offs is to identify what values are at stake in the trade-off and to gain a better understanding of these. What do these values imply and why are these values important? Take the value of freedom in the case of safety belts. Freedom can be construed as the absence of any constraints on the driver; it then basically means that people should be able to do what they want. Freedom can, however, also be valued as a necessary precondition for making one's own considered choices; so conceived freedom carries with it a certain responsibility. In this respect it may be argued that a safety belt that reminds the driver that he has forgotten to use it does not actually impede the freedom of the driver but rather helps him to make responsible choices. It might perhaps even be argued that automatic safety belts can be consistent with this notion of freedom, provided that the driver has freely chosen to use such a system or endorses the legal obligation for such a system, which is not unlikely if freedom is not just the liberty to do what one wants but rather a precondition for autonomous responsible behavior. One may thus think of different conceptualizations of the values at stake and these different conceptualizations may lead to different possible solutions to the value conflict.

Specifying values

A second judgment step would be to argue for specific conceptualizations or specifications of the relevant values. Some conceptualizations might not be tenable because they cannot justify why the value at stake is worthwhile. For example, it may be difficult to argue why freedom, conceived of as the absence of any constraint, is worthwhile. Most of us do not strive for a life without any constraints or commitments because such a life would probably not be very worthwhile. This is not to deny the value of freedom; it suggests that a conceptualization of freedom only in terms of the absence of constraints misses the point of just what is valuable about freedom. Conceptualizations might not only be untenable for such substantial reasons, they may also be inconsistent, or incompatible with some of our other moral beliefs.

Common ground

A third step in judgment is to look for the common ground behind the various values that might help to solve the value conflict. This idea can, for example, be found in Kant's notion of the good will. It is likely that Kant would maintain that the good will can solve all value conflicts, at least in principle. This is probably too optimistic, but that does not reduce the need to look for common ground between values. Even if such common ground cannot always be found, it may be available in specific cases.

6.3.5 Value Sensitive Design

The previous approach treats the occurrence of value conflict merely as a philosophical problem to be solved by philosophical analysis and argument. However, in engineering design value conflicts may also be solved by technical means. That is to say, in engineering it might be possible to develop new, not yet existing, options that solve or at least ease the value trade-off. In a sense, solving value trade-offs by means of new technologies is what lies at the heart of engineering design and technological innovation. Engineering design is able to play this part because most values do not conflict as such, but only in the light of certain technical possibilities and engineering design may be able to change these possibilities. An interesting example is the design of a storm surge barrier in the Eastern Scheldt estuary in the Netherlands discussed earlier.

One approach that takes into account the possibility of solving, or at least easing, value conflicts through engineering design is **Value Sensitive Design**. Value Sensitive Design is an approach that aims at integrating values of ethical importance in a systematic way within engineering design (Friedman, Kahn, and Borning, 2006). The approach aims at integrating three kinds of investigations: conceptual, empirical, and technical:

Value Sensitive Design An approach that aims at integrating values of ethical importance in a systematic way in engineering design.

- *Empirical investigations* aim at understanding the contexts and experiences of the people affected by technological designs. This is relevant to appreciating precisely what values are at stake and how these values are affected by different designs.
- *Conceptual investigations* aim at clarifying the values at stake, and at making trade-offs between the various values. Conceptual investigations in Value Sensitive Design are similar to the kind of investigations described in Section 6.3.4.
- *Technical investigations* analyze designs and their operational principles to assess how well they support particular values, and, conversely, to develop new innovative designs that meet particular morally relevant values particularly well. The second is especially interesting and relevant because it provides the opportunity to develop new technical options that more adequately meet the values of ethical importance than do current options.

As the earlier presented example of the Eastern Scheldt barrier shows, technical investigations may ease value conflicts. Usually, however, technical innovation will not entirely solve value conflicts, so that choices between conflicting values still have to be made. In this respect, innovation through Value Sensitive Design only presents a partial solution to value trade-offs in engineering design.

6.3.6 A comparison of the different methods

We discussed five methods for making a choice between alternatives in the light of design criteria. We saw that each method has its pros and cons: these are summarized in Table 6.4. What is striking is that none of the methods reaches a definite solution for the problem of morally relevant trade-offs in design. Both cost-benefit analysis and multiple criteria analysis suppose the commensurability of values, which might be problematic. Reasoning might help to solve some value conflicts, but probably not in all cases. Similarly technical innovation through Value Sensitive Design often is useful, but in practice it usually does not lead to a definite solution for the problem. Thresholds have the disadvantage that sometimes less is achieved in a given

Table 6.4 Overview of methods for making trade-offs in design

Method	*How are the values weighted?*	*Main advantages*	*Main disadvantages*
Cost-benefit analysis	All values are expressed in monetary terms	• Options are made comparable	• Values are treated as commensurable • May be difficult to adequately express all relevant (moral) concerns in monetary terms
Multiple criteria analysis	Trade-offs between the different values	• Options are made comparable	• Values are treated as commensurable • Result depends on measurement scale
Thresholds	A threshold is set for each value.	• The selected alternatives meet the thresholds • No direct trade-off between the criteria	• Can thresholds be determined independently from each other? • Less achieved than possible
Reasoning	Values are related to each other and possibly traded off through reasoning and judgement.	• Might solve value conflict by reason and judgment	• Not all value conflicts can be solved in this way
Value Sensitive Design	Not applicable	• Can lead to alternatives that are clearly better than all of the present alternatives	• Does not solve the choice problem in many cases

situation than could be the case. Moreover, you still have to weigh the various criteria and values while drawing up the thresholds.

We should not conclude from the above that the choice between alternatives that score differently for various values is random. The methods are useful. However, which method is best will depend on the situation. The discussion of pros and cons can help you to make a choice based on proper reasons. Moreover, it is good to be aware of the shortcomings of the various methods, so that you can try to limit these shortcomings in a concrete situation.

6.4 Regulatory Frameworks: Normal and Radical Design

Engineering design often is redesign. In other words, something is designed that already exists in a comparable shape or based on a comparable working principle, such as a car, a bridge, a chemical plant, or a microchip. Experience with the product has often led to codes and standards for the design of that product and has also sometimes led to legislation from the government. Although some legislation and codes and standards apply to all engineers and technical products, many of the rules are more product-specific. We will call the totality of such product-specific rules the **regulatory framework** for that technology (Van Gorp, 2005; see also Grunwald, 2001). A regulatory framework can be considered a part of morality because it deals with judgments about how to act rightly that are laid down in rules. (In Section 3.2 morality was defined as the totality of opinions, decisions, and actions with which people express what they think is good or right.) These rules are often based on past experience with a technology, like the occurrence of certain types of accidents or the discovery of certain (undesirable) social effects of a technology. In some cases, regulatory frameworks will be partly based on public unrest or discussion. In the case of legislation, rules are usually the result of democratic decision-making.

Regulatory framework The totality of (product-specific) rules that apply to the design and development of a technology.

A regulatory framework can help engineers to make ethically relevant decisions in the design process. In as far as the framework is democratically established, for example through legislation, is also helps to avoid the fallacy of technocracy (see Sections 1.5.2 and 4.5.2). This does not imply, however, that engineers can always just follow the existing regulatory framework without asking some further questions. Even if the framework can be considered a kind of morality it is not necessarily also ethically acceptable. In order to judge whether a regulatory framework can be followed in design one could think of the following set of conditions:[2]

- The framework is complete in that it covers the relevant decisions without neglecting relevant issues;
- The framework is free of contradictions and inconsistencies;
- The framework is unambiguous; it is clear how the framework should be applied to specific situations;
- The framework is morally acceptable;
- The framework is lived by in practice.

The existence of a framework meeting this set of conditions partly depends on the type of design process. Empirical research suggests that regulatory frameworks are more common in normal design than in radical design (Van de Poel and Van Gorp, 2006; Van Gorp, 2005; Van Gorp and Van de Poel, 2008). **Normal design** can be defined as design in which the configuration and working principle of the product remain the same (Vincenti, 1990, p. 209). If that is not the case, then we refer to **radical design**. The **working principle** is the principle on which the working of a piece of equipment is based. The working principle of the propeller engine for an aircraft is different from that of a jet engine, because the thrust of the engine is based on a different physical phenomenon. An important reason for the absence of regulatory frameworks in radical design is that often the rules of the framework are related to

Normal design Design in which the normal configuration and working principle of the product remain the same.

Radical design The opposite of normal design. Design in which either the normal configuration or the working principle (or both) of an existing product is changed.

Working principle The (scientific) principle on which the working of a product is based.

the current working principle or normal configuration. This is especially true for detailed technical codes and standards. For example, the codes used to guarantee the safety of steel LPG tanks cannot simply be transferred to tanks made from synthetic or composite materials. Also test and inspection procedures and norms may become inadequate in cases of radical design as was illustrated in the train wheel example discussed at the beginning of this chapter. Especially in cases of more radical design, or disruptive technologies, the current way of dealing with ethical issues may become obsolete. An example is nano-electronics and the development of increasingly smaller RFID chips (see box).

Case RFID Chips

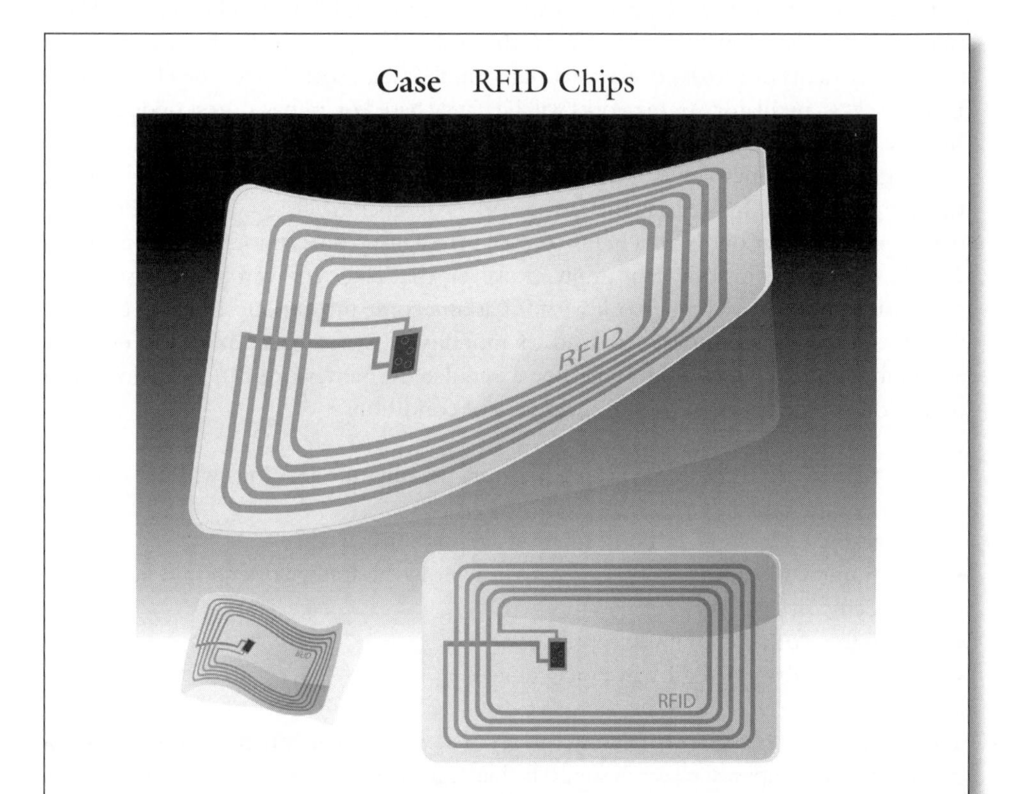

Figure 6.9 A Radio Frequency Identity Chip (RFID). Photo: © Huseyin Bas/ Fotolia.com.

The Radio Frequency Identity Chip (RFID) is a chip or tag consisting of a small integrated circuit and a very small radio antenna. Like bar codes, RFID chips have their own unique identification number. RFID chips are used, among other things, for tracking and tracing of objects, boxes and vehicles in logistic chains. Due to developments in nano-electronics, RFID chips will likely become smaller and may become invisible. Not only objects but also people may be tagged with RFID chips, as is now already the case for some small-scale applications. Eventually, people and objects may so become entangled in an "Internet of things," in which tiny devices exchange information without people noticing.

The philosophers Van den Hoven and Vermaas have argued that RFID chips raise privacy issues that are different from the way privacy is traditionally understood to be endangered by, for example, information technologies. An important metaphor in the traditional debate on privacy is the panopticon (see also Chapter 3 on Jeremy Bentham): a hemispherical prison in which the imprisoned are continuously watched from an authoritative point of view (the dome of the panopticon). While nano-electronics enables continuous surveillance, data storage is not necessarily central (although it can be central). In fact, information may be stored in individual tags, but by locally combining these bits of information, privacy-sensitive information may be revealed.

Many of the current attempts to protect citizen's privacy are focused on restraining the storage and processing of information in central databases, or in constraining the retrieval of information from such databases. With the advance nano-electronics this focus may be too limited. Attention should also be paid to the design of the hardware. Relevant issues for example include the reach of the antenna, the accessibility of the data on the chip for other devices, whether the chips are writable or not, and, if so, by whom.

Source: Van den Hoven and Vermaas (2007).

Even if a regulatory framework is available, such a framework does not always meet the mentioned conditions. Consider the earlier discussed case of coolants. The existing framework, as we have seen, forbad the use of flammable coolants. Although this requirement as such is obviously not unethical, the implication was that coolants were preferred which were more harmful to the environment than the flammable hydrocarbons. On the other hand, it was not obvious that flammable coolants are necessarily unsafe. Modern refrigerators contain only small amounts of coolant and the risk of explosion is minimal as tests showed. Here it should be kept in mind that at the time that the requirement was formulated that coolants should be inflammable, refrigerators used to contain much larger quantities of coolant than nowadays; due to technical innovations that increased the efficiency of refrigerators this amount had been drastically reduced. There may, therefore, be good moral reasons not to follow a regulatory framework even if one is available.

All in all, three types of situations are possible considering the existence of a regulatory framework:

1 A framework may be available that meets all the mentioned conditions. In this situation, engineers can follow the framework in making ethically relevant choices in design.
2 No regulatory framework may be available. This situation is more likely, as we have seen, in radical, innovative design. In this case, engineers themselves have to make a number of ethically relevant choices, for example after discussing the relevant issues with other stakeholders like users, clients, regulators, et cetera. In this

situation, the responsibility of designers for ethical choices in design is larger than usual because they cannot fall back on socially sanctioned rules.

3 It is also possible that a framework is available but that it does not meet one or more of the mentioned conditions. What the best strategy is here will depend on what condition is not met. We focus here on the situation that the current framework is morally unacceptable or at least morally debatable. In that situation, engineers can follow various strategies:

   - Aim at changing the framework. Engineers are involved in the formulation of technical codes and standards or in setting up test procedures and they may thus be able to change these elements of the framework.
   - Inform other parties that formulate other parts of the framework, like the government or a professional association, about what they consider shortcomings or problems within the current framework.
   - Deviate from certain elements of the regulatory framework in the design process. Often regulatory frameworks leave some room for this, although some elements may be mandatory, for example because they are part of legislation.
   - Opt for radical design, in which parts of the framework do not longer apply. Moral reasons can thus be a ground for choosing a radical design. On the other hand, radical design may well lead to unexpected risks because there is less experience involved, as is testified by the innovative wheel design for the German high speed ICE train. This can be a moral reason to opt for normal design.

6.5 Chapter Summary

Design is at the heart of engineering. It is at the design stage that new technologies get shape and that important ethical decisions are made. This chapter has, therefore, investigated the ethical issues raised during design and ways to deal with them.

Engineering design is a systematic process in which certain functions are translated into a blueprint for an artifact, system, or service that can fulfill these functions. The design process can be subdivided in a number of steps, each which raise their peculiar ethical issues:

- *Problem analysis and formulation*. Here ethical issues may arise with respect to the perspective chosen in formulating the design problem, and the formulation and specification of design requirements.
- *Conceptual design*. In this stage, creativity is a major virtue for engineers. Creativity is also morally important, since it enables engineers to come up with creative solutions between possible conflicting moral demands on a design.
- *Simulation* is important to learn about the characteristics and possible consequences of design options. A major issue is here the reliability of simulations, which is usually limited for a variety of reasons.
- *Decision*. In the decision stage, you will often face trade-offs between various morally relevant design requirements. Other important issues are the inclusiveness and the explicitness of the decision-making process.

- *Detail design.* Here ethical issues may arise, for example in relation to the choice between certain materials or the organization of the production process.
- *Prototype development and testing* is important to discover the unexpected effects of the new product when it is used.

Without doubt, some of the most important ethical decisions during design are made in the decision stage. Here the trade-offs between the design requirements may amount to a value conflict, that is, a situation in which the various relevant (moral) values select different options as the best one and there is no clear hierarchy between the values. We have discussed five methods for making decisions in cases of value conflict in design:

- Cost-benefit analysis. The main disadvantage of this method is that, by expressing everything in monetary units, it treats all relevant values as commensurable. Nevertheless the method is systematic and various ethical concerns can be included by adding certain ethical decision criteria.
- Multiple criteria analysis. Like cost-benefit analysis, values are treated here as commensurable, albeit not by expressing everything in terms of money. The method can also be plagued by methodological problems related to the choice of measurement scales.
- Setting thresholds. This method avoids direct trade-offs between the relevant values and may be helpful to set, for example, a minimal level of safety.
- Reasoning about values, which is useful to get a better grip on the values at play, their meaning and possible ways they can be combined or traded off.
- Value Sensitive Design which is a design methodology that can help to solve value conflicts by technical means, that is, by developing new innovative designs.

We have finally discussed to what extent engineers can rely on democratically sanctioned regulatory frameworks in making decisions in engineering design. We have seen that the degree to which such frameworks are available largely depends on whether we deal with redesign of existing products (normal design) or the design of new innovative products (radical design). In radical design, democratically sanctioned rules for making design decisions are usually absent and engineers have a larger responsibility for the design decisions they make.

Study Questions

1. Why is the perspective chosen during the formulation of the design problem morally relevant? Why is it relevant from a utilitarian point of view? And from a Kantian point of view? Give an example that illustrates the moral relevance of the perspective chosen in a design problem.
2. What is the difference between normal and radical design? In what respects is this difference morally relevant?
3. Why is there a danger of technocracy if engineers do not follow the existing regulatory framework in the design of a new technology?

4 Why does multiple criteria analysis presuppose the commensurability of design criteria?
5 Why does the solving of value conflicts by means of new technologies lie at the heart of engineering design and technological innovation? Provide an example illustrating this.
6 Choose a product that is designed in your own discipline. Make an overview of ethical issues that are raised by the design of this product. Go through all the design stages described in Section 6.2
7 In the case described in the text of alternative coolants, the engineers McLinden and Didion eventually chose for HFC 134a.
   a. Do you agree with their choice?
   b. What do you think of the decision procedure (setting thresholds) by which they came to this choice?
   c. Argue your answers.
8 Reread the case at the beginning of Chapter 4 on the design of a suicide barrier for the Golden Gate Bridge.
   a. What values are at stake in the design of the barrier?
   b. Are these values conflicting? If so, how and why?
   c. What would in your view be the best way to deal with this value conflict?
9 In the case of the pesticide 2,4,5-T it could be argued that the designers did not only design a product but also a practice of use. This use practice presupposed rather strict safety rules for the application of the pesticide, which turned out not to be followed in real life. Some people would say that it was the responsibility of the users to use the product as prescribed. Others might say that the designers should have adapted their product to existing or at least realistic use practices.
   a. Who is, in your view, responsible for the dangers that the use of 2,4,5-T had in practice?
   b. Should the designers have designed a product that could be used only in a safe way?
   c. How can this kind of problem be prevented in the future?
10 In many mobile phones and laptop computers, the material tantalum is used.[3] Most of the worldwide tantalum supply comes from legitimate mining operations in Australia, Canada and Brazil, but tantalum is also extracted from the metallic ore coltan that is mined in Congo. Rebel groups exploit coltan mining to raise funds for the civil war that is going on in Congo and in which thousands of people have been killed. Some organizations have alleged that "there is a direct link between the mining of coltan in Congo and the human right abuses." Coltan mining by rebels in Congo also causes environmental degradation. Some manufacturers have shown concerns about the use of coltan from Congo or have declared that they will no longer use coltan from Congo, but such policies are hard to implement because in many cases it is hard to find out where coltan that is traded on the world market is actually coming from. It might well be smuggled to another African country from Congo before it is sold. Banning coltan from such other African countries would, however, affect countries that are legitimately mining coltan and for which the product may be an important source of income.
   a. At which stage of the design process are decisions made about the materials used?
   b. Why is the choice of material ethically relevant in this case?
   c. Do designers have a responsibility to avoid the negative consequences associated with the use of tantalum and coltan?
   d. Should the designers try to avoid the use of tantalum? Would this justify opting for radical design if necessary?

11 In the Fort Pinto case (Section 3.1), an issue was how to trade off safety and economic considerations.
   a. Explain how this trade-off would be made with each of the five methods discussed in Section 6.3.
   b. What are the advantages and disadvantages for each of these methods in this specific case?
   c. Which method (or methods) do you consider most appropriate to trade off safety and economic considerations in this case and why?

Discussion Questions

1 Some people argue that ethical issues in technology arise due to how technologies are used and can, therefore, not be addressed in design. Do you agree? Is design necessarily irrelevant when most ethical issues arise due to how technologies are used?
2 Some philosophers believe that there cannot be incommensurable values because even if values are seemingly incommensurable we actually choose an option and this choice reveals the relative importance of the values. Do you agree with this argument? To assess the argument you can, for example, consider the dilemma in *Sophie's Choice* (see Section 5.3.1): If Sophie chooses one child rather than the other does this show that she loves that child more than the other? Is the only way to show that she loves both children equally to refuse to make a choice (so that both die)?
3 In cost-benefit analysis, human lives are often expressed in money. Do you consider this an acceptable practice? If it is not acceptable, how should we then determine how much money to spend on increasing, for example, human safety?

Notes

1 This and the following three paragraphs are partly drawn from Devon and Van de Poel (2004).
2 The conditions are base on Grunwald (2001). The formulation has been adapted at some places. Grunwald formulates as fourth condition that the people involved should accept the framework.
3 Description is based on www.thestandard.com/article/0.1902.26784.00.html (accessed August 12, 2009).

7

Designing Morality

Peter-Paul Verbeek

Having read this chapter and completed its associated questions, readers should be able to:

- Describe in what sense ethics is also a matter of things;
- Understand the phenomenon of technological mediation;
- Reflect on the moralizing role of technology and on arguments for and against moralizing through technology;
- Integrate considerations of technological mediation in the design process.

Contents

Ethics, Technology, and Engineering: An Introduction, First Edition.
Ibo van de Poel and Lambèr Royakkers.

7.1 Introduction

Case Robert Moses' Racist Overpasses

Figure 7.1 Low overpass. Oversized overturned truck on the BQE under the Brooklyn Bridge. Photo: B. Yanev, NYC DOT.

Robert Moses (1888–1981) was a very influential and also contested urban planner of mid-twentieth century New York. In his 1974 book, *The Power Broker*, biographer Robert Caro argued that Moses also demonstrated racist tendencies. He designed several overpasses over the parkways on Long Island, which were too low to accommodate buses. Only cars could pass below them and for that reason the overpasses complicated access to Jones Beach Island. Only people who could afford a car – and in Moses' days these were generally not Afro-American people – could easily access the beaches now.

This case has become especially famous due to the philosopher of technology Langdon Winner who mentioned it in his article "Do artifacts have politics?" (1980). As Winner notes although the overpasses are extraordinary low, one would normally not be inclined to attach any special meaning to that fact. It turns out, however, that they were deliberately designed to achieve a specific social effect. In the words of Langdon Winner:

> Robert Moses, the master builder of roads, parks, bridges, and other public works of the 1920s to the 1970s in New York, built his overpasses according to specifications

> that would discourage the presence of buses on his parkways. According to … Moses' biographer, Robert A. Caro, the reasons reflect Moses' social class bias and racial prejudice. Automobile-owning whites of "upper" and "comfortable middle" classes … would be free to use the parkways for recreation and commuting. Poor people and blacks, who normally used public transit, were kept off the roads because the twelve-foot tall buses could not handle the overpasses.

Winner's analysis of these low-hanging overpasses have become a paradigmatic example, even though some objections have been raised against it because timetables show that it was actually possible to reach the beach by bus, bypassing the overpasses. The example shows that technological artifacts can be politically or morally charged. This has implications for the ethics of design because it charges engineers with a responsibility to think about the potential moral and political role of artifacts in the design process. The chapter will investigate how engineers can do that.

The chapter starts with arguing that ethics is not just a matter of people but also of things (Section 7.2). To better understand, the moral role of technological artifacts, the notion of "technological mediation" will be elaborated (Section 7.3). After this, the implications of this mediation approach for responsible engineering will be investigated in the Sections 7.4 and 7.5.

7.2 Ethics as a Matter of Things

As the example of the "racist" overpasses shows, the artifacts we deal with in our daily lives help to determine our actions and decisions in myriad ways. Since answering the question how to act is the ethical activity *par excellence* (see Chapter 3), this implies that we should not consider morality as a solely human affair, but also as a matter of *things*. Yet, this new material dimension in ethics raises many questions. Is the conclusion that technologies influence human actions reason enough to actually attribute morality to materiality? And is it morally right to go even one step further and try to explicitly shape this morality of things, by consciously steering human behavior with the help of the material environment?

Langdon Winner's analysis of Moses' Long Island overpasses dates from 1980. In this case, allegedly one man was able to embed his racial ideology within these technological artifacts, thereby racializing their construction and eventual use. A decade later, French philosopher and anthropologist Bruno Latour argued that artifacts are bearers of morality, as they are constantly taking all kinds of moral decisions for people (Latour, 1992). He showed, for example, that the moral decision of how fast one drives is often delegated to a speed bump which tells the driver "slow down before reaching me." Anyone complaining about deteriorating morality, according to Latour, should use their eyes better, as the objects around us are crammed with morality.

The ethics of engineering design is the best place to start analyzing the moral dimension of technological artifacts, since this is also the place where human beings can take responsibility for the moral aspects of their products. In its current form, though, the ethics of engineering design tends to follow a somewhat externalist approach to technology. It mainly focuses on the importance of taking individual responsibility (e.g., "whistle blowing," Section 1.5.3) to prevent technological disasters, and on methods to assess and balance the risks accompanying new technologies. Typical case studies concern technologies which have caused a lot of problems that could have been prevented by responsible actions of engineers, like the exploding space shuttle *Challenger* (Section 1.1), or the Ford Pinto with its rupturing gas tank in crashes over 25 miles per hour (Section 3.1). Case studies like these merely address technologies in terms of their functionality: technologies are designed to do something, and if they fail to do so properly, they were badly designed. What such case studies fail to take into account are the impacts of technologies on our moral decisions and actions, and on the quality of our lives. For analyzing these impacts of technologies, and the various aspects of the role technological artifacts play in their use contexts, the concept of *technological mediation* is a helpful tool. Technological mediation concerns the role of technology in human action, and human experience.

7.3 Technological Mediation

When technologies are used, they always also influence the context in which they fulfill their function. Technological artifacts help to shape human actions and perceptions, and create new practices and ways of living. Cell phones, for example, contribute explicitly to the nature of our communications and interactions. Technologies like ultrasound play active roles in our decisions regarding unborn life. Functionality is too limited a concept for engineering ethics. The impacts of technology transcend functionality: they form a surplus to it, which occurs once the technology is functioning. When technologies fulfill their functions, they also help to shape the actions and experiences of their users. This phenomenon is called **technological mediation**: technologies help to shape the experiences and practices of their users (Latour, 1992; Ihde, 1990; Verbeek, 2005). Technologies are not neutral "intermediaries," that simply connect users with their environment; they are impactful mediators, that help to shape *how* people use technologies, how they experience the world and what they do.

Technological mediation The phenomenon that when technologies fulfill their functions, they also help to shape the actions and perceptions of their users.

For a better understanding of the mediating role of technologies two perspectives on mediation will be discerned: one that focuses on perception and another one on praxis. Each of these perspectives approaches the human-world relationship from a different side. The "experience-oriented" perspective starts from the side of the world, and directs itself at the ways reality can be interpreted and be present for people. The main category here is *perception*. The "praxis-oriented" perspective approaches human–world relations from the human side. Its central question is how human beings act in their world and shape their existence. The main category here is *action*.

7.3.1 Mediation of perception

Mediation of perception The influence of artifacts on human perception, that is, the sensory relationship with reality.

As philosopher Don Ihde has elaborated in his book *Instrumental Glossary Realism*, there are several ways in which technologies can help to shape people's experience of reality (Ihde, 1991). First, when using a technology, users can "incorporate" or "embody" it, as it were. This embodiment relation, for instance, occurs when looking through a pair of glasses; the artifact is not perceived itself, but it helps to perceive the environment.

Technologies can also establish another relationship, however, in which they do not provide access to reality by being "incorporated," but by providing a representation of reality, which requires interpretation. A thermometer, for instance, establishes a relationship between humans and reality in terms of temperature. Reading off a thermometer does not result in a direct sensation of heat or cold, but gives a value which requires interpretation in order to tell something about reality.

When mediating our sensory relationship with reality, technologies transform what we perceive. This transformation of perception, as Ihde argues, always has a **structure of amplification and reduction**. (Ihde, 1991). Mediating technologies amplify specific aspects of reality while reducing other aspects. When looking at a tree with an infrared camera, for instance, most aspects of the tree that are visible for the naked eye get lost, but at the same time a new aspect of the tree becomes visible: one can now see whether it is healthy or not.

Structure of amplification and reduction The fact that mediating technologies amplify specific aspects of (the perception of) reality while reducing other aspects.

This mediating role is not an intrinsic property of technologies themselves, though. Within different use situations, technologies can have a different "identity." The telephone and the typewriter, for instance, were not developed as communication and writing technologies, but as equipment for the blind and the hard of hearing to help them hear and write. In their current use context they are interpreted quite differently. Don Ihde calls this phenomenon **multistability**: a technology can have several "stabilities," depending on the way it is embedded in a use context.

Multistability The phenomenon that a technology can have several "stabilities," depending on the way it is embedded in a use context.

By transforming our perception, technologies help to determine how reality can be present for and interpreted by people; they help to shape what counts as "real." This has important ethical consequences, since it implies that technologies can actively contribute to the moral decisions human beings make. Medical imaging technologies, like MRI and ultrasound, are good examples of this. Obstetrical ultrasound makes visible aspects of a living fetus in the womb, which cannot be seen without them, and which inform us about the health of the unborn child. But the specific way in which ultrasound scanners represent what they "see" helps to shape how the unborn child is perceived and interpreted, and what decisions are made (see box). In this way, technologies fundamentally shape people's experience of disease, pregnancy, or their unborn child.

The very fact of having an ultrasound scan made lets the fetus be present in terms of health and disease, and in terms of our ability to prevent children with this disease from being born.

Case Obstetric Ultrasound

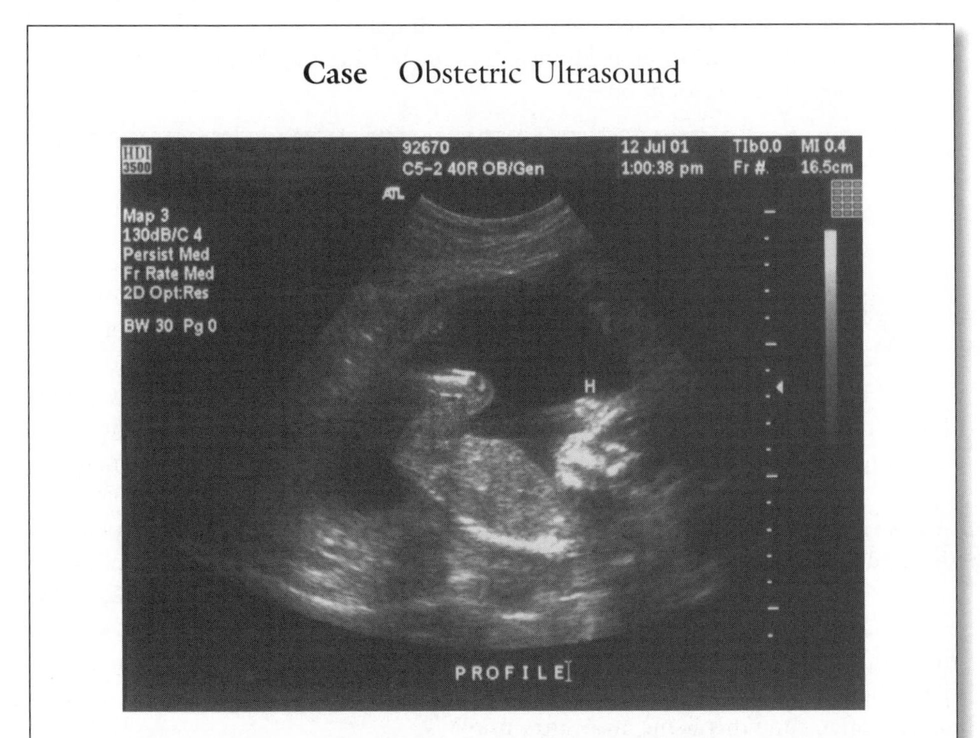

Figure 7.2 Obstetric ultrasound, Photo: © Melking/Fotolia.com.

Ultrasound is not simply a functional means to make visible an unborn child in the womb. It actively helps to shape the way the unborn child is given human experience, and in doing so it informs the choices his or her expecting parents make. Because of the ways in which ultrasound mediates the relations between the fetus and the future parents, it constitutes both fetus and parents in specific ways.

Ultrasound brings about a number of "translations" of the relations between expecting parents and the fetus, while mediating their visual contact. First of all, ultrasound isolates the fetus from the female body. In doing so, it creates a new ontological status of the fetus, as a separate living being rather than forming a unity with his or her mother. This creates the space to make decisions about the fetus apart from the pregnant woman in whose body it is growing.

Second, ultrasound places the fetus in a context of medical norms. It makes visible defects of the neural tube, and makes it possible to measure the thickness of the fetal neck fold, which forms an indication of the risk that the child will suffer from Down's Syndrome. In doing so, ultrasound translates pregnancy into a medical process; the fetus into a possible patient; and congenital defects into preventable suffering. As a result, pregnancy becomes a process of choices: the choice to have tests like neck fold measurements done at all, and the choice what to do if anything is "wrong." Moreover, parents are constituted as decision-makers regarding the life of their unborn child. To be sure, the role of ultrasound is ambivalent here: on the one hand it may encourage abortion, making it possible to prevent suffering; on the other hand it may discourage abortion, enhancing emotional bonds between parents and the unborn child by visualizing "fetal personhood."

7.3.2 Mediation of action

Mediation of action The influence of artifacts on human action.

Script A prescription how to act that is built (designed) into an artifact.

Within the praxis-perspective, the central question is how artifacts mediate people's actions and the way they live their lives. In many ways, the things we use help to shape what we do and how we do things; human actions are not only the result of individual intentions and the social structures in which human beings find themselves, but also of people's material environment (Latour, 1992, 1994). The concept of **script**, as elaborated by Madeleine Akrich and Bruno Latour, is helpful to indicate the influence of artifacts on human actions (Latour, 1992, 1994; Akrich, 1992). Like the script of a movie or a theater play, the script of an artifact prescribes users how to act when they use it. A speed bump, for instance, has the script "slow down when you approach me;" a plastic coffee cup "throw me away after use."

This influence of artifacts on human actions has a specific nature. When scripts are at work, things mediate action as material things, not as immaterial signs. A traffic sign makes people slow down because of what it signifies, not because of its material presence in the relation between humans and world. And we do not discard a plastic coffee cup because its user's manual tells us to do so, but because it simply is physically not able to survive being cleaned several times. The influence of technological artifacts on human actions can be of a non-lingual kind. Things are able to exert influence *as material things*, not only as *signs* or *carriers of meaning*.

As is the case with perception, in the mediation of action *transformations* occur. Within the domain of action these transformations can be indicated as "translations" of "programs of action." When an entity enters a relationship with another entity, the original programs of action of both are translated into a new one. When somebody's action program is to "prepare meals quickly," and this program is added to that of a microwave oven ("heating food quickly"), the action program of the resulting, "composite actor" might be "regularly eating instant meals individually."

In the translation of action, a similar structure can be discerned as in the transformation of perception. Just as in the mediation of perception some aspects of reality are amplified and others are reduced, in the mediation of action one could say that specific actions are *invited*, while others are *inhibited*. The scripts of artifacts suggest specific actions and discourage others.

Invitation-inhibition structure The fact that mediating technology invited specific actions, while other actions are inhibited.

The nature of this **invitation-inhibition structure** is as context-dependent as the amplification-reduction structure of perception. The concept of *multistability* also applies within the context of the mediation of action. The telephone has had a major influence on the separation of people's geographical and social context, by making it possible to maintain social relationships outside our immediate living environment. But it could only have this influence because it is used as a communication technology, not as the hearing aid it was originally supposed to be.

7.4 Moralizing Technology

As elaborated above, many of our actions and interpretations of the world are co-shaped by the technologies we use. Telephones mediate the way we communicate with others, cars help to determine the acceptable distance from home to work, thermometers co-shape our experience of health and disease, and prenatal diagnostic technologies generate difficult questions regarding pregnancy and abortion. This mediating role of technologies also pertains to actions and decisions we usually call "moral" – ranging from the speed we find morally acceptable to our decisions about unborn life. If ethics is about the question "how to act," and technologies help to answer this question, technologies appear to do ethics, or at least help us to do so. Analogously to Winner's claim that artifacts have politics, technology has morality.

In the mediation approach, technologies are analyzed in terms of their mediating roles in relations between humans and reality. The core idea is that technologies, when used, always establish a relation between users and their environment. Technologies do not only enable us to perform actions and have experiences that were scarcely possible before, but in doing so, they also help to shape *how* we act and experience things. Artifacts help to shape human actions, interpretations, and decisions, which would have been different without the artifact, as the case of obstetrical ultrasound illustrated.

Quite often, technologies mediate human actions and experiences without human beings having told them to do so. Some technologies, for instance, are used differently than their designers had envisaged. The first cars – which only made 15 km/h – were used primarily for sports, and for medical purposes; driving at a speed of 15 km/h was considered to create an environment of "thin air," which was supposed be healthy for people with lung diseases. Only after the car was interpreted as a means for long distance transport did it get to play its current, pervasive role in society. And as a means for long-distance transport, it plays an important role in the division between labor and leisure that is part of our everyday lives; in a world without cars, one's social world at work grossly coincides with the social world in one's free time (Baudet, 1986).

In this example, an unexpected mediation came about in a new and initially unexpected use context. Without anyone explicitly intending it, the car helped to shape the way we organize our lives. But unforeseen mediations can also emerge when technologies are used as intended. The very fact that the introduction of cell phones has led to changes in youth culture – such as the fact that young people appear to make ever less appointments with each other, since everyone can call and be called at any time and place – was not intended by the designers of the cell phone, even though it is used here in precisely the context the designers had envisaged.

When mediating the relations between humans and reality, artifacts help to constitute both the objects in reality that are experienced or acted upon and the subjects that are experiencing and acting. This implies that the subjects who act or make decisions about actions are never purely human, but rather a complex blend of humanity and technology. When making a decision about abortion on the basis of technologically mediated knowledge about the chances that the child will suffer from a serious disease, this decision is not "purely" human, but neither is it entirely induced by technology. The very situation of having to make this decision and the very ways in which the decision is made, were co-shaped by technological artifacts. Without these technologies, either there would not be a situation of choice, or the decision would be made on the basis of a different relation to the situation. At the same time, the technologies involved do not *determine* human decisions here. Moral decision-making is a joint effort of human beings and technological artifacts.

7.4.1 Criticizing the moral character of technological artifacts

The idea that artifacts have morality has been severely criticized. A main criticism is that mediation has nothing to do with morality whatsoever. Not only are technological artifacts unable to make moral decisions, but also does technology-induced human behavior not have a moral character. A good example of this criticism are the often-heard negative reactions to explicit behavior-steering technologies like speed limiters in cars. Usually, the resistance against such technologies is supported with two kinds of arguments. First, there is the fear that human freedom is threatened and that democracy is exchanged for technocracy. Should all human actions be guided by technology, the criticism goes, the outcome would be a technocratic society in which moral problems are solved by machines instead of people. Second, there is the charge of immorality or, at best, amorality. Actions that are not the product of our own free will but are induced by technology cannot be described as "moral." And, which is worse, behavior-steering technologies might create a form of moral laziness that is fatal to the moral abilities of citizens.

Yet, these criticisms are deeply problematic. After all, the analyses of technological mediation given above show that human actions are *always* mediated. To phrase it in Latour's words: "Without technological detours, the properly human cannot exist. … Morality is no more human than technology, in the sense that it would originate from an already constituted human who would be master of itself as well as of the universe. Let us say that it traverses the world and, like technology, that it engenders in its wake forms of humanity, choices of subjectivity, modes of objectification, various types of attachment" (Latour, 2002). And this is precisely what opponents of speed limitation forget. Also without speed limiters, the actions of drivers are continually mediated: indeed, as cars can easily exceed speed limits and as our roads are so wide and the bends so gentle as to permit driving fast, we are constantly being invited to further explore the space between the accelerator and the floor. Therefore, giving the inevitable technological mediations a desirable form rather than rejecting outright the idea of a "moralized technology" in fact attests to a sense of responsibility.

This analysis of the moral character of technological artifacts has important implications for engineering ethics and technology design. First of all, the mediation approach to technology makes clear that moral issues regarding technology development comprise more than weighing technological risks and preventing disasters, however important these activities in fact are. What is at stake when technologies are introduced in society are also the ways in which these technologies will mediate human actions and experiences, thus helping to shape our moral decisions and our quality of life. Engineering ethics design, therefore, should also occupy itself with taking responsibility for the future mediating roles of technologies in design.

Moreover, the analysis of technological mediation shows that, even without explicit moral reflection, technology design is inherently a moral activity. By designing artifacts that will inevitably play a mediating role in people's actions and experience, thus helping to shape (moral) decisions and practices, designers "materialize morality"; they are "doing ethics by other means." This conclusion makes it even more urgent to expand the scope of engineering ethics in order to include the moral dimensions of the artifacts themselves, and to try and give shape to these dimensions in a responsible way.

Examples of Moralizing Artifacts

Speed bump: "Lower your driving speed"
Metro tourniquets: "Pay for public transport"
Hotel keys (with large object): "Return your hotel key to the desk"
Door-closer: "Close the door"
Alcohol lock for car (car lock that analyses your breath): "Don't drive drunk"

7.4.2 Taking mediation into ethics

There are two ways to take mediation analyses into engineering ethics and design. First of all, such analyses can be used to develop moral assessments of technologies in terms of their mediating roles in human practices and experiences. Second, the conclusion that artifacts do have a specific form of morality also shifts ethics from the domain of language to that of materiality. When artifacts have moral relevance, ethics cannot only occupy itself with developing conceptual frameworks for moral reflection, but should also engage itself with the actual development of the material environments that help shape moral action and decision-making. Hans Achterhuis has called this the "**moralization of technology**" (Achterhuis, 1995).

Moralization of technology The deliberate development of technologies in order to shape moral action and decision-making.

The first way to take mediation into ethics is closest to common practices in engineering ethics. In fact, it comes down to an augmentation of the current focus on risk assessment and disaster prevention. Rather than focusing on the acceptability and preventability of negative consequences of the introduction of new technologies, it aims to assess the impact of the mediating capacities of technologies-in-design for human practices and experiences. When an action-ethical approach is followed here, moral reflection is directed at the question whether the actions resulting from specific technological mediations can be morally justified. This reflection can take place along duty ethical or consequentialist lines, focusing either on the question if specific norms are met or if the desirable consequences of the technology outweigh the negative ones. In many cases, though, a virtue-ethical approach is at least as fruitful to assess technological mediations. Such an approach focuses on the quality of the *practices* that are introduced by the mediating technologies – the ways in which human beings do things and the kind of life we are living.

Virtue ethics is about the question of "how to live," rather than "what to do in these specific circumstances." Not only the impact of mediation on specific human actions is important then, but also the ways in which mediating technologies help to constitute human beings and the world they are experiencing and in which they are acting. To return to the example of ultrasound again: rather than merely assessing the impact of routine ultrasound scans in obstetrical health care in terms of safety and abortion rates, a virtue-ethical approach would try to assess the quality of the practices that arise around ultrasound scanning, in which the fetus and its expecting parents are constituted in specific ways (as possible patients versus decision-makers) and in specific relations to each other (situations of choice).

The second way to augment engineering ethics with the approach of technological mediation is to not only *assess* mediations, but to also try to help *shape* them. Rather than working from an external standpoint *vis-à-vis* technology, aiming at rejecting or accepting new technologies, engineering ethics then aims to *accompany* technological developments, experimenting with mediations and finding ways to discuss and assess how one could deal with these mediations, and what kinds of living-with-technology are to be preferred. This direction was taken by the Dutch philosopher Hans Achterhuis, who called this the "moralization of technology" (Achterhuis, 1995, 1998). Instead of only moralizing other *people* ("do not shower too long;" "buy a ticket before you enter the subway"), humans should also moralize their *material environment*. To a water-saving showerhead the task could be delegated to see to it that not too much water is used when showering, and to a turnstile the task of making sure that only people with a ticket can enter the train.

Achterhuis' plea for a moralization of technology received severe criticism (cf. Achterhuis, 1998, pp. 28–31). Firstly, autonomy was thought to be attacked when human actions are explicitly and consciously steered with the help of technology. This reduction of autonomy was even perceived as a threat to human dignity; if human actions are not a result from deliberate decisions but from steering technologies, people were thought to be deprived from what makes them human. Here, we hear the echo of Immanuel Kant's ideas that moral decisions have to be the result of autonomous decisions rather than heteronomous influences. If human beings are not acting autonomously, their actions cannot be called "moral." Human beings then simply show a type of behavior that was desired by the designers of the technology, instead of explicitly choosing to act this way. Second, Achterhuis was accused of jettisoning the democratic principles of our society, because his plea for developing behavior-steering technology was considered an implicit propagation of technocracy. When moral issues are solved by the technological activities of designers instead of democratic activities of politicians, these critics hold, not humans but technology will be in control.

Case Cubicle Warrior

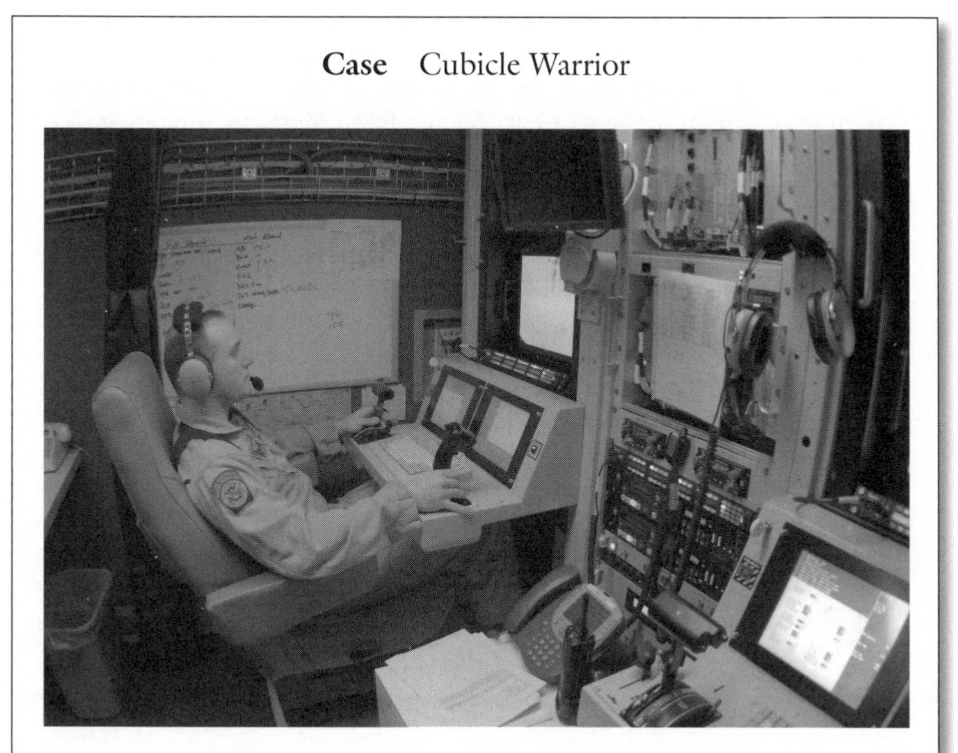

Figure 7.3 Cubicle warrior. Photo: North Dakota National Guard/Senior Master Sgt. David H. Lipp (US Air Force).

The deployment of military robots is growing rapidly. Presently, more than 17 000 military robots are active in the US military. Most of these robots are unarmed, and are mainly used for clearing improvised explosive devices and reconnaissance; however, over the last years the deployment of armed military robots is in the increase. One of the most widely used military robots is the unmanned combat aerial vehicle *Predator*. This unmanned airplane which can remain airborne for 24 hours is currently employed in Afghanistan. The *Predators* can fire Hellfire missiles and are flown by pilots located at the military base in the Nevada desert, thousands of miles away from the battlefield. The *Predators* connect the cubicle warriors (human operators) – who remotely control these military robots behind visual interfaces – with the war zone; they are the eyes of the tele-soldier. These robots can precisely determine a certain target and send the GPS-coordinates and camera images back to the operator. Based on the information projected on his computer screen the cubicle warrior has to decide, for example whether or not to launch a missile. His decision is *mediated* by a computer-aided diagnosis of the war situation. Future military robots will have built into their design ethical constraints, the so-called 'ethical governor' which will suppress unethical lethal behavior. Although these ethical governors are not very sophisticated yet, current research shows some major progress in this development. For example, research has been done – sponsored by the US Army – to create a mathematical decision mechanism consisting of constraints represented as prohibitions and obligations derived directly from the laws of war (Arkin, 2007). Moreover, a future goal is that military robots can refuse orders of a cubicle warrior which according to the ethical governor are illegal or unethical. For example, a military robot might advise a cubicle warrior not to push the button and shoot because the diagnosis of the camera images tells the operator he is about to attack non-combatants, that is, the software of the military robot that diagnoses the war situation provides the cubicle warrior with ethical advice. An ethical governor helps to shape moral decision-making. In other words, the task to see to it that no Rules of Engagement[1] are violated could be delegated to a military robot. A consequence is that humans then simply show a type of behavior that was desired by the designers of the technology instead of explicitly choosing to act this way. This will also be the case with ethical governors, since an ethical governor may form a "moral buffer" between cubicle warriors and their actions, allowing them to tell themselves that the military robot has taken the decision (Cumming, 2006). The consequence of the moralization of military robots is that the decision of a cubicle warrior is not the result of moral reflection, but is mainly determined or even enforced by a military robot (see also Royakkers and Van Est, 2010).

These arguments can be countered, though. First of all, human dignity is not necessarily attacked when limitations of autonomy occur. Our legal constitution implies a major limitation of autonomy, after all, but this does not make it a threat to our dignity. Human behavior is determined in many ways, and autonomy is limited in many ways. Few people will protest against the legal prohibition of murder, so why protest to the material inhibition imposed by a speed bump to drive too fast at places where children are often playing on the pavement? Second, the analysis of technological mediation made clear that technologies *always* help to shape human actions. Therefore, paying explicit attention to the mediating role of technologies should be seen as taking the responsibility that the analysis of technological mediation implies. When technologies are always influencing human actions, we had better try to give this influence a desirable form. Besides, as will become clear below in the example of a Dutch industrial design initiative *Eternally Yours*, the "moralizing" role of technologies does not necessarily have the form of exerting *force* on human beings to act in specific ways. Technologies can also *seduce* people to do certain things; they can invite specific actions without forcefully exacting them.

These counterarguments, however, do not take away the anxiety that a technocracy would come about when technologies are explicitly moralized. It might be true that technologies do not differ from laws in limiting human freedom, but laws come about in a democratic way, and the moralization of technology does not. Yet, this does not justify the conclusion that it is better to refrain from paying explicit attention to technological mediation during the design process. If technologies are not moralized explicitly, after all, the responsibility for technological mediation is left to the designers only. Precisely this would amount to form of technocracy (see Section 1.5.3), and imply the technocratic fallacy (see Section 4.5.2). A better conclusion would be that it is important to find a democratic way to "moralize technology," which is the subject of the next section.

7.5 Designing Mediations

The moralization of technological artifacts is not as easy as it might seem to be. In order to "build in" specific forms of mediation in technologies, designers need to anticipate the future mediating role of the technologies they are designing. And this is a complex task, since there is no direct relationship between the activities of designers and the mediating role of the technologies they are designing. The mediating role of technologies comes about in a complex interplay between technologies and their users.

Technologies, however, can be used in unforeseen ways, and therefore have an unforeseen influence on human actions. The classical energy-saving light bulb is a good example here, having actually resulted in an increased energy consumption since such bulbs often appear to be used in places previously left unlit, such as in the garden or on the façade, thereby cancelling out their economizing effect.[2] Moreover, unintentional and unexpected forms of mediation can arise when technologies do get used in the way their designers intended. A good example is the revolving door which keeps out not only cold air but also wheelchair users. In short, designers play a seminal role in realizing particular forms of mediation, but not the only role. Users with their interpretations and forms of appropriation also have a part to play; and so do technologies, which give rise to unintended and unanticipated forms of mediation.

Designers thus help to shape the mediating roles of technologies, but these roles also depend on the ways in which the technologies are used and on the ways in which the technologies in question allow unforeseen mediations to emerge. Designers cannot simply "inscribe" a desired form of morality into an artifact. The mediating role of technologies is not only the result of the activities of the designers, who inscribe scripts or delegate responsibilities, but also depends on the users, who interpret and appropriate technologies, and on the technologies themselves, which can evoke "emergent" forms of mediation. Figure 7.4 illustrates these complicated relations between technologies, designers, and users in the mediation of actions and interpretations.

The figure makes clear that in all human actions and all interpretations informing moral decisions, there are three entities that "act": 1) the human being performing the action or making the moral decision (in interaction with the technology), but also appropriating the technological artifact in a specific way; 2) the artifact mediating these actions and decisions, sometimes in unforeseen ways; and 3) the designer who – either implicitly or in explicit delegations – gives a specific shape to the artifact used, and thus helps to shape the eventual mediating role of the artifact. Taking responsibility for technological mediation, therefore, comes down to entering into an interaction with the agency of future users and the artifact-in-design, rather than acting as a "prime mover" (cf. Smith, 2003).

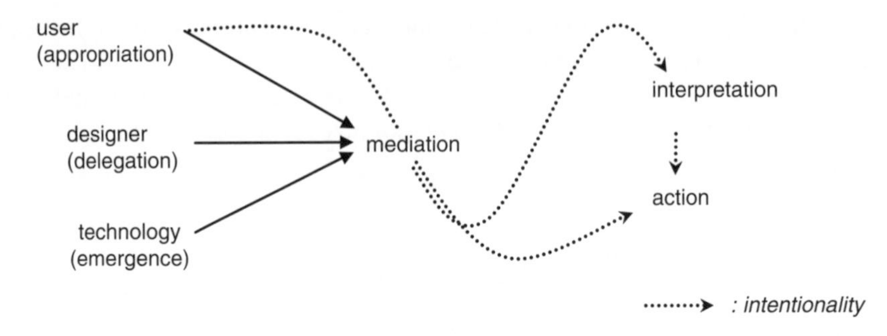

Figure 7.4 Human actions and interpretations informing moral decisions.

The unpredictability of the mediating role of technology that follows from this does not imply, however, that designers are by definition unequipped to deal with it. In order to cope with the unpredictability and complexity of technological mediation, it is important to seek links between the design context and the future use context. Design specifications should be derived not only from the product's intended function but also from an informed prediction of the product's mediating roles and a moral assessment of these roles. A key tool to bringing about this coupling of design context and use context, however trivial it may sound, is the designer's moral imagination. By trying to imagine the ways the technology-in-design could be used and by shaping user operations and interpretations from that perspective, a designer can include the product's mediating role in his or her moral assessment during the design phase. Performing a mediation analysis can be a good basis for making an *informed prediction* of the future mediating role of a technology. An interesting example of **anticipating mediation by imagination** is the work of the Dutch industrial designers collective *Eternally Yours* (see box).

Anticipating mediation by imagination Trying to imagine the ways technology-in-design could be used. This insight is then used to deliberately shape user operations and interpretations.

Case Eternally Yours

Eternally Yours is engaged in eco-design, but in an unorthodox way (cf. Van Hinte, 1997; and Verbeek, 2005). It does not want to address the issue of sustainability only in the usual terms of reducing pollution in production, consumption, and waste. The actual problem, Eternally Yours holds, is that most of our products are thrown away far before actually being worn out. Meeting this problem could be way more effective than reducing pollution in the different stages of products' life-cycles. For this reason, Eternally Yours focuses on developing ways to create product longevity. It does so by investigating how the coming about of attachment between products and their users could be stimulated and enhanced.

In order to stimulate longevity, Eternally Yours seeks to design things that invite people to use and cherish them as long as possible. 'It's time for a new generation of products, that can age slowly and in a dignified way, become our partners in life and support our memories,' as Eternally Yours approvingly quoted the Italian designer Ezio Manzini in its letterhead. Eternally Yours investigates what characteristics of products are able to evoke a bond with their users. According to Eternally Yours, three dimensions can be discerned in the lifespan of products. Things have a technical, an economical, and a psychological lifespan. Products can turn into waste because they simply are broken and cannot be repaired anymore; because they are outdated by newer models that have appeared in the market; and because they do not fit people's preferences and taste anymore. For Eternally Yours, the psychological lifespan is the most important. The crucial question for sustainable design is therefore: how can the psychological lifetime of products be prolonged?

Eternally Yours developed many ideas to answer this question. For instance, it searched for forms and materials that could stimulate longevity. Materials were investigated that do not get unattractive when aging but have "quality of wear." Leather, for instance, is mostly found more beautiful when it has been used for some time, whereas a shiny polished chromium surface looks worn out with the first scratch. An interesting example of a design in this context is the upholstery of a couch that was designed by Sigrid Smits. In the velour that was used for it, a pattern was stitched that is initially invisible. When the couch has been used for a while, the pattern gradually becomes visible. Instead of aging in an unattractive way, this couch renews itself when getting old. Eternally Yours does not only pay attention to materials and product surfaces, however. It also investigated the ways in which services around products can influence their lifespan. The availability of repair- and upgrading services can prevent people from discarding products prematurely.

A second way to formulate an informed prediction of the future mediating role of technologies is a more systematic one. It consists in an augmentation of the existing design methodology of *Constructive Technology Assessment* (CTA, see Section 1.6) in such a way, that it becomes an instrument for a democratically organized moralization of technology, and can be seen as a democratization of the designing process. When a CTA design methodology is followed, not only designers determine what a technology will look like, but all relevant social actors. Following this method, therefore, could take away the fear for technocracy that was discussed above.

Seen from the perspective of technological mediation, however, CTA also has limitations that need to be overcome. CTA primarily focuses on *human* actors, and pays too little attention to the actively mediating role of the *nonhuman* actor that is at the centre of all activity: the technology-in-design. CTA analyzes the complex dynamics of technology development. It bases itself on a notion that technologies are not "given," but the outcome of a process in which many actors are involved. Other interactions between the actors might have resulted in a different technology. By analyzing the dynamics of *technology development*, CTA reveals how technologies emerge from their *design context*, but their role in their *use context* remains unanalyzed. Therefore, organizing a democratic, domination-free discussion between all relevant actors is not enough to lay bare all relevant aspects of the technology in question. The mediating role of the technology-in-design is likely to remain hidden during the entire CTA process if it is not put explicitly and systematically on the agenda. For this reason, participants in the CTA process should not only be invited to integrate assessments of users and pressure groups in product specifications, but also to anticipate possible mediating roles of the technology-in-design.

To be sure, this anticipation of technological mediation introduces new complexities in the design process. Designers, for instance, might have to deal with trade-offs: in some cases, designing a product with specific desirable mediating characteristics might have negative consequences for the usefulness or attractiveness of the product. Introducing automatic speed influencing in cars will make sure that drivers keep to the speed limit, but at the cost of the experience of freedom – which appears to be rather important to some car drivers, judging by the fierce societal resistance against speed limiting measures. Also, when anticipating the mediating role of technologies, prototypes might be developed and rejected because they are likely to bring about undesirable mediations. Dealing with such trade-offs and undesirable spin-offs requires a separate moral decision-making process (see also Section 6.2.4 on trade-offs).

Technology design appears to entail more than inventing functional products. The perspective of technological mediation reveals that designing should be regarded as a form of materializing morality. This implies that the ethics of engineering design should take more seriously the moral charge of technological products, and rethink the moral responsibility of designers accordingly.

7.6 Chapter Summary

The analyses of technological mediation have major implications for the ethics of engineering design. The insight that technologies inevitably play a mediating role in the actions of users makes the work of designers an inherently moral activity. Ethics is about the question how to act, and technologies appear to be able to give material answers to this question by inviting or even exacting specific forms of action when they are used. This implies that technological mediation could play an important role in the ethics of engineering design. Designers should not only focus on the functionality of technologies but also on their mediating roles. The fact that technologies always mediate human actions charges designers with the responsibility to anticipate these mediating roles.

This anticipation is a complex task, however, since the mediating role of technologies is not entirely predictable. But even though the future cannot be predicted with full accuracy, ways do exist to develop well-informed and rationally grounded conjectures. In order to cope with the uncertainty regarding the future role of technologies in their use contexts, designers should try to bridge the gap between the context of use and the context of design.

One way to do so is by carrying out a "mediation analysis" with the help of the designer's imagination, which can be facilitated by the vocabulary developed in this chapter. Such an analysis will not allow designers to predict entirely how the technology they are designing will actually be used, but it will help to identify possible use practices, and the forms of mediation that might emerge with it.

Designers could also make use of an augmented form of constructive technology assessment, in which the connection between design and use is not only made in imagination but also in practice. In this case, a mediation analysis is carried out not by the designer individually, but by all stakeholders together, who engage in a democratically organized debate in order to decide how to feed back the outcomes of this analysis into the design process. Following this method could take away part of the fear that deliberately designing behavior-steering technology would lead to technocracy, since the inevitable mediating role of technology is made subject to democratic decision-making here.

Study Questions

1. In what two ways can technologies mediate our perception of the world? Give an example of both ways.
2. In what way is the transformation of reality by an electronic microscope a structure of amplification and reduction? What aspects of reality are amplified? What aspects are reduced?
3. What is meant by the multistability of technology? Give an example of an everyday technology that illustrates the multistability of technology.
4. Which two criticisms have been raised against the idea that technology has morality? Do you consider these criticisms convincing? Why (not)?
5. a. What reasons are there to consider the phenomenon of technological mediation a *moral* phenomenon?
   b. In what two ways can mediation analysis be taken into ethics? Give an example of both ways.
6. a. What is meant by the moralization of technology? Give an example.
   b. What objections can be raised against moralization by technology?
   c. How can technologies be moralized in a democratic way?
7. Explain how it is possible to anticipate the mediating role of technology through imagination.
8. Reread the case on the Golden Gate Bridge at the beginning of Chapter 4. Give an analysis of this case in terms of "scripts." How do the design of the bridge and the context of use contain a structure of invitation and inhibition with respect to committing suicide? In what respects do they "invite" and in what respects "inhibit" suicide? How would the addition of a suicide barrier change the script? Can you think of certain suggestions for the design of the barrier on basis of the analysis of the script?
9. Give a mediation analysis of Intelligent Speed Adaptation (ISA) in cars. ISA is a technology that uses GPS to detect where a vehicle finds itself, in order to limit the speed of the vehicle to the speed limit that applies at that specific place. Elaborate how the mediating roles of these technologies have moral significance. Discuss whether or not you think that ISA is an acceptable technology. If possible, also elaborate if and how this technology could be (re)designed to make it more acceptable.
10. Give a mediation analysis of technologies for the identification of hereditary breast cancer. Elaborate how the mediating roles of these technologies have moral significance. Discuss whether or not you consider these technologies acceptable. If possible, also elaborate if and how such technologies could be (re)designed to make them more acceptable.

Discussion Questions

1 Do you agree that obstetric ultrasound changes the relation between parents and fetus? Do you think that it changes the decisions parents make with respect to, for example, abortion? Can you think of other forms of "representing the unborn child" that would have other effects?
2 Suppose that people could be genetically modified in such a way that they automatically behave ethically. Would you consider such a form of genetic manipulation morally desirable? Is there any difference between this technological intervention and the examples of moralization by technology discussed in this chapter? Do you consider this scenario a realistic or useful thought experiment to think about the desirability of moralizing technology?

Notes

This chapter is based on Verbeek (2006a, 2006b, and 2008).

1 Rules of Engagements compromise directives issued by competent military authorities that delineate both the circumstances and the restraints under which combat with opposing forces is joined.
2 Steg (1999); Weegink (1996). For the current generation of energy-saving light bulbs, which are based on LED technology, the situation will probably be different.

8

Ethical Aspects of Technical Risks

Having read this chapter and completed its associated questions, readers should be able to:

- Discuss why engineers are responsible for safety and how they can apply this responsibility in engineering practice;
- Describe the main approaches to risk assessment;
- Describe the main ethical considerations for judging the moral acceptability of risks and apply these to concrete cases;
- Argue why risks that are of similar magnitude are not necessarily equally acceptable;
- Identify ethical issues in risk communication and to judge different ways of dealing with them;
- Explain what is meant with engineering as a societal experiment and to reflect on the conditions under which such experiments are morally acceptable.

Contents

Ethics, Technology, and Engineering: An Introduction, First Edition.
Ibo van de Poel and Lambèr Royakkers.

8.1 Introduction

Case A Coal Mining

Between 1900 and 1975, about 600 million tons of coal were mined in the province of Limburg in the Netherlands (Pöttgens, 1988). In an area of about 220 km^2 the surface dropped about 2.5 m on average. In some locations there was a drop of more than 10 m. The coal mining operations were damaging the houses. Hundreds of millions in damages were paid by the companies mining the coal. The risk of damage to the houses was known in advance, but it thought to balance out against the benefits of the coal mining.

Case B DC-10 Disaster

On March 3, 1974 the freight door of a DC-10 opened during flight (Eddy, Potter, and Page, 1976). As a result, the plane crashed killing 346 people. The risk was known beforehand, at least to some of the people involved. On June 12, 1972 a similar accident had almost occurred. Already towards the end of the 1960s the possibility of this type of accident had been anticipated as a result of tests. Also the *Failure Mode and Effect Analysis* (FMEA) of the freight doors revealed the possibility of this type of accident. One important reason why nothing had been undertaken to reduce or avoid the risks was because there was an

Figure 8.1 Tacoma Narrows Bridge. Photo: © Siramstrong / Fotolia.com.

ongoing conflict between the supplier responsible for the door (Convair) and the plane manufacturer (Douglas). Neither party wanted to weaken its legal position. Neither therefore wanted to take the first step in the direction of adapting the design, even though each was aware of the shortcomings. The company's management ignored a memo written by a Convair engineer in which the shortcomings of the door design were outlined.

Case C Tacoma Narrows Bridge Collapse

In 1940 in the United States, the Tacoma Narrows Bridge collapsed (Petroski, 1982, ch. 13). It was an innovative bridge design, but the bridge started to vibrate when it was hit by side winds. The bridge finally collapsed when it was closed for safety reasons. The fact that the bridge could start to vibrate had not been anticipated by the designers. From experiences with previous bridges they had wrongly concluded that narrower suspension bridges could be built.

Case D Asbestos

Asbestos is a product that started to come into large scale use at the beginning of the twentieth century. Due to a number of positive characteristics, such as heat resistance, durability, and good insulation properties, it was applied in a large number of products. However, during the course of time asbestos proved to have some extremely harmful side effects. Inhaling asbestos fibers can lead to asbestos-related diseases such as asbestosis and mesothelioma (cancer of the lung and stomach lining), which can be lethal. According to some estimates, as many as 10 000 people in the US and 4000 people in the UK die yearly due to asbestos related diseases.[1] These diseases only become manifest after several decades. The use of asbestos has been banned in the meantime in many countries.

Case E Greenhouse Effect

Carbon dioxide (CO_2) is one of the greenhouse gasses responsible for the heating up of the earth's atmosphere. According to certain calculations without the greenhouse effect the temperature on earth would be –18°C thus probably making human life on earth impossible. Since the end of the nineteenth century human-instigated CO_2 production has increased exponentially resulting in an intensified greenhouse effect. Though there has been controversy over the existence of intensified greenhouse effect, most scientists now agree that the earth is indeed warming up with potentially large-scale and catastrophic consequences.

These five cases demonstrate that to a certain extent hazards are inherent to technology. The first two examples present hazards that were known beforehand. In case A those involved consciously took the risks of coal mining because of the expected advantages. In case B, with the DC-10, the risks were – in retrospect – regarded as unacceptable by most of the managers and engineers involved. However, the relevant risks were not removed or diminished for the reasons given in the example. Cases C and D reveal that the negative effects of technology do not only emanate from known risks but also from unknown hazards. Sometimes such hazards have to do with the fact that certain technology can fail in a way that had not been foreseen beforehand, such as in the Tacoma Narrows Bridge case. Sometimes they are due to unforeseen side effects of technology, as with asbestos. Case E shows that the existence of certain hazards can be controversial. Even if most scientists now agree that the intensified greenhouse effect exists and that the earth is warming up, for a long time this was controversial. There is still no agreement on what exactly the consequences of an intensified greenhouse effect will be. Case E is also different from the others in the sense that it is not precisely clear how this hazard can be attributed to a specific technology. More so than with other hazards it is a hazard that arises in the use phase of technology rather than in the design or production phase.[2] Furthermore, less so than in a number of other cases, it is hard to ascribe the possible hazard to one specific technology. It is connected more with the large-scale use of a wide range of technologies.[3]

In this chapter, we discuss the moral issues that are raised by the risks and hazards of technologies, and how engineers can deal with those issues. We will discuss the responsibility of engineering for safety (Section 8.3) and the current methods for assessing risks (Section 8.4). Section 8.5 discusses the moral acceptability of technological risks and the next section focuses on communication of risks. Then, we will pay attention to situations of uncertainty and ignorance (Section 8.7). Finally, in Section 8.8 some conclusions are drawn regarding the responsibility of engineers. However, to begin with we shall define some of the key terms.

8.2 Definitions of Central Terms

We speak of a **hazard** if a technology, or its use, can cause damage or otherwise undesirable effects. The term *risk* is a specification of the term hazard. It is an attempt to name or specify the phenomenon of hazard, which is often done in quantitative terms. In this chapter we shall mainly concentrate on safety risks (risks of events in which there can be fatalities or injured) and health risks (risks in which the health of people is endangered). We shall not consider environmental risks or social risks. These are given more attention in the chapter on sustainability (Chapter 10).

> **Hazard** Possible damage or otherwise undesirable effect.

The same hazard can be expressed as a risk in various ways. In this chapter, the term **risk** will be defined as the product of the probability of an undesirable event and the effect of that event, unless stated otherwise. This probability is often taken to be the relative frequency of an undesirable event, such as "once every ten years." The effect is often expressed as the number of fatalities. With this definition the term "risk" is a measure for the expected number of fatalities per time unit. Other definitions of risk are used too, like:

> **Risk** A risk is a specification of a hazard. The most often used definition of risk is the product of the probability of an undesirable event and the effect of that event.

- The probability of an undesirable event taking place.
- The maximum negative effect of an undesirable event.

Safety The condition that refers to a situation in which the risks have been reduced as far as reasonably feasible and desirable.

Acceptable risk A risk that is morally acceptable. The following considerations are relevant for deciding whether a risk is morally acceptable: (1) the degree of informed consent with the risk; (2) the degree to which the benefits of a risky activity weigh up against the disadvantages and risks; (3) the availability of alternatives with a lower risk; and (4) the degree to which risks and advantages are justly distributed.

Safety is sometimes defined as the absence of risk and hazards. Usually, a technological product cannot be made absolutely safe in this sense. Safety therefore also often refers to the situation in which the risks have been reduced in as far that is reasonably feasible and desirable. So conceived, safety is related to the notion of **acceptable risk**. We will discuss the acceptability of risks in more detail in Section 8.5.

For a number of reasons, it is not always possible to predict the hazards of a technology beforehand and to express them reliably as risks. One reason is the *complexity* of causal relations between potential harmful agents and specific undesirable effects. Complexity may be due to such factors as interactions between different substances or between substances and specific environments, long delay times and intervening variables. The impossibility of expressing hazards in risks may also be due to **uncertainty**, that is, a lack of knowledge. Uncertainty may, in turn, be caused by a number of underlying factors, like modeling errors, indeterminacy, and the drawing of system boundaries (Renn, 2005, p. 30). In a more circumscribed sense, the notion of uncertainty is often used to refer to situations where we know the type of consequences, but cannot meaningfully attribute probabilities to the occurrence of such consequences (Felt et al., 2007, p. 36). In cases of uncertainty, we can therefore not calculate the risks. Sometimes, we do not even know that something can go wrong, that there is a hazard. In such cases, the term **ignorance** is often used (Felt et al., 2007, p. 36). What is typical of ignorance is that we do not know what we do not know. Therefore it is extremely hard, if not impossible, to anticipate the consequences of ignorance because often we do not know what we have to be prepared for.

Uncertainty A lack of knowledge. Refers to situations in which we know the type of consequences, but cannot meaningfully attribute probabilities to the occurrence of such consequences

Ignorance Lack of knowledge. Refers to the situation in which we do not know what we do not know.

The impossibility of expressing hazards in risks may also be due to **ambiguity**. Ambiguity refers to the fact that different interpretations or meanings may be given to the measurement, characterization, aggregation, and evaluation of hazards. The International Risk Governance Council (IRGC) distinguishes between interpretive ambiguity – referring to different interpretations of scientific data (for example, how to extrapolate dose-response relations to low doses for which no data are available) – and normative ambiguity – referring to disagreement about the relevant (moral) values and their relative importance (Renn, 2005).

Ambiguity The property that different interpretations or meanings can be given to a term.

8.3 The Engineer's Responsibility for Safety

From where does the engineer's responsibility for safety come? Many codes of conduct for engineers attribute a responsibility for safety to engineers (see Chapter 2). Besides that, there are legal obligations concerning the safety of products or technical codes and standards, in which safety often plays an important role. Nevertheless, the law and codes of conduct are not sufficient a moral argument to establish that engineers are responsible for safety. To this purpose we must explore the ethical frameworks that were dealt with in Chapter 3.

Consequentialism, duty ethics, and virtue ethics all provide arguments why engineers should strive for safe products; the exact arguments differ for each ethical framework. Consequentialism states that engineers must strive for good consequences: safe products definitely fall into that category. The desirability to design safe products is sometimes described as "do no harm." This can be defended in terms of consequentialism or utilitarianism with the freedom principle of Mill (see Section 3.7.2). It is a kind of minimum standard that applies to striving for good consequences. In duty ethics the notion "you should not harm anyone" can be seen as a general norm. This can be defended via the universality principle of Kant. Imagine you were allowed to harm others. It would imply that others would be allowed to harm you too according to this universality principle. It is impossible for people to want this, because it would mean giving away all the safety that people wish to have. People would have the right to harm you without you being able to hold them responsible. In virtue ethics, care for the users or, more in general, for people who suffer the consequences of your design, is an important virtue. Striving for safe products, therefore, is an important virtue.

During the design process, engineers can follow different strategies for ensuring safe products, such as:

1 **Inherently safe design**: avoid hazards instead of coping with them for example by replacing substances, mechanisms, and reactions that are hazardous by less hazardous ones.
2 **Safety factors**: constructions are usually made stronger than the load they probably have to bear. Adding a safety factor to the expected load or maximum load is an explicit way of doing this.
3 Negative feedback: For cases that a device fails or an operator loses control, a **negative feedback mechanism** can be built in that causes the device to shutdown. An example is the dead man's handle that stops the train when the driver falls asleep or looses consciousness.
4 **Multiple independent safety barriers**: A chain of safety barriers can be designed that operate

Inherently safe design An approach to safe design that avoids hazards instead of coping with them, for example by replacing substances, mechanisms and reactions that are hazardous by less hazardous ones.

Safety factor A factor or ratio by which an installation is made safer than is needed to withstand either the expected or the maximum (expected) load.

Negative feedback mechanism A mechanism that if a device fails or an operator loses control assures that the (dangerous) device shuts down.

Multiple independent safety barriers A chain of safety barriers that operate independently of each other so that if one fails the others do not necessarily also fail.

independently so that if the first fails the others still help to prevent or minimize the effects. This can, for example, be achieved through redundancy in design (see box). Also emergency escapes can be quite useful (Hansson, 2007a).

Redundant Design

The failure of a component or sub-system can often be compensated by producing systems with redundant designs. Nuclear reactors have redundant systems to sustain electricity (to operate pumps etc.) and to cool the reactor core (see, e.g., Mostert, 1982). If one system drops out, one can in principle fall back on the redundant system so that the reactor can continue to operate safely. The O-rings of the *Challenger* were designed redundantly too. There were two O-rings for each connection, so that if the first were to fail the second would compensate. The redundancy of the O-rings was one of the arguments given to allow the flight to continue on the evening before the fatal flight (Vaughan, 1996). Airplanes often have redundant systems too, for example, for the control of the plane.

For the system as a whole it can also be useful to have some fall-back options. Take for example a back-up system for an electronic databank or spare capacity reserves if an electricity plant shuts down. The latter can be very important in preventing large areas of the grid from dropping out.

What is important is that these strategies do not only address known risks but also to some extent uncertainties to avoid the ostrich's fallacy (see Section 4.5.2). Negative feedback mechanisms may also, for example, be effective if the causes of a certain accident are not foreseen or are unknown. In the case of the ICE train discussed in Chapter 6, for example, none of the mentioned strategies for safe design was employed, while any of these strategies might either have decreased the probability of the accident or the consequences of it, even if the exact risk was unknown (see Brumsen, 2006):

1 Compounded wheels may be said to be inherently more dangerous than monobloc wheels because they introduce new potential failure mechanisms. Even if these mechanisms were probably not (exactly) known beforehand, it seems obvious that the new wheel design was inherently less safe.
2 The design and inspection lacked an adequate margin of safety. For example, it turned out that the minimally required diameter of the wheels was probably set too low.
3 The train lacked negative feedback mechanisms in case of an accident. Some trains for example have sensors that monitor wheel breaks and that can also be used to stop the train automatically. Most trains also have an emergency break that can be used by passengers. The emergency brake of the ICE could only be handled by the conductor, who refused to employ it despite warnings from passengers who had heard the wheel breaking. After the wheel had broken, the train drove for another

2 minutes without derailing. An adequate feedback mechanism might have prevented the disaster or would at least have decreased the number of fatalities.

4 The rail track could have formed an independent safety barrier if it had been especially designed for high speed trains as is the case for the French TGV. Such tracks do not have switches and have curves that are adjusted to high speeds. For example, a French TGV that derailed on December 21, 1993 at a speed of 300 kilometers an hour, drove on just outside the rails for several kilometers causing only two minor injuries. In the case of the German ICE disaster, the train changed tracks only two hundreds meter before the bridge into which it eventually ran, which was probably a major factor in the severity of the accident.

8.4 Risk Assessment

To judge whether certain hazards are acceptable, an attempt is usually first made to map them and express them as risks. This takes place by carrying out so-called **risk assessments**. In engineering there are many types and methods for risk assessment. The exact methods differ from one engineering domain to the other. We shall not attempt to give an overview of all the methods for risk assessment used in engineering, but limit ourselves to a general overview.

Risk assessment A systematic investigation in which the risks of a technology of an activity are mapped and expressed quantitatively in a certain risk measure.

A risk assessment usually consists of four steps:

1 Release assessment
2 Exposure assessment
3 Consequence assessment
4 Risk estimation. (Covello and Merkhofer, 1993)

Release assessment

Releases are any physical effects that can lead to harm and that originate in a technical installation. Examples are shock waves, radiation, and the spread of hazardous substances. In general, we can distinguish between two kinds of releases: incidental and continuous. Incidental releases are usually unintended and are due to, for example, an explosion in a chemical plant or an accident with a nuclear power plant. Such releases can often cause immediate and major harm. Continuous releases are often anticipated and may be accepted as side-effects of, for example, production processes. Continuous releases do not necessarily or always lead to exposure or harm.

In the case of incidental releases, an important step is the detection of so-called **failure modes** and accident scenarios. These are series of events that lead to the failure of the installation or to an accident. The probability of the occurrence of such scenarios is calculated too. This takes place in two ways.[4] In the first method, the probability of certain accidents occurring is calculated using statistical data about accidents in the past. If such statistical data are absent event trees and fault trees are often

Failure mode Series of events that may lead to the failure of an installation.

Event tree Tree of events in which one starts with a certain event and considers what events will follow.

Fault tree Tree of events in which we move backwards from an unwanted event (a fault) to the events that could lead to the undesirable event.

used to calculate the probability of an accident. For **event trees** we start with a certain event and consider what events will follow. For **fault trees** we move backwards from an unwanted event (a fault) to the events that preceded and could have led to the undesirable event. To each event in the event or fault tree a probability value is attached on the basis of failure data concerning components. Next, the probability of a specific accident scenario is calculated.

Exposure assessment

In this step the aim is to predict the exposure of vulnerable subjects like human beings to certain releases. Exposure assessment usually describes what vulnerable subjects (human beings, animals, the environment) are exposed to a certain release, through what mechanisms (for example, inhalation of toxic substances by humans), and the intensity, frequency, and duration of the exposure.

Consequence assessment

In the third step the focus is on determining the relationship between exposure and harmful consequences. In some risk assessments, the analysis is limited to acute harm or to the number of direct fatalities. In other cases, long-term effects on health or the environment are also considered. An important part of this step is usually the determining of dose-response relationships. Such relationships can be established through tests on animals, epidemiology and models (Covello and Merkhofer, 1993, pp. 127–178).

Animal tests Tests for determining dose-response relationships by exposing animals to various dosages and assessing their response.

In the case of **animal tests** the harmful effects are tested by exposing animals to dosages. Different animals are given different dosages. Next to that there is a control group. By comparing the groups a dose-response relationship can be determined for the type of animal involved. The idea behind this is that it tells us something about the dose-response relationship in humans.

Epidemiological research Research in which population data is used to find out what the relationship is between the occurrence of certain diseases or certain mental deviations and certain factors that may cause these deviations.

In **epidemiological research** we use population data to find out what the relationship is between the occurrence of certain diseases or mental deviations and certain factors that may cause these deviations. The advantage of epidemiological research is that we have no need to translate the effects on animals to the effects on humans. Epidemiological research has its disadvantages too, however. First, it can only occur after the fact, when certain health effects have already occurred. Second, it requires reliable statistical data. This often requires extensive empirical research. Often the time and money for this are lacking. Third, only statistical correlations are usually established. Demonstrating a statistical correlation however is not enough to prove the existence of a causal relationship. A nice example is that an empirical study found that married people eat statistically significant less candy than unmarried people.[5] An analysis of the data showed, however, that both being married and eating less candy were strongly correlated to the underlying variable age. Older people are more likely to be married and are more likely to eat less candy. So it would have been wrong to conclude in this case that being married is a cause of eating less candy. To establish a cause we thus need to exclude all other possible causal factors, which is often difficult in practice.

Many **models for dose-response relationships** have a hypothetical and descriptive nature. They presuppose a certain relationship between dose and effect, but they hardly or do not explain how a dose of a harmful substance leads to consequences.

Models for dose-response relationships Models that presuppose or predict a certain relationship between dose and response.

Risk estimation

In the fourth step the risk is determined and presented using the results obtained earlier. In this step we determine in what measure the risk is expressed. This can be done using the number of expected fatalities per time unit, for example, or the reduced lifespan of people that work or live in the neighborhood of an installation.

8.4.1 The reliability of risk assessments

In many cases, risk assessments only have limited reliability. This is because the results often depend on the original assumptions made, as the box on the estimated risks of dioxin shows. In connection with this it is striking that many risk assessments do not give an estimate of the accuracy and reliability of the final result. One may well wonder whether it would not be more responsible to list uncertainty intervals for results, or to state explicitly under which conditions results apply.

Case The Risks of Dioxin

In 1978, a test was carried out with rats to determine the health effects of dioxin. Rats that were given 100 000 picograms of dioxin per kilogram of body weight per day developed cancer significantly more often than the control group. At 10 000 picogram/kg per day there was only a small increase in cancer and at 1000 picogram/kg per day no effects could be measured.

On the basis of these data, regulatory bodies in the United States and Canada made very different assessments concerning which concentrations are acceptable. In the United States, the assumption was made that animal tests are not sensitive enough to measure the health effects at low doses. Moreover, a more or less linear relationship was assumed between dose and effect. As a result, they concluded that a dose of 0.006 picogram/kg per day in humans would lead to an individual risk of less than 1 in 1 million. That means one fatality per million people exposed. In Canada, it was assumed that dioxin was not an initiator of cancer but a promoter. In contrast with cancer initiators, cancer promoters are assumed to have a no-effect level, so that no harmful effect can be found below a given level. Applying this model led to the conclusion that around 1 to 10 picogram/kg per day would be safe for people. On the basis of these conclusions the acceptable exposure to dioxin would be about one thousand times higher in Canada than in the United States.

Source: Covello and Merkhofer (1993, pp. 177–178).

One relevant issue is also the degree of evidence that is needed to establish a risk during a risk assessment on the basis of, for example, epidemiological data. In establishing a risk on the basis of a body of empirical data one might make two kinds of mistakes. One can establish a risk when there is actually no risk (a so-called **type I error**) or one can mistakenly conclude that there is no risk while there actually is a risk (a so-called **type II error**). Science traditionally aims at avoiding type I errors because one usually does not want to assume a hypothesis as knowledge unless there is strong evidence for it. Several authors have argued that in the specific context of risk assessment it is often more important to avoid type II errors (Cranor, 1990; Shrader-Frechette, 1991). The reason for this is that risk assessment not just aims at establishing scientific truth but also has a practical aim, that is, to provide the knowledge on the basis of which decisions can be made about whether it is necessarily to protect the public against certain risks. It might be worse not to protect the public against a risk than to take unnecessary precautions against a risk that turns out not to exist.

Type I error The mistake of assuming that a scientific statement is true while it actually is false. Applied to risk assessment: The mistake that one assumes a risk when there is actually no risk.

Type II error The mistake of assuming that a scientific statement is false while it actually is true. Applied to risk assessment: The mistake that one assumes that there is no risk while there actually is a risk.

8.5 When are Risks Acceptable?

Some engineers and scientists believe that if the risks of two different activities are the same according to risk assessments, the activities are equally acceptable. In other words, if one activity is acceptable the other (which has the same risk) must be acceptable too. This argument is flawed for a number of reasons. First, the question whether the *risks* of technology A are acceptable is not the same question as the question whether technology A is acceptable. This will become clear when we consider the ethical objections to human cloning (see box).

Ethical Objections to Cloning

Philosopher of technology Tsjalling Swierstra has reconstructed which arguments played a role in the social discussion on human cloning. He made the following list of arguments:

Cloning:

1 undermines the uniqueness of humans;
2 is contrary to human dignity;
3 leads to psychosocial problems in the cloned child;
4 suffers from numerous scientific and technical risks;
5 will lead to misuse and has unforeseen and undesirable consequences;
6 is unnatural;
7 is based on an overreaching desire for manipulation and leads to the abject instrumentalization of people;
8 will lead to undesirable changes in our self-image.

Source: Based on Swierstra (2000, p. 42).

Objection 4 is already formulated in terms of risks. The third and fifth objection can easily be reformulated in terms of risks. However, objections like cloning undermine the uniqueness of humans (the first) and can lead to an undesirable change in our self-image (the eighth) cannot so easily be understood in terms of risks. This is because these kinds of objections cannot be justified on the basis of consequentialism; they stem from a duty ethics or virtue ethics approach. Especially the argument that cloning leads to a kind of human or kind of society that is not virtuous and therefore is undesirable, to which the second and seventh objection refer, is clearly based on virtue ethics (Swierstra, 2000). The question whether cloning is ethically acceptable is thus more encompassing than the question whether the risks of the technology are acceptable.

Even if we restrict our analysis to the acceptability of risks, it is a fallacy to conclude that if the magnitude of the risks of two technologies is the same these risks are equally acceptable. A number of reasons why the conclusion that equally large risks are equally acceptable is flawed are given in the box.

Why the Magnitude of the Calculated Risk Does Not Tell Us Everything about the Acceptability of the Risk

There are a number of arguments why equally large risks are not necessarily equally ethically acceptable:

As not all risk assessments are equally *reliable*, the results of risk assessments are not easily comparable. The critical question belonging to such a comparative judgment "Are the estimations of risks reliable?" is highly relevant in order to avoid a fallacy (see Chapter 4). Take for example the comparison between the risks of nuclear power plants and traffic risks. Traffic risks are usually calculated using a large number of statistical data based on years of experience. In the case of nuclear power plants, the risks cannot be calculated using statistical data and all sorts of assumptions have to be made to estimate the risk in question. That is why the estimations of traffic risks are usually more reliable than the estimates of the risks for nuclear power plants.

Risks are often *multi-dimensional*, while only one dimension is used in the comparison of risks in many cases. This dimension often is the number of expected fatalities per time unit. Fatalities can occur both through accidents in traffic and because of accidents in nuclear power plants. In many other aspects, however, the risks in question are not that easy to compare. Take, for example, the lasting risk of nuclear waste.

It is not obvious that *a small probability of a major accident is as acceptable as a large probability of a small accident*, even if the product of probability and effect is the same. In this respect the risks of traffic and of nuclear power plants differ. The probability of having an accident in traffic is far higher than the probability of a nuclear accident at a power plant. However, there are far more fatalities with nuclear accidents compared to traffic accidents. Accidents in which multiple deaths occur – even if the total risk is the same as that of another accident – are often considered less acceptable, because the degree of social disruption is much higher. Whether or not this is a good argument is a matter of debate.

The acceptability of a risk partly depends on the degree to which people *voluntarily* take a risk or consent to a risk.[6] Still, the distinction between voluntary and imposed risk is not always clear-cut. To what extent are traffic risks voluntary if someone has to travel a lot for work? Nevertheless, traffic risks are more voluntarily taken than the risks of a nuclear power plant being built near you without any consultation. The risks of skiing are more voluntary than traffic risks.

Risks as such are not acceptable, but they can be acceptable because risky activities bring certain *benefits*. Instead of assessing isolated risks, it is, therefore, a better idea to assess risky activities. This way it is possible to weigh the risks and the benefits. If an activity does not result in any advantage in someone's eyes then it is reasonable that he or she will reject the activity if it involves a risk, whatever small.

In extension of the above, it makes sense to weigh *different options* – to achieve the same goals – when the acceptability of risks is being assessed. This means that the acceptability of nuclear energy as a technology for generating energy not only depends on the question whether the advantages of nuclear energy weigh against the risks (and other negative effects), but also depends on the question whether other technologies – like wind energy – are more attractive.

The acceptability of certain risks also depends on how justly the risks and advantages of a specific risky activity are *distributed*. Take, for example, a chemical plant that is being built in a ghetto in a Third World country. If the advantages like employment, profit, and useful products do not go to the people living in the ghetto, the risk may be unacceptable.

Source: Based on Otway and Von Winterfeldt (1982), Slovic, Fischhoff, and Lichtenstein (1990), Shrader-Frechette (1991), and Stern and Feinberg (1996).

The above arguments refute the proposition that immediate conclusions can be drawn from risk assessments about the acceptability of risks. Risk assessments are nevertheless an important source of information for judging the acceptability of risks. Besides that, ethical considerations play a role too. Though there is no full agreement and an exhaustive enumeration is impossible (see, e.g., Lave, 1984; Shrader-Frechette, 1991; Harris, Pritchard, and Rabins, 2005; Hansson, 2003), we can at least mention the following four ethical considerations. We will elaborate on these points in the sections below:

1 the degree of informed consent with the risk;
2 the degree to which the benefits of a risky activity weigh up against the disadvantages and risks;
3 the availability of alternatives with a lower risk; and.
4 the degree to which risks and advantages are justly distributed.

8.5.1 Informed consent

Risks are more acceptable if those who run the risk consent to the risk in question. Some posit that risks are only acceptable if those running the risk have agreed to the risk after having received complete information concerning the risk. This principle is known as **informed consent**. The principle of informed consent originally stems from medical practice; it is closely related to ideas from normative ethics (see box).

Informed consent: Principle that states that activities (experiments, risks) are acceptable if people have freely consented to them after being fully informed about the (potential) risks and benefits of these activities (experiments, risks).

Informed Consent and Normative Ethics

The principle of informed consent can easily be justified on the basis of ideas from normative ethics, as discussed in Chapter 3. It is a good match for Mill's freedom principle, which states everyone is free to lead his or her own life as long as it does not harm others. Informed consent is aimed at creating conditions through which people can act according to the freedom principle. This principle posits that risks are only acceptable if people have chosen for them freely.

Besides the above, informed consent closely ties in with Kant's second formulation of the categorical imperative: "Act as to treat humanity, whether in your own person or in that of any other, in every case as an end, never as means only." As we saw in Section 3.8, this means that we must respect the moral autonomy of others to reach their own choice. In other words, informed consent is aimed at creating the conditions under which people can make an autonomous choice. People must decide for themselves whether a risk is acceptable or not.

There are different ideas about how the principle of informed consent should be applied in technology. One idea is to allow this to occur through the economic market. In such a scenario, it is assumed people will decide for themselves which risks they wish to take concerning the purchase of certain risky or dangerous products. The choice of many individuals results, through a kind of invisible hand, in an optimal risk level. However, it is doubtful whether market transactions lead to informed consent in practice. First, consumers are often insufficiently or incompletely aware of the risks of technical products. So we cannot speak of real consent with the risks involved. People, for example, buy cell phones without being aware of any possible health risks. In such a case we cannot say that people have consented to those health risks. Second, technical products often introduce risks that affect people other than the buyers or sellers of the products in question without their consent. One example is paint with organic solvents that contribute to environmental problems such as smog. On top of this, the economic market does not behave as ideally as many economists would have us believe. The choice of consumers is often limited because of the existence of monopolies, for example. In a number of cases, safer technologies simply do not reach the market even if they are technically feasible.

The Ford Pinto and Informed Consent

In Chapter 3 we discussed the Ford Pinto case. We saw that Ford decided not to alter the design of the Pinto, because the benefits to society from such an alteration supposedly did not weigh against the costs to society.

What is striking about the way, in which Ford dealt with this problem, is that the company believed that it had to make the choice whether reducing the risk weighed up against the extra costs. There is, however, another way to deal with this problem: leave the choice to the consumer. Ford could have given consumers the choice to have an improved tank installed at limited costs, or to settle for the original design. In this way the consumer could make his/her own choice and the extra risk would have been voluntary.

Informed consent can also be applied to technology by asking everybody who is potentially suffering from a risk for his or her consent. Unless everybody agrees that the risk be taken, the risk is considered unacceptable. A major disadvantage of this approach is that it gives almost unlimited veto power to individuals and will in many cases lead to a situation in which no risk is accepted at all, eventually making everybody worse off. One could try to avoid this situation from occurring by introducing the possibility of offering compensation for certain risks or by introducing the possibility of trading risks against each other.

8.5.2 Do the advantages outweigh the risks?

An important reason why risks can be ethically acceptable is that risky activities can have advantages. More generally we could argue that risky activities and thus the risks that are linked to these activities are acceptable if the benefits of the activities outweigh the costs. These ideas are in agreement with consequentialism of which utilitarianism is a specific type (see Chapter 3). Instruments have been developed based on such utilitarian arguments to determine the most desirable level of risk for a technology. One example of this is **risk-cost-benefit analysis** (Fischoff, Lichtenstein, and Slovic, 1981; Lave, 1984; Shrader-Frechette, 1985, 1991). This is a variant of the regular cost-benefit analysis (see Section 6.3.1). In risk-cost-benefit analysis, the social costs for risk reduction are weighed against the social benefits offered by risk reduction. The optimal risk level of a certain technology or product is where the net social benefit – the social benefits minus the social costs – is as high as possible. The basic ethical ideal behind this is that we should strive for the greatest happiness for the greatest number (Bentham). Such risk-cost-benefit analyses are often carried out by engineers.

Risk-cost-benefit analysis This is a variant of regular cost-benefit analysis. The social costs for risk reduction are weighed against the social benefits offered by risk reduction, so achieving an optimal level of risk in which the social benefits are highest.

There are two important objections to risk-cost-benefit analysis. First, such an analysis may commit the fallacy of pricing (see Section 4.5.2): it is not always possible to express all the relevant costs, benefits, and risks in money in a comparable fashion. People, especially, tend to have reservations about expressing the value of human life in monetary terms (as happened in the Ford Pinto case). Second, little attention is paid to informed consent and to the just distribution of costs and benefits, although these are important ethical considerations too. The second objection can in part be met by taking certain moral principles into account in risk-cost-benefit analysis.[7] A risk is only accepted, for example, if no one suffers from it – after compensation for those who might – or if it proves that those who were worst off are now better off.[8]

8.5.3 The availability of alternatives

The acceptability of the risks of a technology also depends on the availability of alternatives with lower risks. Suppose that two technologies, say A and B, introduce the same risks. Then we may commit the sheer size fallacy (see Section 4.5.2) if we leave out the question whether an alternative is available. Now also suppose that for technology A an alternative is available with lower risk and no major other disadvantages, while for technology B such an alternative is absent. In this case, we may well conclude that the risks of technology A are unacceptable (because an alternative with lower risks is available) while the risks of technology B are acceptable (since no alternative is available and the technology is in other respects acceptable).

The importance of the availability of alternatives is also reflected in some environmental laws, like the Integrated Pollution Prevention and Control Directive of the European Union and the Clean Air Act and Clean Water Act in the US, which all use the notion of **best available technology** or a comparable notion. The idea is that environmental emissions should be reduced to the degree that is possible with the best available technology. This

Best available technology As an approach to acceptable risk (or acceptable environmental emissions), best available technology refers to an approach that does not prescribe a specific technology but uses the best available technological alternative as yardstick for what is acceptable.

approach does not prescribe a specific technology but uses the best available technological alternative as yardstick.

Best Available Technology

The European Union defines best available technology as follows:[9]

> "best available techniques" means the most effective and advanced stage in the development of activities and their methods of operation which indicate the practical suitability of particular techniques for providing in principle the basis for emission limit values designed to prevent and, where that is not practicable, generally to reduce emissions and the impact on the environment as a whole:
>
> a. "techniques" shall include both the technology used and the way in which the installation is designed, built, maintained, operated and decommissioned;
> b. "available techniques" means those developed on a scale which allows implementation in the relevant industrial sector, under economically and technically viable conditions, taking into consideration the costs and advantages, whether or not the techniques are used or produced inside the Member State in question, as long as they are reasonably accessible to the operator;
> c. "best" means most effective in achieving a high general level of protection of the environment as a whole.

8.5.4 Are risks and benefits justly distributed?

An important ethical consideration in accepting risks is the degree to which risks and benefits of risky activities are justly distributed. It is, for example, not just if certain groups of people always have to carry the load of certain activities, while other groups reap the benefits. Some people believe that everyone should be treated equally with regards to risks. This argument can be supported by Kant's first categorical imperative, which states that you should act only according to the maxim whereby you can at the same time will that it should become a universal law (see Section 3.8). This implies an equality principle.

Equal treatment concerning risks can be achieved by setting standards. Standards treat everybody equally, which does not mean that everyone runs the same risk. It means that the maximum permissible risk is equal for everybody in principle. Though standardization can be defended by appealing to the ethical principle of equity, there are ethical objections to be raised as well. A first possible objection is that little account is given of the pros and cons of risky activities. This utilitarian objection is clarified by means of Figure 8.2 (Derby and Keeney, 1990). Say that technology 1 is characterized by a curve through points A and B in Figure 8.2. To allow technology 1 to meet the standards, considerable costs are involved (point A). These high costs only reduce the risk slightly compared to point B. Some utilitarians will therefore find point B more desirable than meeting the standard against high costs in point A.

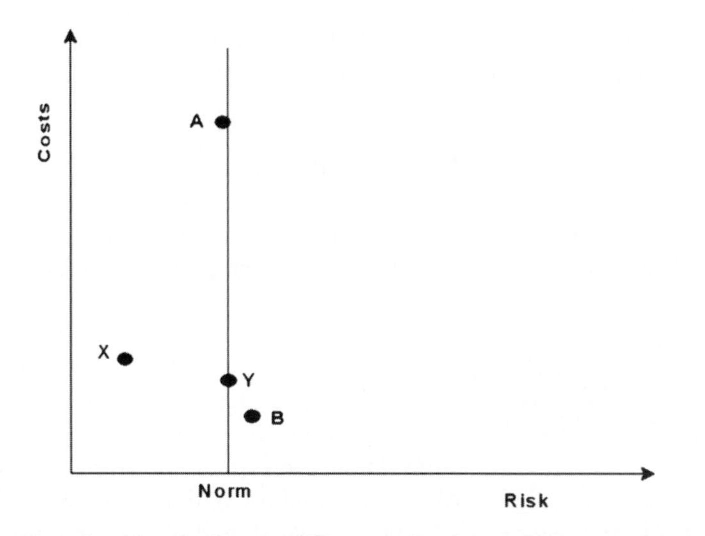

Figure 8.2 Costs for risk reduction for different technologies. The points A and B represent technology 1. Points X and Y represent technology 2. From Derby and Keeney (1990).

What is also possible is that standardization leads to higher risks than the situation in which costs considerations are taken into account. Take technology 2, for example, which is characterized by a curve through points X and Y in Figure 8.2. Technology 2 meets the standard (point Y), but a much safer product can be designed against little additional cost (point X). Some utilitarians will find point X a much more desirable result than point Y.

A second possible ethical objection against standardization is that it is paternalistic (see Section 1.5.2 on paternalism). People do not get to choose which risks they find acceptable – the regulator, often being the government, does that for them instead. This objection is especially applicable to **personal risks**, that is, risks that only affect the buyer, user, salesperson, or producer. Market regulation for such products can be an option provided certain conditions are met, such as the availability of full information about risks and the freedom of choice between different products, because this leads to informed consent. For **collective risks**, that is, risks that affect larger groups, for example, floods, market regulation does not work. There can be no informed consent through market regulation because the risks affect other people besides the user and producer. These kinds of situations demand a collective decision about what acceptable risks are. Such a collective decision can – but need not – result in standardization.[10]

Personal risks Risks that only affect an individual and not a collective. For example, the risk of smoking. The relevant distinction with collective risk is whether the individual can stop or avert the risks for him or her individually. We can individually decide not to smoke but cannot individually prevent flooding for ourselves.

Collective risks Risks that affect a collective of people and not just individuals, like the risks of flooding.

One can also wonder whether an equal treatment of people concerning risk standards leads to a just division of risks. This is doubtful because the degree to which people benefit from a risky activity differs. An equal distribution of risks is, for example, not justified if only a limited group has the benefits of a specific risky activity. Justice and equality are related, but are not the same.

8.6 Risk Communication

According to some professional codes of conduct, engineers must inform the public about risks and hazards (see Chapter 2). In some cases specialists are used, who are called **risk communicators**. Risk communication raises a number of ethical questions (Morgan and Lave, 1990; Valenti and Wilkins, 1995; Jungermann, 1996; Johnson, 1999). A first question is whether risk communication should only inform or also (try to) persuade. Can the government discourage smoking, or should it only inform about the risks of smoking? Another question is whether people should always be informed about risks even if it is not always in their best interests or if it is likely that they will interpret the information the wrong way. As a risk communicator, should you inform people of the risk of burglary if people have to leave their homes as quickly as possible because of the safety risk as a result of a coming hurricane?

Risk communicators Specialists that inform, or advise how to inform, the public about risks and hazards.

In both examples, the contrast between duty ethics and consequentialism plays a role in the background. From the perspective of duty ethics, the consequences of risk communication are not relevant for the question "What is responsible risk communication?" Risk communication must first be honest (do not lie and always tell the complete truth). Next to that, it must respect the freedom of choice and autonomy of people and hence not be paternalistic (see also Section 1.5.2 on paternalism). Here, the principle of informed consent is of importance too. It implies that you must not try to convince people but only inform them. From the perspective of consequentialism, the considerations and conclusions would be quite different in some cases. Risk communication is judged by means of the goodness of the consequences. Attempts to convince people by means of risk communication or withholding certain information can be morally right if it results in good consequences.

In risk communication ethical questions about the amount of information you give and how you present and structure your information also arise. How detailed should the information be for example? If a risk assessment was carried out, should you just give the result or should you talk about the uncertainty margin too? Should you explain how the risk assessment was carried out, so that people can check how reliable it is? From psychological research, we know that the way in which risks are presented has a great influence on the way an audience interprets these risks (see, for example, Martin and Schinzinger, 1996, pp. 134–136). It is even possible for people to take opposite decisions on the basis of the same information framed differently. The risk measure used can also influence how people interpret risks. Ethical judgments are often hidden in the use of a certain risk measure. In the risk measure "number of deaths per time unit," the assumption made is that each death has the same value. If the risk is expressed as a decrease in life expectancy, the implicit assumption is that the death of a young person is worse than the death of an old person. This is the case because the death of a young person weighs more heavily in this risk measure than the death of an old person.

Risk Perception

From psychological research on risk perception among the public, it appears a large number of factors plays a role on how people perceive risks:[11]

- The (perceived) voluntariness of the risk;
- The expected benefits of the activity or technology that causes the risk and the distribution of these effects;
- The maximal occurring negative effects and the possible controllability of these effects (degree of social disruption);
- The situation the risk is related to. Many people estimate risks related to work as smaller compared to other situations;
- The proximity and visibility of risks. Close risks are usually experienced as being greater than risks further away, or those that are less visible or imaginable;
- The way risks are presented and the risk measure that is used.

Source: Based on Slovic, Fischhoff, and Lichtenstein (1990) and Martin and Schinzinger (1996, pp. 134–137).

Given the importance of the way in which risks are presented upon how they are perceived, the question to consider is what types of presentation are acceptable. It is important to realize that it usually is impossible to present risks neutrally. It makes a difference whether you express the maximum dosage of dioxin per day in picograms, milligrams, or kilograms. The latter presentation – maybe unintentionally – gives the impression that the risk is far smaller than in the first case. Nearly always, a certain method of presenting will intentionally or unintentionally make a certain interpretation more probable than another. That is not to say that you should consciously strive for a particular interpretation, but you should consider how you can present data in the most honest and best way.

8.7 Dealing with Uncertainty and Ignorance

Up to now, we have presupposed that it is possible to predict and express risks related to the hazards of technologies to a certain extent. But what happens if that is not the case? Consider the supposed health issues surrounding cell phones and the possible negative health and environment effects of growing and consuming genetically manipulated crops. In some of these cases, risks are calculated, but the question is whether all possible hazards have been assessed in a reliable way. A crucial question thus arises: Is it acceptable to introduce a new technology with potential hazards into society when there is scientific controversy or uncertainty about these hazards? This is an ethical question because, on the one hand, it is desirable to protect society from hazards, while, on the other hand, outright forbidding a technology may also be undesirable. Much of the current ethical debate on this question somehow focuses on the precautionary principle.

8.7.1 The precautionary principle

Precautionary principle Principle that prescribes how to deal with threats that are uncertain and/or cannot be scientifically established. In it most general form the precautionary principle has the following general format: If there is (1) a threat, which is (2) uncertain, then (3) some kind of action (4) is mandatory. This definition has four dimensions: (1) the threat dimension; (2) the uncertainty dimension; (3) the action dimension; and (4) the prescription dimension.

The **precautionary principle** was initially proposed to deal with environmental problems (see box). In Chapter 10, we will discuss how the precautionary principle applies to environmental issues and sustainability. The principle can, however, also be applied to unknown risks and we will discuss that application here. The principle is mainly suitable for situations in which we cannot fully express hazards as risks because we have insufficient scientific knowledge. In general the precautionary principle states that precautionary measures must be taken if there are indications of a certain hazards, despite the fact that the hazards cannot be completely scientifically proven.

The Precautionary Principle

The precautionary principle originates from the Rio Declaration, the closing statement of the first conference of the United Nations on sustainable development, which was held in Rio de Janeiro in 1992:

> In order to protect the environment, the precautionary approach shall be widely applied by States according to their capabilities. Where there are threats of serious or irreversible damage, lack of full scientific certainty shall not be used as a reason for postponing cost-effective measures to prevent environmental degradation.[12]

Irreversible damage is usually understood in terms of environmental resources than cannot be replaced or restored. In the above formulation, the precautionary is primarily an *argumentative* principle: it indicates that certain reasons are invalid for arguing against environmental measures. The principle has also been formulated as a prescriptive principle that prescribes certain actions. A well-known formulation here is the so-called Wingspread Statement:

> When an activity raises threats to the environment or human health, precautionary measures should be taken, even if some cause-and-effect relationships are not fully established scientifically. (Raffensperger and Tickner, 1999, pp. 354–355)

The precautionary principle is hotly debated. Some philosophers and legal scholars have argued that the principle is basically a form of practical rationality or what Aristotle called practical wisdom (Adorno, 2004; Hansson, 2009). Others have argued that the principle is incoherent because "it forbids the very measures it requires" (Sunstein, 2005, p. 366). Partly the controversy seems to be based on different understandings of the "precautionary principle." Those who argue that the principle is basically a form of practical wisdom see the principle as an open ended principle that can be specified in various ways (see the main texts on the four dimensions of the precautionary principle that can be further specified). The main thrust of the principle is, according to them, that decisions are not only based on known risks but also on what we have called uncertainty above. As we have seen, engineers in fact already do that in safety engineering. Those who argue that the principle is incoherent have in mind a strong version of the principle that forbids any activities that potentially raise risks that have not yet been established scientifically. Since such potential risks are often inherent both to doing something and refraining from that something, they consider the principle as incoherent.

The precautionary principle has also led to debates between the United States and Europe in the World Trade Organization (WTO), for example on genetically modified organisms (GMOs). With reference to the precautionary principle, the European Union introduced a de facto moratorium on new GMOs in 1998. In 2003, this was replaced by labeling requirement for genetically modified food. The US has opposed these measures because there was no conclusive scientific evidence of the risks of GMOs and the measures were seen as trade barriers. In 2009 the WTO ruled that the de facto moratorium was illegal under the WTO rules.[13]

Per Sandin has argued that the precautionary principle as a prescriptive principle contains four dimensions (Sandin, 1999): If there is (1) a threat, which is (2) uncertain, then (3) some kind of action (4) is mandatory.

The four dimensions in this formulation are:

1 the threat dimension;
2 the uncertainty dimension;
3 the action dimension; and
4 the prescription dimension, expressed here in the phrase "is mandatory."

One technology to which the precautionary principle has been applied is nanotechnology, in particular nanoparticles (see box). Application of the precautionary principle to this technology seems to imply that the potential hazards of nanoparticles should first be properly assessed before they are introduced on a large scale in society. Although this is a sensible strategy, one could raise the question whether the strict separation that this strategy proposes between toxicity testing and introduction into society is tenable. The point is that in many cases reducing complexity and uncertainty is not possible without the introduction of the new technology into society. It is often only *after* introduction into society that the hazards of certain technologies can be properly assessed as we will see below.

Case Nanoparticles

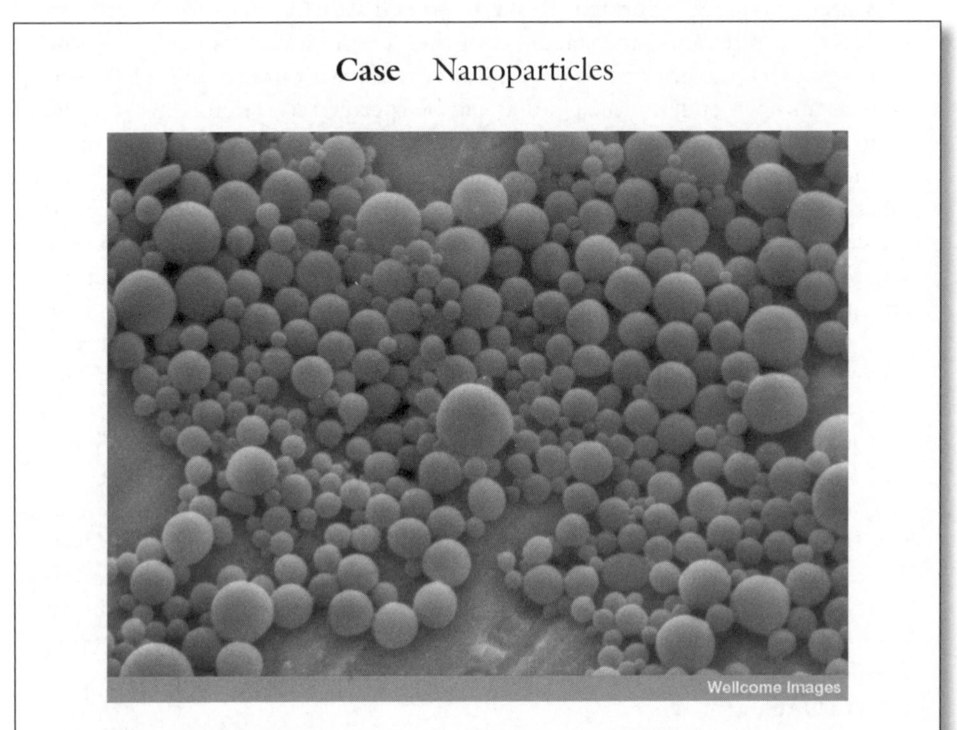

Figure 8.3 Nanoparticles. Photo: Annie Cavanagh, Wellcome Images.

Nanoparticles are very small particles with a size in the range of 1 to 100 nanometer (10^{-9} meter). Such small-scale particles often have quite different mechanical, optical, magnetic and electronic properties than the bulk material they are made of. Nanotechnology therefore makes it possible to create products with new characteristics and functions. Currently, nanoparticles are already used for sporting goods, tires, stain-resistant clothing, shoe polish, sunscreens, cosmetics, electronics, and self-cleaning windows.

Although some of the new properties of nanoparticles may prove very useful and economically important, they also imply a potential hazard because the toxicological properties of nanoparticles may be different from that of the bulk material. Currently, the toxicity of nanoparticles can be determined with tests on cultured cells or isolated organs (in vitro), with animal tests (in vivo), or with theoretical models based on knowledge about the effects of other small particles (e.g., ultrafine particles) on the human body (The Royal Society & The Royal Academy of Engineering, 2004, and Oberdörster, Oberdörster, and Oberdörster, 2005). However, the current knowledge of potential hazards of synthetic non-biodegradable nanoparticles is limited. The philosophers John Weckert and James Moor have therefore proposed to apply the precautionary principle to the risks of nanoparticles (Weckert and Moor, 2007). They propose the following more precise formulation of the principle:

> If an action A poses a credible threat P of causing some serious harm E, then apply an appropriate remedy R to reduce the possibility of E. (Weckert and Moor, 2007, p. 144)

P causing harm E is here the threat component (1), "credible" refers to the uncertainty component (2), remedy R is the action component (3) and "apply" refers to the prescription component (4). Weckert and Moor argue that their version of the precautionary principle applied to nanoparticles "will require at least a concerted attempt at establishing the risks to health and the environment and perhaps slowing the development of products until the threats have been properly assessed" (Weckert and Moor, 2007, p. 144). The Health Council of the Netherlands comes to a similar conclusion in applying the precautionary principle to non-biodegradable nanoparticles: before such particles are taken into production and are brought onto the market on a large scale, their toxicological properties should be properly investigated (Health Council of the Netherlands, 2006, p. 15).

8.7.2 Engineering as a societal experiment

Like science, engineering has an experimental nature. In science, hypotheses are deduced from theories, which are then tested in experiments. A hypothesis can be confirmed or falsified through the experiment. Analogous to this, newly designed products can be seen as hypotheses for properly functioning products. The hypothesis that they are properly functioning is usually tested through simulations and experiments in the laboratory, in small-scale field tests, or in clinical trials (in the case of medicine). Although such simulations, experiments, tests, and trials are very important in engineering, they do not always provide complete and reliable knowledge of the functioning of technological products and the hazards and risks involved (see also Section 6.2.3). For a variety of reasons, it is often not possible to completely predict the possible hazards of new technologies before they enter society (Krohn and Weyer, 1994):

1 For carrying out a risk assessment, one first needs to identify possible hazards and failure mechanisms. It is, however, well possible that certain hazards are overlooked. In laboratory experiments as well, certain hazards may not surface, so that they become only apparent after the technology has been introduced into society. Examples of partly unknown failure mechanisms are Tacomo Narrows Bridge (Section 8.1) and the German ICE train (Section 6.1).
2 Laboratory and field tests are not always representative of the circumstances in which the product eventually has to function. You need to know what circumstances are relevant in actual practice and which are irrelevant for performing

a good test. This knowledge may only become available after a product has been introduced into society. An example is the case of 2,4,5-T discussed in Section 6.2.6.
3 Risks may be due to long-term cumulative effects of substances, sometimes in interaction with other substances. Examples are the possible effects of DDT, dioxin, and radioactive radiation. Cumulative or interactive long-term effects can hardly be studied in the laboratory; so that the testing actually takes place during the use of the product in society.
4 Natural and socio-technical systems may be characterized by recursive and non-linear dynamics. Even if a recursive natural system is deterministic, due to such dynamics future developments and possible hazards may be impossible to predict. The only way to find out the effects may be to actually introduce a new technology or substance into society.

The above implies that at least in some cases society has become the laboratory for engineering experiments. This means that it is ultimately during actual implementation of a technology in society that its functioning and possible hazards and risks are tested. However, unlike traditional scientific experiments, **societal experiments** are usually difficult to terminate if something goes wrong. Moreover, in societal experiments with new technologies, the consequences can be much larger and can have an impact on third parties. If an elementary particle in a particle accelerator does not behave as expected, it is simply an interesting fact. If a nanoparticle turns out to be toxic, after it has been introduced into society, it is a potential disaster next to an interesting technical fact. Societal experiments in technology are nearly always large scale, can have irreversible negative consequences, and usually involve people as experimental subjects. In other words, they are experiments with major social consequences. Despite, or maybe because of, the mentioned characteristics of societal engineering experiments, it is still hardly recognized that society has become the laboratory for new technologies. As the European Expert Group on Science and Governance writes:

Societal experiments We speak of the introduction of new technology in society as a societal experiment if the (final) testing of possible hazards and risks of a technology and its functioning take place by the actual implementation of a technology in society.

> [W]e are in an unavoidably experimental state. Yet this is usually deleted from public view and public negotiation. If citizens are routinely being enrolled without negotiation as experimental subjects, in experiments which are not called by name, then some serious ethical and social issues would have to be addressed. Even if no simple or accessible solutions exist to this problem, if our concern is public trust, surely a minimal requirement is that we acknowledge the public predicament. (Felt et al., 2007, p. 68)

A crucial question then is under what conditions it is acceptable to carry out societal experiments with new technologies. One important principle that has been proposed to judge this is informed consent (see also Section 8.5.1). Since World War II, informed consent has become the leading principle for experiments involving human subjects (Whitbeck, 1998a). Before that date, the leading moral principle was whether the experimenter would be willing to subject him- or herself to the experiment (Whitbeck, 1998a). For medical experiments involving human subjects, informed consent is usually legally required. An example is medical experiments on human subjects that are regularly advertised in local media. For other experiments it might not be legally required but can still be considered an important moral yardstick.

The engineering ethicists Mike Martin and Roland Schinzinger have proposed to apply the principle of informed consent also to societal experiments in engineering (Martin and Schinzinger, 1996). Such an application raises at least three issues. One issue is whether it makes sense to ask people to consent to *uncertain* hazards or even to hazards about which the experimenters are *ignorant*. What does consenting to an experiment which has an unknown risk amount to? It seems that it would imply having to accept all risks that emerge from the experiment because unknown risks potentially cover *any* risk. It is hard to see how people could rationally accept such experimental conditions. But if they do not give their informed consent (which seems the only rational thing to do), any societal experiment involving ignorance would be unacceptable.

This brings us to the second issue: is the principle not too restrictive? As soon as one individual objects to a certain societal experiment, it should be abandoned even if this experiment might bring large benefits to the rest of society. This seems unfair, at least in some cases, for example if the actual hazard for the person objecting to the experiment is small and the social benefits are large. Of course, societal experiments in engineering can be carried out on a smaller scale than the whole of society, but still informed consent may be sometimes too restrictive a condition if large potential benefits are involved. Even for medical experiments, doubts have been raised whether the principle of informed consent does not unjustly exclude experiments from which large segments of society would profit (Hansson, 2004b).

A third issue is how to deal with people who are indirectly involved in the experiment but are not able to give their informed consent. One specific example is future generations. For example, the introduction of nuclear energy in society amounts to a societal experiment that involves future generations because nuclear waste that remains radioactive (with current technology) for thousands of years is generated; even if this waste is stored or disposed as well as possible, it will introduce uncertain and possibly unknown hazards to future generations. Some might conclude that the principle of informed consent shows that the introduction of nuclear energy is thus unacceptable, at least as long as it generates hazards for future generations. This judgment may, however, be a bit too quick as other sources of energy, especially fossil fuels, also generate hazards for future generations due to their contribution to the greenhouse effect or may not be able in the near future to meet the world energy demand (sustainable energy sources like solar and wind energy). One way to deal with this issue may be to introduce the notion of **hypothetical consent**, in contrast to actual consent: under which conditions would it be safe to assume that future generations consent to an experiment that involves hazards for them?

Hypothetical consent Hypothetical consent refers to a form of informed consent in which people do not actually consent to something but are hypothetically supposed to consent if certain conditions are met, for example that it would be rational for them to consent or in their own interest.

The mentioned issues raise serious doubts whether it is desirable to apply the principle of informed consent to societal experiments with technology. It might, nevertheless, be possible to reformulate the principle to address these issues. Alternatively, if one rejects informed consent as a leading principle for societal experiments, one needs to propose an alternative approach that addresses at least the main underlying moral concern that people may be subjected to societal experiments without knowing and without the ability to have a say, that is, without respect for their moral autonomy. More specifically, one could think of a set of principles, for example along the following lines:

- Experimental subjects are to be informed about the experiment, its set-up, risks and potential hazards, uncertainties and ignorance, and expected benefits.
- Societal experiments should be approved by democratically legitimized bodies. This can for example be parliament but also a governmental body that is controlled by parliament or the government.
- Experimental subjects should have a reasonable say in the set-up, carrying out, and (rules for) stopping of the experiment.
- Experimental subjects that are especially vulnerable to the hazards involved in the experiment should either not be subject to the experiment or be additionally protected.
- The experiment should entail a fair distribution of risks and benefits among different groups and among different generations.

8.8 Chapter Summary

As an engineer, you have a moral responsibility to ensure the safety of the technologies you design. Safety should here not be understood as the absence of risk but rather as the reduction of hazards and risks to an acceptable level. In this chapter, a number of methods have been reviewed that engineers can apply to live by this responsibility in professional practice. You can employ a number of different strategies for safe design including inherently safe design, adding safety factors to your design, adding negative feedback mechanisms, and providing multiple independent safety barriers. In addition, risks can be assessed through risk assessment, a systematic process consisting of four steps: 1) release assessment; 2) exposure assessment; 3) consequence assessment; and 4) risk estimation. On the basis of such risk estimations, one can reflect on the acceptability of certain risks. However, the question whether a risk is acceptable depends on more than just the magnitude of the risk. More specifically, the following ethical considerations were identified that are important to judge the acceptability of risks:

1 the degree of informed consent with the risk;
2 the degree to which the benefits of a risky activity weigh up against the disadvantages and risks;
3 the availability of alternatives with a lower risk; and
4 the degree to which risks and advantages are justly distributed.

Reflection on the acceptability of risks not only demands technical and scientific expertise, but ethical expertise too. Engineers do not have this expertise to any greater extent than other people. That is why the question whether a risk is acceptable cannot usually be answered by engineers alone. This limits the responsibility of engineers on the one hand, but it gives them additional responsibility on the other, that is, they must properly inform others of the risks (risk communication) and involve them in decisions concerning the acceptability of risks.

Engineering often takes place under conditions of partial ignorance – in circumstances in which not all the risks of technology can be foreseen beforehand. To deal with potential hazards in such cases, one might employ the precautionary principle. In its most general formulation, this principle says that: If there is (1) a threat, which is (2) uncertain, then (3) some kind of action (4) is mandatory. This principle can be made more concrete in engineering in several ways. It can for example call for additional safety measures (along the lines discussed above) but also call for more risk assessment before a technology is introduced into society. Although the precautionary principle is useful, it cannot take away all the uncertainty with which the introduction of a new technology into society is accompanied. In that respect, the introduction of technology into society amounts to a societal experiment and the question is under what conditions such experiments are ethically acceptable. An often mentioned criterion here is informed consent, although it may be doubted whether this criterion can be usefully applied to societal experiments. Nevertheless, it is important that societal experiments somehow respect the moral autonomy of potential victims of such experiments.

At first sight, the role of uncertainty and ignorance in engineering seems to restrict the responsibility of engineers. As we saw in Section 1.3, knowledge of the consequences is a condition for responsibility. The fact that technology development always involves uncertainty and ignorance therefore diminishes the responsibility of engineers. On the other hand, science teaches us that there always is ignorance and unknown risks, and thus there is a special responsibility for engineers because they can indicate where there is uncertainty and ignorance and what the potential hazards may be. So engineers must not only communicate about what they know but also about what they do not know – a task that many engineers find very difficult. Nevertheless, this competence can be viewed as an important virtue for engineers.

Study Questions

1 Why engineers do have a responsibility for safety?
2 What is the difference between uncertainty and ignorance?
3 Give five arguments why the argument "if technology x with risk r is acceptable then a technology y with the same risk r is also acceptable" is not sound.
4 What is the difference between type I errors and type II errors? What type of error is worse in your view during a risk assessment? Argue you answer.
5 What is the difference between personal and collective risks? Is the risk of nuclear energy an individual or collective risk? Could this risk be dealt with by informed consent and, if so, how?
6 Why is an equal distribution of risks not always just? Give an example to illustrate your answer.
7 What is meant by engineering as a societal experiment? Can you give an example of the introduction of a technology in society that is clearly experimental in this sense? Argue your answer.

8 Mention a technology of which the risks are acceptable while the technology itself is unacceptable. Can you also think of an example of a technology that is acceptable while its risks are unacceptable?

9 Consider the following situation. A country is preparing for the outbreak of a rare disease.[14] If the disease arrives in the country and if it is not abated 600 people will die. To abate the disease, the following abatement programs are available:
- Program A that saves 200 people;
- Program B in which there is a probability of 1/3 that 600 people are saved and 2/3 than nobody is saved;
- Program C in which 400 people die; and
- Program D in which there is a probability of 1/3 that nobody dies and 2/3 that 600 people die.
  a. Are programs A and C different in terms of expected fatalities and people saved? And programs B and D?
  b. Which program you think will be preferred by most people?
  c. Is it possible to present the risks and advantages of the programs neutrally? If so, how? If not, what would be the best way to present the risks and advantages of the various programs?

10 In a risk assessment of genetically modified corn, it is argued that: "Since we have not been able to show scientifically that there are adverse health effects, genetically modified corn does not pose a health risk."
  a. Is this argument sound?
  b. Suppose that the precautionary principle is applied to the introduction of genetically modified corn as consumer food. What would that then imply?

11 As an engineer you are responsible for the safety of a new train tunnel. The tunnel consists of twin train tunnels. Every x meters a connection will be established between both train tunnels, so that in cases of an accident (for example the outbreak of fire), train passengers can more easily escape. This will reduce the number of expected fatalities and injuries as the result of accidents. You need to make a design decision about the desirable value of x. Possible values for x are 100, 250 or 500 meters. Relevant data are given in Tables 8.1 and 8.2.

Table 8.1

Connection every x meter	*Average number of fatalities per accident*	*Average number of injuries per accident*	*Additional costs per year (construction and maintenance)*
x = 50	5	100	500 000 Euros
x = 100	10	200	300 000 Euros
x = 250	20	200	200 000 Euros
No connections	40	500	No additional costs

Table 8.2

Probability of an accident (per year)	0.01
Average number of passenger per train	400
Average number of trains per year	2500

Table 8.3

Fatality	500 000 Euros
Injured	50 000 Euros

A relevant design standard prescribes that the probability of an individual train passenger being fatality injured due to an accident should be lower than 10^{-7} per trip. This design standard is not legally binding. In a handbook, you have found the data for the societal costs of fatalities and injuries as result of a train accident (Table 8.3).

a. What is the maximum distance x if the design standard is applied?
b. How large should x be on basis of the ethical framework of classical utilitarianism?
c. How should in your view a decision be made about the desirable distance x?

Discussion Questions

1 Who should in your view decide about the acceptability of risks? Engineers? Politicians? Company managers? The public? Argue your answer and discuss what your view would imply for the responsibility of engineers with respect to safety.
2 Do you consider informed consent a good principle for deciding about the acceptability of technological risks and hazards? Argue why or why not. If you do not consider informed consent a good principle, indicate how the moral autonomy of possible victims should then be protected. Or is this moral autonomy not important in your view?
3 Should risk communicators take into account the effect of their information on the public or should they solely try to ensure that people interpret the information in the right way and can make their own decision?
4 Do you agree that the precautionary principle is incoherent because it forbids the very measures it requires? Explain why you think that the principle is coherent or incoherent and what this (in)coherence implies for the acceptability of the principle.

Notes

1 http://reports.ewg.org/reports/asbestos/facts/fact1.php; www.hse.gov.uk/statistics/causdis/asbfaq.htm#hseopinion (accessed November 13, 2009).
2 Which does not mean to say that design and production adaptations are unimportant when it comes to striving to reduce CO_2 emissions.
3 Although it is obviously so that one technology contributes much more to CO_2 emissions than another.
4 A third method that is sometimes used is to ask experts to estimate the risks.
5 The example is taken from Simon (1974).
6 It has been estimated that people are willing to accept voluntary risks that are up to 1000 times larger than involuntary risks. See, for example, Starr (1990).
7 For examples, we refer to Kneese, Ben-David, and Schulze (1983), and Shrader-Frechette (1985).
8 The First principle is known as the Pareto Principle (cf. Zandvoort, 2000); the second one as the Difference Principle (Rawls, 1971).

9 Article 2.12 Directive 2008/1/EC of the European Parliament and of the Council of January 15, 2008 concerning integrated pollution prevention and control (Codified version) (accessed January 14, 2009).

10 It can also be based on informed consent of all involved but is then likely to sustain the status quo, which is often concerned morally problematic.

11 Some of these factors, especially the first three, are closely related to the acceptability of risks. This implies that people might well implicitly use a definition of risk that maintains a certain relationship between the magnitude of a risk and its acceptability.

12 www.gdrc.org/u-gov/precaution-7.html (accessed September 29, 2009).

13 www.euractiv.com/en/trade/eu-gmo-ban-illegal-wto-rules/article-155197 (accessed September 29, 2009).

14 The example is based on Tversky and Kahneman (1981, p. 453). See also Martin and Schinzinger (1996, p. 134).

9

The Distribution of Responsibility in Engineering

Having read this chapter and completed its associated questions, readers should be able to:

- Describe the problem of many hands and explain how it applies to engineering;
- Judge responsibility distributions by the moral fairness and by the effectiveness requirement;
- Explain the difference between moral responsibility and legal liability;
- Distinguish different notions of legal liability and discuss their pros and cons;
- Describe the different models for allocating responsibility in organizations, to discuss their pros and cons, and to apply them;
- Describe how engineering designs may affect the distribution of responsibility;

Contents

Ethics, Technology, and Engineering: An Introduction, First Edition.
Ibo van de Poel and Lambèr Royakkers.

9.1 Introduction

Case *Herald of Free Enterprise*

Figure 9.1 *Herald of Free Enterprise* capsized. Photo: © Press Association.

On March 6, 1987 the roll-on/roll-off passenger and freight ferry the *Herald of Free Enterprise* capsized just outside the Zeebrugge harbor. Water rapidly filled the ship. Of the almost six hundred people on board 383 were eventually saved, 189 bodies were recovered and four people were registered missing. The main cause of the disaster was the fact that the inner and outer bow doors were open when the ship set sail. The doors were sometimes deliberately left open for a while to allow all the car exhaust fumes to escape. That was done to prevent the passengers from feeling ill and getting headaches.

It was the job of the assistant boatswain to close the doors, but he had fallen asleep. The first officer was to check whether the doors had indeed been closed and should report to the captain. However, the first officer was also expected to assist the captain on the bridge when setting sail. The absence of warning lights made it impossible to see from the bridge whether the bow doors were closed.

On at least two previous occasions, similar negligence had led to ships setting sail with their bow doors open but without disastrous results.[1] In the case of the *Herald*, as is often the case, it was human error that preceded the disaster, but it was the design that contributed to the occurrence of the disaster in the first place. Roll-on/roll-off ships are inherently unstable when water enters a deck.

It is likely that the maritime engineers who designed roll-on/roll-off ferries like the *Herald* were aware of the inherent instability of such ships. Moreover, there were, and are, simple technical solutions if one wants to prevent rapid capsizing when water enters a deck.[2] Bulkheads created on the decks could easily impede the water and prevent rapid capsizing.[3] However, such bulkheads increase the loading time of roll-on/roll-off ferries and this increase in loading time, in turn, implies an increase in transportation costs. Moreover, bulkheads decrease the efficient use of space so reducing the transportation capacity of the ship. In Northwest Europe, shipping companies are in sharp competition with trains and planes, therefore they do not want to face increasing costs or longer loading times.

The official investigation into the sinking of the *Herald of Free Enterprise* commenced on April 27, 1987. The investigation was carried out by the Admiralty High Court, an investigation council of the British Supreme Court. The Admiralty High Court does not have the power to legally prosecute but it can recommend legal prosecution to ordinary courts of law. During the process much attention was given to the role of the assistant boatswain, Stanley, to the first navigating officer, Sabel, to captain Lewry and to Kirby, the shore captain. Kirby was directly responsible for drawing up the instructions to be followed on board, including the safety instructions. The role of the shipowner, to be precise of several of its directors, was also examined. The outcome of the investigation was that the ship's captain, Lewry, was suspended for a year and the first navigating officer, Sabel, for two years. The assistant boatswain, Stanley, got off free and the shipowner was officially reprimanded. The Admiralty High Court's written assessment of the shipowner was devastating. It identified a "disease of sloppiness" and negligence at every level of the corporation's hierarchy.

Despite all these resolute claims those responsible for the disaster were not immediately legally prosecuted. It was only after quite some time and pressure on the part of the Belgian authorities and the families of the victims that proceedings began in England. In 1989 it became clear that the shipping company was going to be accused of "corporate manslaughter," in other words, of deaths caused by a company rather than by an individual or individuals. By that time Townsend Thoresen had been taken over by the shipping company P&O. In September 1990 the case was taken to the Central Criminal Court at the Old Bailey in London. Eight people were accused. In addition to the shipping company, the assistant boatswain Stanley, the first officer Sabel and the captains Lewry and Kirby there were three directors who were accused: Develin, Ayres, and Alcindor.

When the case began the judge ordered the jury to forget everything they knew about the case. Even the Admiralty High Court's report, with its devastating criticism of the shipping company, should not be used. The prosecutor tried to prove that a number of mistakes had been made by the shipping company and the ship's crew alike and that it was "obvious" that sailing out of port with the ship's bow doors open would capsize the vessel. Most of the witnesses – experienced seamen included – did not find that this was so "obvious." The judge also maintained that there was insufficient evidence to support a verdict possibly to be voiced by the jury to the effect that there was an "obvious" connection between the open doors (cause) and the capsizing of the *Herald* (effect). On October 19, 1989, partly under pressure of the judge, the jury decided that there was insufficient evidence and so all the defendants were acquitted.

Source: This case description is mainly based on Van Gorp and Van de Poel (2001) and Baeyens (1992).

The *Herald of Free Enterprise* illustrates a number of issues with respect to responsibility in engineering. First, it shows how difficult it may be to pinpoint responsibility and blame in cases in which many people are involved in an activity and in which many causes contributed to a disaster. This is known as the problem of many hands. Second, it shows that even if we may have good reasons to hold someone morally responsible (blameworthy) for his/her actions that person might not be legally guilty or liable. So there is a difference between moral responsibility and legal liability. Third, the case raises the question how we can best organize active responsibility in complex organizations in order to avoid disasters as with the *Herald of Free Enterprise.*

In this chapter, we will first discuss the so-called problem of many hands in Section 9.2. Dealing with the problem of many hands requires attention for the distribution of responsibility in engineering. In this chapter, we will discuss three ways in which responsibilities are actually distributed in engineering, that is, through 1) the law, 2) organizational models for responsibility, and 3) technological design. We will argue that in each case the resulting responsibility distribution can be evaluated in terms of moral fairness (are the appropriate persons held responsible?) and in terms of effectiveness (does the responsibility distribution contribute to avoiding harm and to achieving beneficial results?).

9.2 The Problem of Many Hands[4]

Up until this point, we have focused on how individual engineers can behave responsibly. The social consequences of technology are, however, the result of the interaction between the actions of many different actors. Apart from engineers, this includes users, governments, companies, managers, and the like. One might assume that if all of the actors would behave individually responsibly, the overall result would be beneficial for society. This assumption does not always hold water, as we have seen in the Challenger case (Section 1.1): the fragmentation of decision-making led to different parts of the organization focusing purely on their areas of responsibility, and thus not feeling responsible for safety as a whole. We see a similar pattern in the case of the *Herald of Free Enterprise.*

The problem of many hands typically describes the problem where a lot of people are involved in an activity, like a complex engineering project, therefore making it difficult to identify where the responsibility for a particular outcome lies (Thompson, 1980; Bovens, 1998). In part this is a practical problem. It is often difficult in complex organizations or engineering projects to identify and prove who was responsible for what. Especially for outsiders it is usually very difficult, if not impossible, to know who contributed to, or could have prevented a certain action, who knew or could have known what, et cetera. This is especially problematic if one wants responsibility to have juridical implications, because the law requires evidence of irresponsible behavior and this evidence has to meet a certain standard of proof.

The problem of many hands is also a moral problem. This is so because it may turn out that nobody can reasonably be held morally responsible for an engineering disaster. This is morally problematic for at least two independent reasons. The first is that many people, including victims, members of the public, and also the engineering community, may find it morally unsatisfactory that if an engineering disaster occurs nobody can be held responsible. Of course, the search for somebody to blame may be misunderstood, but at least in some situations it seems reasonable to say that someone should bear responsibility. In fact, some philosophers have introduced the notion of **collective responsibility** to deal with the intuition that there is more to responsibility in complex cases than just the sum of the responsibilities of the individuals considered in isolation. Intuitively, we may say that a collective is responsible in cases where, had it been an action performed by one person, he or she would have been held responsible. This addresses the intuition that for outsiders it should not make a difference whether a complex engineering project was undertaken by one person or by a large number of persons in a division of labor. The second reason for attributing responsibility is the desire to learn from mistakes, and to do better in the future (Fahlquist, 2006a, 2006b). If nobody is held morally responsible for a disaster, this may not happen.[5]

Collective responsibility The responsibility of a collective of people.

We can now characterize the **problem of many hands** as the occurrence of the situation in which the collective can reasonably be held morally responsible for an outcome, while none of the individuals can reasonably be held responsible for that outcome. On this definition, the case of the *Herald of Free Enterprise* is probably not a problem of many hands because it is likely that at least some of the individuals involved meet the conditions for individual moral responsibility, although it may be difficult to distribute the share of moral responsibility in a fair way. To illustrate the problem of many hands we will therefore look at another example.

Problem of many hands The occurrence of the situation in which the collective can reasonably be held morally responsible for an outcome, while none of the individuals can be reasonably held responsible for that outcome.

9.2.1 The CitiCorp building

To illustrate the problem of many hands in engineering we will return to a case described in Section 3.9.3: the design and construction of the Citibank Headquarters in midtown New York. As we saw there this 59-story building was completed in 1977 and was designed by LeMessurier, a renowned structural engineer. In 1978, LeMessurier learned due to a series of serendipitous events that the tower's steel frame was structurally deficient. In Chapter 3 we focused on LeMessurier's behavior after this discovery. Here we focus on the situation in 1977 before LeMessurier discovered the flaw in the building. Apparently, the building was structurally deficient at that time, although nobody knew that. Who is to be held responsible for this structural deficiency?

To answer this question, we start by briefly sketching the main causes of the structural deficiency of the building. We then focus on the three main actors that causally contributed to this structural deficiency: 1) LeMessurier who designed the building; 2) the contractor who during construction decided to replace the welded joints by bolted joints; and 3) the employee at LeMessurier's firm who approved this change but did not inform LeMessurier about it (the "approver"). We will argue that none of these actors can reasonably be held responsible for the building being structurally deficient in 1977 and that this leads to a problem of many hands.

The structural deficiency of the CitiCorp building was mainly caused by a combination of two facts.[6] One was the peculiar design of the building, the other the change from welded to bolded joints. It was the combination of these two facts that made the building structurally deficient. Each of these facts considered in isolation did not jeopardize the structural strength of the building. The design was peculiar because the first floor was several stories above ground, with the ground support of the building being four pillars placed in between the four corners of the structure rather than at the corners themselves. The reason for this construction was that there had been a church on the building site and it had been agreed that this church would be reconstructed beneath the building after its completion. However, as LeMessurier found out in 1978 the combination of the peculiar design and the bolded connections made the structure vulnerable to high winds that strike the building diagonally at a 45 degree angle. Based on the New York weather records, a storm with a probability of occurrence once every 16 years (a so-called "16-year storm") would be sufficient to cause total structural failure.[7]

Now that we have some insight in the causes of the structural deficiency of the Citicorp building, let us look whether each of the three mentioned actors can reasonably be held responsible: LeMessurier, the contractor, and the approver. In doing so, we will apply the conditions for individual moral responsibility that were presented in Section 1.3. An individual is thus morally responsible for the structural deficiency if:

1 he did something wrong;
2 a causal connection is present between the wrong-doing and the structural deficiency;
3 he could have known that the building was structurally deficient; and
4 he acted freely.

The fourth condition is fulfilled for all three persons: they were not forced to act in a certain way. The second condition is also met: each of them made a causal contribution to the structural deficiency, for example, by changing the design from welded to bolded joints (the contractor), by approving the design change (the approver) and by choosing this particular design (LeMessurier).

The crucial responsibility conditions here are, therefore, the knowledge condition (the person could have known the deficiency) and the wrong-doing condition. If we apply these conditions, the following picture arises. LeMessurier cannot reasonably be held responsible in 1977, because he then did not know of the change from welded to bolted joints, which was crucial to foresee the structural deficiency of the building. The contractor, of course, knew about the change and probably also about the peculiar design but it seems reasonable to say that the contractor could not have known that the combination of these two factors would lead to structural deficiency. There are two reasons why the contractor could or should not have known this. First, in normal circumstances it would not have been a problem to change from welded to bolted joints. Second, the contractor, not being a structural engineer like LeMessurier, lacked the knowledge and expertise that was required to foresee this particular structural deficiency. Moreover, the contractor asked for and received approval for the change from the approver and, therefore, was not at fault. Consequently, the contractor cannot reasonably be held responsible.

What about the approver? Could or should he have foreseen the structural deficiency before approving the change? And if so, did he act wrongly in approving the change? Here are some reasons why the approver cannot reasonably be held responsible. According to Morgenstern, LeMessurier argued that the

> choice of bolted joints was technically sound and professionally correct. Even the failure ... to flag him [LeMessurier] on the design change was justifiable; had every decision on the site in Manhattan waited for approval from Cambridge, the building would never have been finished. Most important, modern skyscrapers are so strong that catastrophic collapse is not considered a realistic prospect; when engineers seek to limit a building's sway [the purpose for having welded joints], they do so for the tenants' comfort. (Morgenstern, 1995)

Furthermore, it even took LeMessurier several weeks in 1978 after hearing about the change in joints and being asked by a student about the structural strength to find out the vulnerability to 45-degree winds. Even if LeMessurier could, and possibly should, have foreseen the structural deficiency if he had known about the change from bolted to welded joints (which he did not), it seems reasonable to assume that the approver could not have foreseen the structural deficiency. The reason for that is that the approver is likely to have had considerably less experience and knowledge about the rationale for the design compared to LeMessurier. Hence, it is not reasonable to hold the approver responsible. It then turns out that none of the actors can reasonably be held responsible. To show that this is a problem of many hands, we also need to show that the collective can reasonably be held responsible in this case, a task to which we turn now.

In the CitiCorp case, we can define the collective as LeMessurier, the contractor, and the approver together. We assume that these three people can cooperate and share information. To attribute responsibility reasonably to the collective some conditions need to apply. For the moment, we will assume that these conditions are similar to the ones applying to individuals, that is, the collective acted freely, made a causal contribution, could have known it and was doing wrong.[8] It seems obvious that the collective acted freely, as each of the individuals acted freely. The collective also made a causal contribution to the structural deficiency. It is less clear whether the collective also meets the knowledge and wrong-doing conditions.

An important argument why the collective meets the knowledge condition is that if they had shared their knowledge and expertise they could have known that the building was structurally deficient. LeMessurier in fact drew this conclusion in 1978 after being informed about the change from welded to bolded joints.

Is the wrong-doing condition also met? From a consequentialist point of view, it obviously is: structural failure once in 16 years is unacceptably high; no engineer would contest that. One possible counter-argument is that the building still met the New York City building code because that code only requires taking into account 90-degree winds and not 45-degree winds and the building was only structurally deficient for the latter. Nevertheless, the effect of quartering winds was known long before the 1970s – the city's building code of 1899 already required to take all possible directions into account, although some later codes did not (Kremer, 2002). Moreover, wrong-doing is not confined to breaching the code. Engineers are expected to live up to a standard of reasonable care. According to Pritchard, "What counts as reasonable care is a function of both what the public can reasonably expect and what experienced, competent engineers regard as acceptable practice" (Pritchard, 2009). In this case, given the innovative design of the structure, it seems to require taking into account 45-degree winds (cf. Kremer, 2002). It thus seems reasonable to hold the collective in 1977 responsible for the structural deficiency of the CitiCorp building.

9.2.2 Causes of the problem of many hands

In the Citicorp case, the problem of many hands is primarily due to the distribution of information over the various actors. Due to the way information was distributed, neither LeMessurier nor the contractor nor the approver could reasonably have known that the actual built construction was structurally deficient. Still, at the collective level, the structural deficiency could reasonably have been foreseen (and the other responsibility conditions are also met). This reveals a more general cause of the problem of many hands: the distribution of information. The crucial point is that applying the knowledge condition to each of the individuals in isolation might yield a different result than applying it to the entire group of actors at once. This is why we might sometimes judge that none of the individuals could reasonably foresee a certain harm, while at the collective level that same harm is foreseeable.

The conflict between applying the responsibility conditions to the individuals and to the collective can also occur for other conditions, like the wrong-doing condition. An example is the responsibility of individual car drivers to the greenhouse effect. Individual car drivers by using their car emit concentrations of greenhouse gases that are – considered in isolation – completely harmless (assuming that there is a level below which greenhouse gas emissions have no effect); all car drivers together, however, introduce a considerable risk for future generations. What is essential about this example is that while none of the individuals is doing something wrong or is at fault, at the collective level there is obviously harm done, so it would be natural to assume that there is also wrong-doing.

The problem of many hands can also arise due to a combination of conditions for responsibility. For example, an employee of a company who knows of a defect in a product may – due to the hierarchical nature of the organization and the specific procedures within the organization – lack the freedom to repair the defect or to warn customers about it. His superior may have the freedom to act but maybe could not have known about the defect. The above suggests that the CitiCorp example is not an exception but that the problem of many hands is likely to occur regularly in engineering (and elsewhere).

9.2.3 Distributing responsibility

The notion of collective responsibility is helpful to articulate the moral intuition that under certain conditions people should be held responsible for disasters in complex engineering projects even if none of the individuals meet all the conditions for blameworthiness. However, it is not immediately clear what ascribing responsibility to the collective implies for the individuals who together form that collective. This requires attention to the **distribution of responsibility** among the members of a collective. But how should we distribute responsibility? In answering this question we should keep in mind that there are at least two reasons for ascribing responsibility. One is that we consider it morally important to hold people responsible for their actions and the consequences of these actions if certain conditions are met. In Section 1.3, we discussed the conditions that need to apply for holding people fairly responsible: wrong-doing, causal contribution, foreseeability, and freedom of action. We will call this the **moral fairness requirement**. The moral fairness requirement can also be applied to active responsibility: in that case, we will take it to mean that people should only be ascribed a certain active responsibility if they are able to live up to that responsibility. This, among other things, means that they should have the means and authority to fulfill their active responsibility.

Distribution of responsibility The ascription or apportioning of (individual) responsibilities to various actors.

Moral fairness requirement The requirement that a distribution of responsibility should be fair (just). In case of passive responsibility, this can be interpreted as that a person should only be held responsible if that person can be reasonably held responsible according the following conditions: wrong-doing; causal contribution; foreseeability; and freedom of action. In terms of active responsibility it can be interpreted as implying that persons should only be allocated responsibilities that they can live by.

The other reason why we ascribe responsibility is that we want to avoid harm and stimulate desirable outcomes. For utilitarians, this aim is the only aim of responsibility ascriptions. The distribution of responsibility that has the best consequences, that is, is effective in preventing harm, is the morally required distribution of responsibility. We will call this the **effectiveness requirement**. The effectiveness requirement seems especially relevant for active responsibility because then nothing has gone wrong yet, but it is also relevant for passive responsibility because it is

Effectiveness requirement The moral requirement that states that responsibility should be so distributed that the distribution has the best consequences, that is, is effective in preventing harm (and in achieving positive consequences).

desirable that people learn from their mistakes and are deterred from doing certain things and both aims presuppose assuming responsibility for what went wrong.

We will assume that an ideal distribution of responsibility is both morally fair and effective. The problem of many hands shows that it is sometimes hard to meet both requirements at once. In cases like the CitiCorp case, it seems morally unfair to hold one of the actors responsible for the structural deficiency. Yet this distribution of responsibility, or rather the absence of it, does not seem very effective in avoiding harm. How can we reconcile the requirements of fairness and effectiveness? We do not have a clear-cut answer to this question. Instead we will discuss below a number of mechanisms for distributing responsibility and their moral fairness and effectiveness. These mechanisms are the law (Section 9.3), organizational models for distributing responsibility (Section 9.4) and technological designs (Section 9.5).

9.3 Responsibility and the Law

Responsibility is not only a moral concept, but also a legal concept. The way the notion of responsibility is used in the law is however different from how it is used in ethics. We will therefore use the term **liability** to refer to legal responsibility. In what respects is liability different from moral responsibility? First, the conditions or basis by which someone is held liable are often different from the conditions by which someone is held morally responsible. The conditions for liability are laid down in the law and may differ for different types of actions, for different types of consequences and in different countries. For moral responsibility, usually the conditions set out in Section 1.3 are used. This difference means that it may well be possible for a person to be morally responsible for an action while he or she is not liable as we saw in the case of the *Herald of Free Enterprise*. Also the opposite may occur: a person may be liable without being morally responsible. Secondly, liability is established in an official and well-regulated procedure in court. It requires a verdict by a judge or a jury and the liability conditions must be proven to apply in a formal juridical sense. Moral responsibility can be established more informally. Third, liability usually implies the obligation to pay a fine or to repay damages, while this is not necessarily an implication of moral responsibility. Fourth, liability always applies after something undesirable has occurred, while responsibility is relevant both after the fact as well as before something undesirable has occurred (active responsibility; see Section 1.4).[9] In the case of the *Herald*, it could for example be argued that the engineers had an active responsibility to include certain safety measures like bulkheads in their design, even if they are not legally liable when they do not include such measures. The key differences between moral and legal responsibility are summarized in Table 9.1.

Liability Legal responsibility: backward-looking responsibility according to the law. Usually related to the obligation to pay a fine or repair or repay damages.

Even if liability and moral responsibility are different notions, one might make an attempt to make them as similar as is feasible. One could for example base liability on the same conditions as passive moral responsibility. An argument in favor of translating moral responsibility into liability is that if morally irresponsible behavior never leads

Table 9.1 Key differences between moral responsibility and legal liability

Moral responsibility	*Legal liability*
Moral blameworthiness based on conditions of wrong-doing, causality, freedom and foreseeability.	Based on conditions formulated in the law
Can be established more informally; you can also consider whether you are yourself responsible	Established in well-regulated procedure in court; juridical proof of conditions required
Not necessarily connected to punishment or compensation	Usually implies the obligation to pay a fine or to repay damages
Backward-looking and forward-looking	Backward-looking

to punishment on the grounds of legal liability then there remains little incentive to act morally so that few people will be encouraged to do that. However, one may doubt the assumption that people are inclined to behave immorally unless they are punished. An argument in favor of basing liability upon moral responsibility is the consideration that it would be undesirable to have immoral laws. However, laws that deviate in some respects from morality are not necessarily immoral. One reason for this is that not everything that is legally allowed is also morally allowed (see Section 4.5.1 about the fallacy of confusion of law and ethics). In most countries, adultery is not forbidden by the law but that does not make adultery morally allowed in these countries. Rather the law is silent on it.

Both arguments thus do not necessarily lead to the conclusion that moral responsibility and liability *must* coincide. One might equally well say that the law, and therefore also liability, could never apply to all cases of moral responsibility. Therefore even if one tries to translate moral responsibility into liability, this will never completely succeed. In some cases it may even be desirable to make someone liable even if his or her moral responsibility is debatable. Holding people liable, even if they cannot reasonably be held morally responsible, may make them more cautious, and may prevent negative effects and so help to solve the problem of many hands. In other words, the ascribing of liability can be based on considerations of effectiveness rather than on considerations of moral fairness. Below we will discuss some possibilities and limitations of liability as tool for preventing undesirable consequences of technology.

9.3.1 Liability versus regulation

Liability is one of the legal tools that can be used to deal with the social consequences of technology. It is, however, not the only possible tool. The other main legal tool is **regulation**. Regulation can forbid the development, production or use of certain technological products, but more often it formulates a set of the boundary conditions for the design, production, and use of technologies. In Section 6.4,

Regulation A legal tool that can forbid the development, production, or use of certain technological products, but more often it formulates a set of the boundary conditions for the design, production, and use of technologies.

we have seen that such regulations together form a regulative framework for the design of a technology. If such a regulative framework meets certain conditions, it can be considered an adequate way of dealing with the ethical issues raised by the design of a technology as we have seen. However, such regulative frameworks are usually absent in the case of radical, innovative design. One reason for this is that regulation is usually based on our current knowledge of a technology and its consequences and on past experiences with that technology. Regulation is therefore often not able to deal with innovation. As a consequence, regulation will either have to forbid certain innovations or will lag behind the technological developments.

Given the large economic and social benefits of innovation, most of today's governments refrain from regulation that forbids certain innovations outright. The consequence is that regulation tends to lag behind technological development and its consequences. This is primarily a problem of lack of knowledge and experience, but it is further aggravated by the fact that legal regulation is a long and cumbersome process. Even if certain negative consequences of a technology are discovered it may take years before they are adequately addressed in new regulations. In such situations, liability may provide an attractive alternative legal framework for dealing with the social consequences of technology.

Liability does not require the government to foresee the consequences of new technology but rather makes the ones developing those technologies, usually companies and the engineers employed by them, legally liable for those consequences under certain conditions. One might argue that this places the responsibility where it can be met best: in the hands of the ones developing technology. They have the best knowledge of new innovations and their possible effects and are in the best position to avoid certain disadvantages. Moreover, a scheme of liability would stimulate them to employ their (active) moral responsibility. A next question then is: what is the best form of liability to stimulate this?

9.3.2 Negligence versus strict liability

The conditions that must be met in order for a person to be liable depends on the law and may, therefore, differ from country to country. Nevertheless, in large parts of the Western world, the main condition for liability is **negligence**. In order to claim negligence, proof must be given of:

1 A duty owed. This is usually a **duty of care**, which is the legal obligation that individuals adhere to a reasonable standard of care while performing any acts that could foreseeably harm others. Duties of care typically arise in particular relationships such as between parent and child or between landlord and tenant. Also the relation between engineer and the public defines such a duty of care. For engineers, the standard

Negligence Not living by certain duties. Negligence is often a main condition for legal liability. In order to show negligence for the law, usually proof must be given of a duty owed, a breach of that duty, an injury or damage, and a causal connection between the breach and the injury or damage.

Duty of care The legal obligation to adhere to a reasonable standard of care when performing any acts that could foreseeably harm others.

of care both depends on what the public can reasonably expect from engineers and what is common practice in engineering. The duty of care is thus based on the sometimes implicit moral responsibilities of engineers and need not be made explicit in the law;
2. A breach of that duty;
3. An injury or damage; and
4. A causal connection between the breach and the injury or damage.

Negligence does not require that the defendant actually foresaw the damage but that a reasonable person in the position of the defendant could have foreseen the damage.

In contrast to negligence, **strict liability** does not require the defendant to be negligent in order to be liable. It is usually enough that the defendant engaged in a risky activity and that this activity caused the damage done. Technological innovation is obviously a risky activity in the sense that it might produce unknown hazards to society. So innovation is a possible candidate for strict liability. In fact, the US and the countries of the European Union recognize **product liability**, which makes a manufacturer liable for defects in a product, without the need to proof that that manufacturer acted negligently.

Strict liability A form of liability that does not require the defendant to be negligent.

Product liability Liability of manufacturers for defects in a product, without the need to proof that those manufacturers acted negligently.

EU Council Directive 85/374/EEC for Product Liability

Article 1: The producer shall be liable for damage caused by a defect in his product.

Article 4: The injured person shall be required to prove the damage, the defect and the causal relationship between defect and damage.

Article 6.1 A product is defective when it does not provide the safety which a person is entitled to expect, taking all circumstances into account …

Article 7 The producer shall not be liable as a result of this Directive if he proves: … (e) that the state of scientific and technical knowledge at the time when he put the product into circulation was not such as to enable the existence of the defect to be discovered …

Article 15.1 (b): Each Member State may by way of derogation from Article 7 (e), maintain or … provide in this legislation that the producer shall be liable even if he proves that the state of scientific and technical knowledge at the time when he put the product into circulation was not such as to enable the existence of a defect to be discovered.

One reason for applying strict liability to technological products is that it motivates engineers and the other people involved in innovation to be very careful, for example, by investigating possible hazards and taking precautions. Strict liability will therefore probably result in a higher level of safety. It may in fact be the only way to meet Mill's freedom principle (Section 3.7.2) or the principle of informed consent (Section 8.5.1) for technological risks (see Zandvoort, 2000). Both principles forbid subjecting people to (unknown) risks unless they have consented to the risks or the hazardous activity.

Strict liability, however, also has disadvantages. First, strict liability may well slow down the pace of innovation. This is often considered undesirable because innovation is an important source of social and technological progress in today's society. Against this, it may be argued that strict liability does not outlaw innovation, but only requires careful innovation. Second, it seems morally unfair to hold people liable when they are not at fault or could not have foreseen the damage. If strict liability is applied, engineers or the corporations for which they work may well be liable for the hazards of a technology, while they are not morally responsible (if moral responsibility is understood in terms of the conditions discussed in Chapter 1). On the other hand it also seems unfair to the potential victims that they have to bear the damage: they could have done even less than the engineers to prevent the damage. In fact, one of the motivations for the European Union to introduce product liability is that strict liability (also called "liability without fault") in their view would result in a fairer distribution of the risks and benefits of technological innovation:

> liability without fault on the part of the producer is the sole means of adequately solving the problem, peculiar to our age of increasing technicality, of a fair apportionment of the risks inherent in modern technological production.[10]

Nevertheless, the EU directive for product liability makes an exception for defects that could not have been foreseen given the state of scientific and technical knowledge at the time the product was put into circulation (see box).[11] Such unforeseeable risks are called **developments risks** and most schemes of product liability make an exception for development risks.

Development risks In the context of product liability: Risks that could not have been foreseen given the state of scientific and technical knowledge at the time the product was put into circulation.

The exception for development risks is based on two considerations. One is that otherwise innovation would be too much hampered. The other is the concern that it would be morally unfair to make manufacturers responsible for damage they cannot foresee. Although both considerations are reasonable, it is questionable whether they provide a conclusive argument for excluding development risk from liability. First, excluding development risks from liability seems to put a bonus on not developing scientific and technological knowledge about the potential harms of new products. As soon as such knowledge is available, liability may apply. This is obviously an undesirable effect from the point of view of effectiveness because it might mean that society is unnecessary subject to certain risks. Second, even if it may be impossible to predict all the dangers of a new technology beforehand – and there are good reasons to suppose so as we saw in Chapter 8 – it is not obvious that the victims of the yet unknown risks should bear the damage rather than the ones having introduced the product or society as a whole. The fairness argument of the EU for product liability seems to apply to the development risk as well.

9.3.3 Corporate liability

Not only individual persons, but also corporations can be held liable for the law. In such cases, the corporation is treated as a legal person. This is called **corporate liability**. An example is found in the case of the *Herald of Free Enterprise* in which the shipping company was prosecuted for manslaughter. A main advantage of corporate liability is that victims or the government do not need to find out which individuals in a company were responsible for, for example, a defect in a product but that they can simply sue the company as a legal person. If the company as a legal person is convicted it is bound to pay a fine or compensate the damage done. Despite this advantage, corporate liability also has a number of disadvantages and limitations. First, corporations, unlike natural persons, do not possess a conscience. They have "no soul to damn and no body to kick."[12] They cannot be put in prison for example. Therefore, legal instruments that are reasonably effective when applied to natural persons are not necessarily effective when it comes to corporations.

Corporate liability Liability of a company (corporation) when it is treated as a legal person.

Second, most modern corporations are characterized by **limited liability** (Kraakman et al., 2004, pp. 8–9). This means that the shareholders are liable for the corporation's debts and obligations up to the value of their shares. Corporations may, however, well inflict more damage than the total value of their shares. This additional damage is then to be borne by the victims. This point is further aggravated by the fact that companies, unlike natural persons, can disappear by being split up, mergers, or bankruptcy. Bankruptcy, mergers, or splitting up are sometimes a deliberate strategy employed by corporate officials to avoid or limit liability claims.

Limited liability The principle that the liability of shareholders for the corporation's debts and obligations is limited to the value of their shares.

Third, both the moral fairness and the effectiveness of corporate liability to a large extent depend on how the liability of the corporation is translated to individuals within the organization. A liability claim on a company may, for example, result in the dismissal of employees who did not partake in the damage for which compensation is claimed. On the other hand, managers who played a major role in the damage done may emerge unscathed. This is especially the case if their behavior was not illegal and if no individual liability on their part can be shown. This point draws attention to the allocation of responsibility in organizations, a theme that will be discussed next.

9.4 Responsibility in Organizations[13]

Most modern organizations are characterized by a division of tasks and roles. This has implications for who can be held responsible for what in organizations. In this section, we will discuss three different models for distributing responsibility in organizations:

1 The hierarchical model where those highest in the organization's hierarchy are held responsible;
2 The collective model in which each member of the organization is held to be jointly and severally responsible for the acts of the organization as a whole; and
3 The individual model in which each member of the organization is held responsible in relation to his or her contribution.

The models are in the first place intended to establish who is passively responsible for undesirable consequences. Which model is actually applicable depends to an important degree on the formal and actual organizational form of an organization. In the case of the *Herald of Free Enterprise*, some individuals – like the assistant boatswain who had fallen asleep – were prosecuted due to their (alleged) contribution to the disaster (individual model), others – like directors Develin, Ayres and Alcindor – were primarily prosecuted because they were highest in hierarchy (hierarchical model). We might also pose the normative question of which model could "best" be applied, not only for allocating passive responsibility (after something has happened) but also for the distribution of active responsibility. Two considerations are then, again, important. First, whether the model is morally fair in how it allocates responsibility and, second, whether, it is effective in avoiding undesirable behavior. Below, we will discuss these issues for the three models.

Hierarchical responsibility

Hierarchical responsibility model The model in which only the organization's top level of personnel is held responsible for the actions of (people in) the organization.

In the case of **hierarchical responsibility model** it is only the organization's top level of personnel that is responsible for the actions of the organization. The hierarchical model is attractive because of its relative simplicity and clarity. In present-day practice, though, the hierarchical responsibility model is not always effective in preventing undesirable consequences. A main reason for this is the fact that the managers of organizations may be, to an extent, outsiders within their own organization. In practice it is often very difficult for executives within an organization to get hold of the right information in time or to effectively steer the behavior of lower organizational units. Nevertheless, the knowledge that they will later have to account for damage done by the organization may motivate mangers to gather the necessary knowledge and to better steer the organization. The hierarchical responsibility model also seems somewhat morally unfair. If managers are not well informed about what is going on within their organization and are only able to steer to a limited degree, how can they then fairly be held responsible for undesirable activities? Therefore, the allocation of responsibility along strict hierarchical lines may lead to moral objections.

Collective responsibility model The model in which every member of a collective body is held responsible for the actions of the other members of that same collective body (and for the responsibility of the collective).

Collective responsibility

With the **collective responsibility model** every member of a collective body is responsible for the actions of the other members of that same collective body. The collective responsibility model is not very attractive to large organizations, because it is not possible to allocate responsibility in differing degrees to individual members of the collective. Everyone is responsible to an equal degree for the actions of the collective body. Individual differences in being at fault or being able to prevent certain damage cannot be accounted for in this model. This is often seen as morally unacceptable. Another disadvantage of the collective responsibility model is that no one in particular tends to feel morally responsible for the consequences of the activities of the organization as a whole. In fact, everyone is held equally responsible for the actions of the whole, whether people as individuals have contributed to that or not. In such a situation it quickly becomes attractive to let someone else burn her of his fingers, especially in large organizations. The collective responsibility model would only seem to be applicable in a number of more exceptional cases. One important condition when it comes to introducing the model is that the members of the collective must be able to effectively influence each other.[14] This demands small-scaleness and equality between the members of the collective. In large organizations this condition is usually not met. One possibility would be to strive towards achieving smaller organizations with greater solidarity.

Individual responsibility

In the **individual responsibility model** each individual is held responsible insofar as he or she meets the conditions for individual responsibility as discussed in Section 1.3. A main advantage in this model is that it is morally fair. The model also might seem effective because it encourages individuals to behave responsibly. Nevertheless, as we have seen, individual moral responsibility may lead to the problems of many hands: the organization may collectively bring about undesirable consequences for which no individual can be held responsible. This is clearly a disadvantage of this model. One need, however, not conclude from this that the individual model should never be applied. One also could try to avoid the problem of many hands through certain organizational measures, like the better sharing of information or empowering individual employees so that they can live by their individual responsibility. The latter can for example be achieved by a good company policy for internal whistle-blowing so that employees can raise issues without the fear of being dismissed (see also Section 2.3.4).

Individual responsibility model The model in which each individual is held responsible insofar as he or she meets the conditions for individual responsibility.

Conclusion

Obviously, none of the models discussed is ideal in terms of moral fairness and effectiveness. There is no single answer to the question how responsibility can best be distributed in organizations. It should be noted that which model can be best applied in a particular case partly depends on how the organization in question is actually organized (and on the legal status of the organization). If an organization is, for example organized along strictly hierarchical lines, it will be both morally unfair and ineffective to apply the collective responsibility model. The relation, however, also works in the other direction: if a certain responsibility model is judged most desirable in a particular situation or for a particular task, attempts can be made to make the organization fit the responsibility model. The box discusses a situation in which a design team has to be set up and a choice has to be made for a certain responsibility model and hence for certain organizational set-up of the design team.

How Should the Responsibility for Safety be Distributed?

Suppose that you are working as an engineer for a company producing cars. The head of the R&D department asks you to set up a design team for the design of a new type of truck. One of the issues you will have to take into account when setting up the design team is how to distribute the responsibility for the safety of the truck. In this case, the question is not how to allocate responsibility for safety given a certain organizational set-up, but rather how to best allocate this responsibility and of, next, finding an organizational set-up that matches that allocation. (Obviously, in your considerations other concerns than the responsibility for safety will play a role, but we will leave aside such concerns for the moment.) In this case, you could start with formulating the desiderata for an allocation of responsibility for safety. One could think, for example, of the following desiderata:

1 All individuals (or groups) to which a certain responsibility is allocated should be able to live up to those responsibilities.
2 For outsiders, that is, people who are not members of the design team, it should be clear whom to address if there is a concern or question about the safety of the truck.
3 The distribution of responsibilities should be effective in the sense of resulting in a safe truck.

Different models or combinations of models for allocating responsibility could be considered for achieving this. For example:

1 The hierarchical model, in which the leader of the design team is responsible for safety;
2 The collective model, in which all design team members are equally responsible for the safety of the truck;
3 The individual model in which each member of the design team is responsible relative to his or her individual contribution;
4 The responsibility for safety can also be allocated to a special safety official. Many organizations, for example, have special HSE (Health, Safety, and Environmental) officers for this type of concern.

When we evaluate these models with the mentioned desiderata, each seems to have specific pros and cons. The hierarchical model may do good on desideratum 2 (a clear address for outsiders), but it might be questionable whether the design team leader is able to oversee and steer all safety-related decisions, so that desiderata 1 and 3 are not fully met. The individual model will motivate all design team members to take safety seriously. However, guaranteeing the safety of the entire truck may require an integral approach that takes into account the interaction between different parts of the truck that are designed by individual engineers (or sub teams). On the individual model, nobody may be responsible for this interaction, which may have disastrous results for safety, for which it is very difficult to hold any of the individual design team members responsible; hence the individual model may not meet desiderata 2 (address) and 3 (effectiveness). On the collective model, the whole team (each individual member) can be addressed for failure to approach safety in an integral way. It is, however, questionable whether this model is effective (desideratum 3) and fair (desideratum 1) because the model presupposes that all design team members know and understand what the other members are doing and are able to influence that, which may be very difficult to attain in practice. For such reasons, one could choose for appointing a safety official who is responsible for an integral approach (the fourth model), possibly in combination with the individual model, so that each team member is also responsible for his or her own contribution.

In the philosophical literature on responsibility, various authors have pleaded for reinforcement of the individual model (Bovens, 1998). It should be noted that if the individual model is chosen, this would require changes in the way most organizations are currently organized. The individual model requires that within organizations people have the freedom to operate in actively responsible ways.[15] In some organizations this freedom may be limited. Moreover, the law often allocates liability to the organization as a whole (corporate liability) or to the owners or managers, rather than to individual employees. Again, this is a reflection of the way most organizations are currently organized and of the legal status of employees. As we have seen in Section 2.3.4, freedom of speech is, for example, not guaranteed within organizations to the same degree as it is in the relation between individuals and the state. Reinforcing the individual model may thus require major organizational and legal changes. In addition, the individual model has the great disadvantage that no one in particular is responsible for the collective consequences of individual actions. In all the other models some kind of provision is made for this, even though these other models have their own drawbacks. We may then conclude that a combination of the models, tailored to the specific requirements of the situation, will often be the best option.

9.5 Responsibility Distributions and Technological Designs[16]

Not only do the law and organizations influence the distribution of responsibility, but engineering design does too. An example is the automatic pilot in an airplane. The automatic pilot takes over a number of actions from the pilots and consequently also takes over parts of their task. Design decisions about which tasks to allocate to the automatic pilot and which to human pilots are usually made with an eye to effectiveness in terms of safety and costs. What is less often taken into account, is that such design decisions also affect the passive responsibility for errors, for example in case of an accident. For example, if the automatic pilot is designed in a way that it can only be turned on and off during take-off and landing, the human pilots do not longer have the freedom to correct the plane in case of a calamity during flight and can no longer be held responsible if such a calamity results in a disaster because freedom to act is one of the conditions for responsibility. In such cases, the designers (or producers) of the automatic pilot may rather be the ones that are to be held responsible. Sometimes, however, they may also not meet the responsibility conditions, so that nobody can be held responsible, and a problem of many hands occurs. This may be considered as an undesirable effect of the way the automatic pilot was designed in the first place, even if the most effective design, that is, the design that results, for example, in the lowest number of accidents, was chosen.

Another important, though often overlooked, issue is that if certain tasks are allocated to humans through design decisions, it should be ascertained that the conditions exist or can be created under which those individuals can responsibly carry out those tasks. Human pilots, for example, need information to be able to steer a plane. (The knowledge condition is one of the conditions for responsibility.) The system thus should provide them the right information on time. Note that even if this is the case, the pilots are dependent on the system, and on the system designers, for getting reliable information; pilots cannot simply look out of the window of the plane and estimate the flight altitude.

An example of a situation in which a mismatch occurred between the responsibilities allocated to certain humans and the actual conditions under which these responsibilities have to be acted upon, is the controversy over the safety of the pesticide 2,4,5-T (see Section 6.2.6). The developers of this pesticide and the scientists testing its safety assumed that the farmers would use the product according to prescriptions. These prescriptions, however, seemed to ignore the actual conditions under which farmers had to work; conditions which make it very hard, if not impossible, to live up to the responsibility allocated to them.

Technological design may not only allocate responsibilities to individuals, as in the case of the automatic pilot or the pesticide 2,4,5-T, but may also imply more complex divisions of labor and responsibility. Some technologies, for example, require a certain social structure to function properly. The Greek philosopher Plato in the *Republic* already argued that to navigate a ship one needs one and only one captain and that the crew needs to obey the captain. In other words, navigating a ship requires a hierarchical social structure. In a similar fashion, the contemporary philosopher of technology Langdon Winner has argued that the atomic bomb requires an authoritarian social structure: "[T]he atom bomb is an inherently political artefact. As long as it exists at all, its lethal properties demand that it can be controlled by a centralized, rigidly hierarchical chain of command closed to all influences that might make its working unpredictable. The internal social system of the bomb must be authoritarian; there is no other way" (Winner, 1980, p. 131). Winner might exaggerate the extent to which an authoritarian structure is required, but that the atomic bomb requires some social structure of control to prevent certain misuse seems undeniable. Technologies might therefore further or require an allocating of tasks and responsibilities along hierarchical lines. They may also further other, more complex, divisions of labor and responsibility as in the case of the V-chip (see box).

Case V-chip

Figure 9.2 Former US President Bill Clinton holds up a V-chip during ceremonies where he signed the Telecommunications Reform Act at the Library of Congress in Washington, DC February 8, 1996. Photo: Paul J. Richards/AFP/Getty Images.

The V-chip is an electronic device that can be built into televisions to block television program that are violent or otherwise unsuitable to children (FCC, 2009). In the USA, the V-chip is required for all televisions of 13 inch and larger since January 2000. The V-chip functions as follows. The television stations broadcast a rating as part of the program. The parents program the V-chip by setting a threshold rating. All programs above the rating are then blocked by the V-chip.

In order for the V-chip to function properly, it requires a uniform rating system and organizations doing the rating. The latter can be the TV station, the program makers, the government or an independent review board. In the USA, the National Association of Broadcasters, the National Cable Television Association, and the Motion Picture Association of America have established a ratings system known as "TV Parental Guidelines." The program makers or television stations give ratings to the programs. A TV Parental Guidelines Monitoring Board monitors the application of the rating system and deals with complaints.

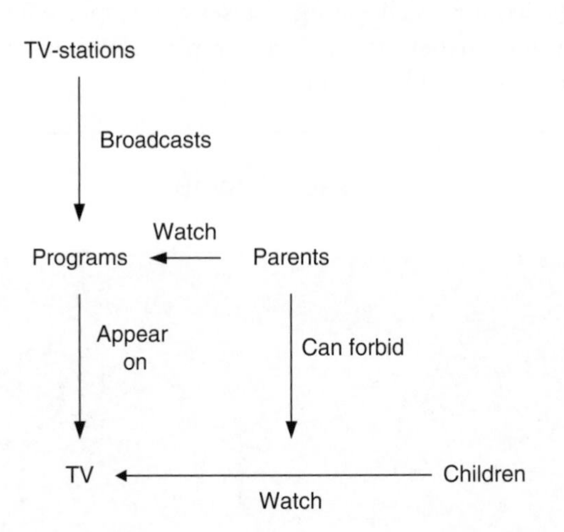

Figure 9.3 Traditional divisions of labor with respect to violence on television.

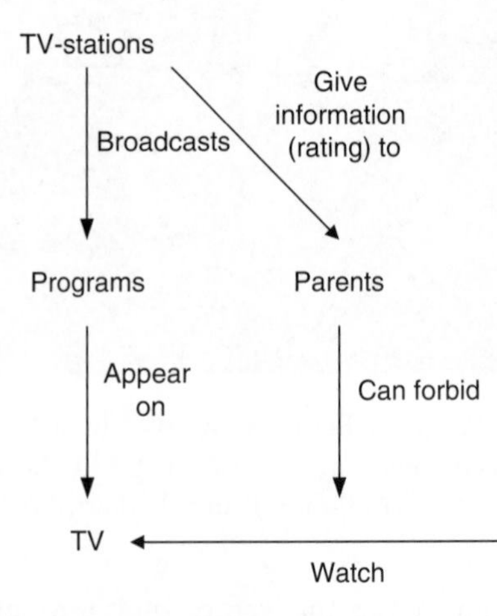

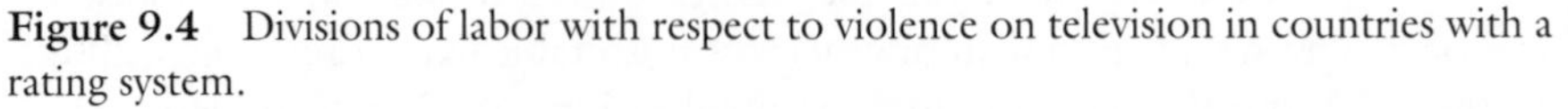

Figure 9.4 Divisions of labor with respect to violence on television in countries with a rating system.

If one compares the V-chip with the previous situation, or the situation existing outside the USA, some interesting shifts in the division of tasks become clear (see Figure 9.3). In the traditional situation, the parents decided directly what their children saw on television. They did so presumably by switching on or off the TV. There were in most countries no formal or legal restrictions for TV programs although there were some moral and aesthetic constraints, for example of "good taste."

The second situation is one in which the programs contain a rating (Figure 9.4), which may be helpful for parents to decide which programs their children are allowed to see and which not. This system operates in many European countries. In this system the parents still mainly decide what their

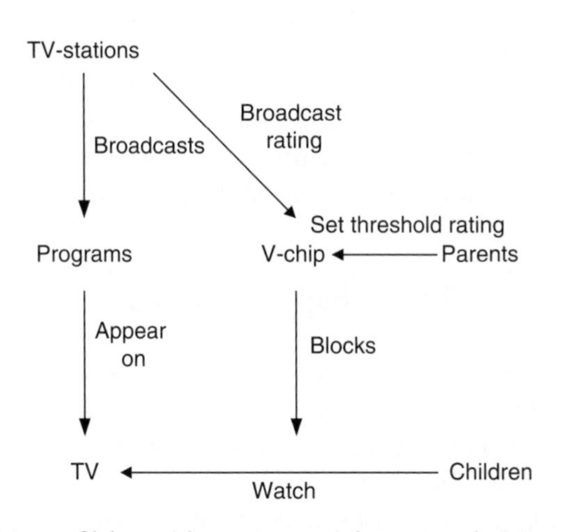

Figure 9.5 Divisions of labor with respect to violence on television with implementation of the V-chip.

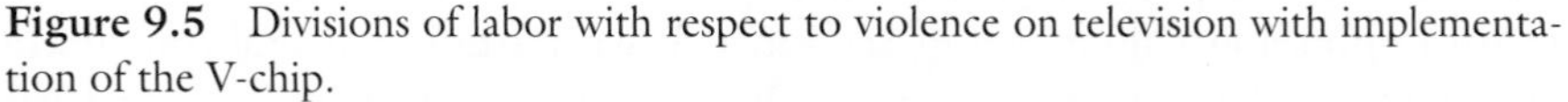

children watch, although the judging of the programs have partly been taken over by others, who are applying the ratings to programs.

With the V-chip the actual role of the parents has further diminished. If they choose to use the V-chip, they only have to set the rating they find acceptable for their children. Note, however, that in doing so they presume – at least tacitly – not only that the rating system is applied properly but also that the rating scheme coincides with their own norms and values. Actually, it might be the case that parents would judge some programs as unacceptable which are still rated as acceptable and, at the meantime, they might consider some programs acceptable or even useful which are rated unacceptable.

The V-chip thus allocates a number of tasks and responsibilities to a variety of actors. We could judge this resulting responsibility distribution in terms of effectiveness and moral fairness. For effectiveness, a major question is what the consequences of the responsibility distribution are. For example, do fewer children watch violent TV programs than without the V-chip? How does the V-chip affect the programs that are broadcasted? It seems that TV stations and program makers get an incentive to make responsible programs but the system might also be an excuse to make tasteless television because everybody is free to block programs they don't like with the V-chip.

With respect to moral fairness, one question is whether the actors involved can reasonably live up to the responsibilities that are allocated to them. Can parents be expected to program the V-chip? Can TV Stations rate programs? On both counts, the answer seems yes. A further question is whether these parties are the appropriate parties to assume these responsibilities. For example, if one would argue that parents should be the ones who are primarily responsible for what their children watch on TV a crucial question is: Does the V-chip diminish the responsibility of parents (because other parties assume a role too) or does it enable the parents to assume more responsibility than before (because they now have better means to control what their children watch on television)? One could also wonder whether it is appropriate that TV stations rate programs and so influence what children watch on TV. Some people would probably argue that this is nor a proper task of TV stations in a liberal society and that these decisions should be entirely left to the parents.

9.6 Chapter Summary

Even if all the individuals involved in technological development act responsibly, the overall effect of their actions are not necessarily benign. A main reason for this is the problem of many hands. This is the situation whereby a collective can reasonably be held responsible for an outcome (such as the negative consequences of technological development) while none of the individuals involved meet the conditions for responsibility discussed in Section 1.3 (wrong-doing, causal contribution, foreseeability, and freedom of action). Overcoming the problem of many hands requires paying attention to the distribution of responsibility. In this chapter, we have discussed three mechanisms for distributing responsibility: 1) the law; 2) organizational models for allocating responsibility; and 3) technological development. In all these cases, the resulting responsibility distribution can be assessed in terms of moral fairness (are the appropriate persons held responsible?) and effectiveness (does the responsibility distribution contribute to avoiding harm and to achieving desirable effects?).

The law makes certain people liable for certain actions and outcomes. Liability is the legal counterpart of moral responsibility. One can, however, be morally responsible without being legally liable and the other way around. Liability can also be used as a tool to prevent possible negative consequences of technology. It is then an alternative to regulation. Unlike regulation, liability does not require that the government foresees the consequences of new technology. Rather it makes the ones developing those technologies, usually companies and the engineers employed by them, legally liable for those consequences under certain conditions. What these conditions comprise of depends, among others things, on whether liability is based on negligence or is strict. Negligence requires a causal connection between the breach of a duty owed and injury or damage. Strict liability does not require the breach of a duty; it is enough to show that the defendant undertook a risky activity that caused the damage. An example of strict liability that is relevant for engineering is product liability. Product liability, however, excludes developments risks, that is, risks that could not have been reasonably foreseen at the time the technology was developed.

We have discussed three models for distributing responsibility in organizations: the hierarchical; the collective; and the individual. Each of the models has its own particular advantages and disadvantages in terms of fairness and effectiveness. None of the models provides a general solution to the problem of many hands. In peculiar circumstances, the models might, however, be usefully applied or combined. Also, engineering design influences the allocation of responsibility in technology. Differently designed technologies may provide users with different degrees of freedom and knowledge and may, hence, influence their responsibility because freedom of action and foreseeability are preconditions for responsibility. Technologies may also allocate certain tasks to certain actors and so influence the allocation of responsibility.

Even if there is often not one responsibility distribution that is obviously the most attractive, additional attention on how to distribute responsibility might help to avoid or at least soften the problem of many hands. In the end, paying attention to the question of how to distribute responsibility is a responsibility to be taken up by individuals, but it is a responsibility that is easily overlooked if one focuses on individual responsibility only.

Study Questions

1. Explain why the problem of many hands is a moral problem.
2. In what ways do you think corporations may be moral agents? How do they differ from human agents?
3. What are the disadvantages of corporate liability?
4. Strict liability
   a. What is strict liability?
   b. On which ethical principle(s) is strict liability based?
   c. Do you think strict liability is ever justified?
5. What is the difference between responsibility and liability?
6. What are the three models for allocating responsibility in organizations, and describe these models.
7. Looking back, which model for allocating responsibility could be best applied in the LeMessurier case for safety?
8. Explain that technological design can influence the allocation of responsibility.
9. Do you think that TV stations have the responsibility to rate programs with respect to the V-chip? Explain your answer.
10. Consider the following additional information on the safety of roll-on/roll-off ferries[17]

When it comes to formulating legal safety requirements, it is the *International Maritime Organization (IMO)* that has an important part to play. This international organization is responsible for adopting legislation for ships. IMO knew as early as 1981 that if water entered the car decks of roll-on/roll-off ships, they could be lost in a rapid capsize (Van Poortvliet, 1999, p. 52). The IMO did not adjust its regulations at the time to solve this problem, while a simple solution was available.

Because legislation adopted by the IMO needs to be implemented by *governments*, only governments accepting an IMO convention have to implement it. When making a convention it is, therefore, important to make it acceptable for as many governments as possible, otherwise only a small percentage of all fleets will be obliged to abide by the convention. A shipping company can decide to sail under the flag of another country which has not ratified an IMO convention, if complying with the convention costs a lot of money. So there is a certain amount of pressure on the IMO not to issue safety requirements that are too tight.

Apart from the IMO, *insurance and classification companies* also have a part to play in the formulation of safety requirements. For hull insurance bought by operating companies from insurance companies such as Lloyd's of London, a ship needs to be classified. Classification organizations are private organizations that have to monitor compliance with legislation during construction and the certification of sea worthiness during a ship's lifetime. Only the equipment and the construction are taken into account by the classification organizations, not passenger safety.

There is little incentive for *shipping companies* to ask for, or for *shipyards* to design ships, that are even safer than required by IMO conventions and hull insurance regulations. When disasters occur, the investigation that follows usually concludes that it was a human error that led to the disaster. Little attention is given to the design of the ship as long as on completion it complies with regulations.

a. Discuss for each of the following actors whether they are responsible (blameworthy) for the inherent instability of roll-on/roll-off ferries. Use the conditions for responsibility discussed in Section 1.3.
   - Maritime engineers designing these ferries
   - The IMO
   - Governments
   - Insurance companies
   - Classification organizations
   - Shipping companies
   - Shipyards
b. Is this a problem of many hands?
c. How could the active responsibility for increasing the safety of roll-on/roll-off ferries best be allocated?

Discussion Questions

1 In the aftermath of technological disasters like that of the *Herald of Free Enterprise*, the Challenger (Chapter 1) and the High Speed Train in Germany (Chapter 6), there is often a lot of attention on who is to blame. It could, however, be argued that for engineers the main concern is not blame but how to prevent such disasters in the future. In other words: one should not focus on backward-looking responsibility and blameworthiness but rather on forward-looking responsibility or active responsibility. Do you agree? Do the two perspectives exclude each other or are they somehow connected? Would the problem of many hands still be a problem if one focuses on forward-looking responsibility?
2 The text mentions two requirements for distributing responsibility: moral fairness and effectiveness. Do you consider one of these requirements more important than the other? How should conflicts between both requirements be dealt with?
3 Should legal liability in your view be based on moral responsibility as much as possible or not? What other considerations may be relevant for legal liability apart from moral responsibility (if any)? Would your view have consequences for currently existing forms of legal liability as they are discussed in this chapter?
4 Engineers often try to increase the safety of technological systems by technological devices (for example, automatic pilot, automatic shut-down of system, completely automated process control). What does this imply for the responsibility of the operators

of these systems? Do you think that this makes those systems safer overall? Do you consider increasing safety by safety devices a desirable development or should safety be dealt with in another way?

Notes

1 *MV Herald of Free Enterprise Report of Court No. 8074 Formal Investigation*. London: Crown, 1987.
2 Platform Ethiek en Techniek TU Delft, Werkconferentie Ethische aspecten van de Ingenieurswetenschappen 19 April 1996, Delft, The Netherlands: 1996.
3 Platform Ethiek en Techniek TU Delft, Werkconferentie Ethische aspecten van de Ingenieurswetenschappen 19 April 1996, Delft, The Netherlands: 1996.
4 This section is based on and partly drawn from Van de Poel et al. (manuscript).
5 It is sometimes argued that pinpointing responsibility may hamper learning and openness about incidents and near-accidents, because the focus is on blame instead of on openness and on learning. This may be true if the focus is on juridical responsibility, the paying of damage and/or the public blaming of the culprit. However, willingness to learn from accidents seems to imply the acceptance of at least some *moral* responsibility.
6 There was in fact a third factor: people from LeMessurier's team had defined the diagonal wind braces as trusses instead of columns so that no safety factor applied. The result was a smaller number of joints, which increased the structural deficiency. We leave this out because the building would also have been structurally deficient without this mistake; although the probability of failure would probably have been lower than once every 16 years, it would still have been unacceptably high.
7 The building was designed with an electric damper that, if functioning, would reduce the probability of failure to once every 55 years. That damper might however fail due to a power failure during a heavy storm.
8 Applying these conditions to the collective raises the important philosophical question whether the collective can act and be held responsible. We here side-step this problem.
9 The law also implies sometimes forward-looking responsibility (e.g., the responsibility of parents for their children), but here we focus on liability, which seems to occur after the fact.
10 Council Directive 85/374/EEC.
11 Countries are free not to incorporate this exception in their national law, but most EU countries have followed the directive in this.
12 Attributed to Edward Thurlow, 1st Baron Thurlow (December 9, 1731–September 12, 1806).
13 This section is based on Bovens (1998).
14 Bovens gives four conditions under which the collective responsibility model can be usefully applied (1998, p. 103):
    - The collective must be characterized by a high degree of de facto solidarity.
    - Efficient, professional supervision from outside is not feasible.
    - Those who are held responsible for the conduct of other members or the collective should be aware beforehand that such a model of responsibility will be employed.

- Those who are held responsible should have the chance to exercise a certain degree of influence on the eventual outcome.

15 Here it has been presumed that the model can only be introduced if the opportunities are created for individuals within the organization to operate in responsible ways.

16 This section is based on and partly draws from Van de Poel (2007).

17 Text is based on and partly drawn from Van Gorp and Van de Poel (2001).

10

Sustainability, Ethics, and Technology

Michiel Brumsen

Having read this chapter and completed its associated questions, readers should be able to:

- Distinguish between anthropocentrism and biocentrism;
- Describe what environmental problems are;
- Reflect on the notion of "sustainable development" and how it can be justified and operationalized;
- Reflect on the achievability of a sustainable society;
- Integrate considerations of sustainability in the design process.

Contents

Ethics, Technology, and Engineering: An Introduction, First Edition.
Ibo van de Poel and Lambèr Royakkers.

10.1 Introduction

Case Biofuels

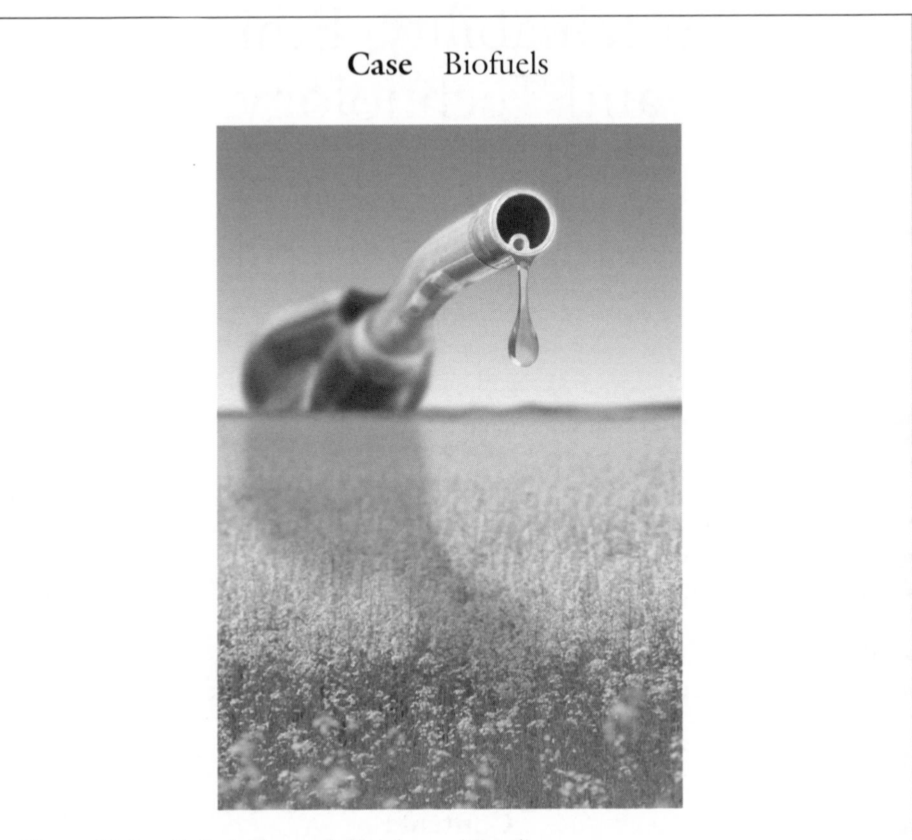

Figure 10.1 Biofuel. Photo: © Hajohoos / Fotolia.com.

How can we continue to provide the energy for our transport needs? According to a theory known as "Peak Oil," the production of oil will decline from now on. At the meantime, our transportation needs seem to continue growing. This scenario implies that we cannot continue to rely on oil for transportation. An additional reason for aiming at a reduced or no reliance on oil for transportation is the substantial contribution that the use of oil makes to the greenhouse effect through car emissions.

One proposed approach to this problem is to develop and use biofuels. Biofuel is solid, liquid, or gaseous fuel obtained from relatively recently lifeless or living biological material. The main difference with fossil fuels is that the latter are derived from long dead biological material. The use of biofuel appears to have two obvious advantages. First, we will not run out of fuels since we can always grow more. Second, the plants that are used to make biofuels extract the greenhouse gas CO_2 from the atmosphere, so reducing the greenhouse effect. The proverbial two birds with one stone.

But as with all things, if something sounds too good to be true ... it usually is. To start with its contribution to global warming. The reasoning above ignores the fact that farming costs energy because of the machinery used, and because of the fertilizers that need to be produced. Turning crops into biofuels also consumes energy. Given that this is energy spent "outside" the cycle of letting the plants grow and using the end products as fuel, it is obvious that the contribution to global warming is not as minimal as it would first seem. Perhaps more important are the unwanted side effects of using crops for producing fuel. Higher demand generally means higher prices. Crops such as sugarcane, wheat, and corn are likely to become more expensive which could well mean that those that are already struggling to feed themselves will no longer manage.[1]

Sometimes it is claimed that second generation biofuels solve these problems. Second generation biofuels are based on non-food crops. However, producing such crops still requires land, water, and fertilizer, which may become short in supply or the prices of which will drastically increase. Therefore, second generation biofuels are likely to contribute to an indirect increase in food prices. Moreover, the drive to produce large amounts of such crops is likely to reinforce already existing negative trends such as deforestation (to obtain arable land) and reliance on monocultures in agriculture.

Third generation biofuels – producing fuels by means of bacteria or algae – are being researched and developed in answer to these concerns. However, while they fare substantially better on the above points than first- or second-generation biofuels, they are currently prohibitively expensive to produce.

Source: Based on Naylor et al. (2007); Inderwildi and King (2009); Zah et al. (2007).

A number of questions emerge from the above. Do we have an obligation to prevent future environmental problems, even if they mostly affect future generations? Isn't it better to leave every generation to solve the problems that are relevant to their lifetime? If we do have such an obligation, then why is that? Moreover, the discussion about biofuels clearly shows that there is a trade-off between our obligations to future generations and the present one: to what extent would we be justified in endangering the food provision to the poorest people alive now, in our efforts to minimize the impact to future generations of the fuel used for transportation? These are some of the questions that we shall address in this chapter.

Climate change, deforestation, air pollution, soil contamination, the availability of clean water, and overfishing are just a few examples of environmental problems that have increasingly drawn the attention and concern of both governments and individuals over the past few years. In many cases, technology is part of the cause of these problems but often it is also potentially a part of the solution. In philosophical ethics, these problems have led to attention for what is called "environmental ethics." We will, therefore, begin this chapter with a short discussion of what we are to understand by "environmental ethics" (Section 10.2). We shall then discuss the kinds of environmental problems we can distinguish (Section 10.3). After that, we shall consider the term sustainability, how we can justify sustainable development from a moral standpoint, and finally which kinds of sustainability can be defended (Section 10.4). Striving for sustainable development is not taken as a non-negotiable viewpoint. The questions discussed here summarize a response to the slogan "Sustainability, we should all be doing it!" The question is: should we, and under which conditions, and what exactly does it entail? Following that, we shall consider whether a sustainable society is at all possible (Section 10.5). In Section 10.6 we shall look specifically at the role engineers can play in this.

10.2 Environmental Ethics?

Ethics is first and foremost involved with interaction between people. So even though you may have come across the notion that people have a certain responsibility regarding the environment, is this in fact morally justifiable?

In the first instance two different answers can be given to this. First, you could say that responsibility for the environment is derived from responsibility for humankind and society. For if we pollute the environment at will and exhaust our resources – burn up all our fossil fuels now – there may be severe consequences for people now and in the future. In environmental ethics this kind of justification is known as **anthropocentrism** (Achterberg, 1994); it is a position that states that the environment only has an instrumental value, that is, the value of its use by us (Baxter, 1974). Note that within this position it is possible to argue that for reasons of self-interest we must take far-reaching measures in the long run to protect the environment. A healthy environment is necessary to provide the essential basic needs, such as food, clean air, and clean water. The second answer argues that the environment has a value of its own (an intrinsic value) and, therefore, should be considered in moral arguments. This is known as **biocentrism**. So even if the environment (or part of it) does not have any use value, it still has moral value. The notion of intrinsic value stems from environmental ethics and can be found in many policy documents on protection of the environment. This does justice to the moral intuition that being worthy of protection does not depend on use value. We find this perfectly plausible for works of art, so why not for the environment?

Anthropocentrism The philosophical view that the environment has only instrumental value, that is, only value for humans and not in itself.

Biocentrism The viewpoint that the environment has intrinsic value (value of its own).

If the environment has moral value, be it instrumental or intrinsic, it deserves to be protected. From this it follows that people in general are responsible for the environment. Moreover, engineers have a special responsibility, because technology has both positive and negative effects on the environment and because they have the power of expertise as argued in the first chapter. Engineers recognize this responsibility, as we saw in the professional codes in Chapter 2. This also has to do with the fact that there is much consensus in society concerning our environmental problems and that environmental care is of importance, and moreover that technology can play an important role in this.

10.3 Environmental Problems

There are many kinds of environmental problems, and solving them first requires their precise definition. In general three kinds of environmental problems are distinguished:

1 We speak of **pollution** if something is added to the environment. Examples of this are the various ways air, water, and soil absorb "foreign" matter making them less suitable for supporting life.
2 If something is taken away from the environment, we speak of **exhaustion**. An example is the use of **non-renewable resources** like fossil fuels, ore, and materials like tropical wood too. We call a resource non-renewable if it can only be consumed without the possibility of producing more of it now or in the future by for example growing plants. **Renewable resources** are resources of which more can be produced. So while fossil fuel is a non-renewable source, biofuel is a renewable source. A resource can also be renewable but easily depleted, as is the case with over-fishing. Once too much fish of a certain kind has been caught it may take years, or even become impossible, to restore the fish population.
3 We refer to **degradation** if there are structural changes to the environment. Examples of this are soil erosion, a decline in biodiversity, the buildup of greenhouse gases, and the gap in the ozone layer. Note that degradation often occurs in conjunction with adding to or taking away from the environment, so the categories are not mutually exclusive.

Pollution Environmental problems in which something undesirable or damaging is added to the environment.

Exhaustion A type of environmental problem in which something valuable is removed from the environment that cannot, or at least not easily, be renewed.

Non-renewable resources Natural resources that cannot be renewed or reproduced. An example is fossil fuel.

Renewable resources Natural resources that can be renewed or reproduced.

Degradation Structural damage to the environment. An example is soil erosion.

The consensus mentioned earlier is hard to find when we are confronted with questions like: what is the seriousness of the present situation and how do we expect it to develop if we do not take any action? How should we weigh environmental values against other ones, such as economic values? And what should we protect or maintain: a healthy and pleasant environment (what does that entail?), or should we strive for unspoiled nature?

We do know that the continuity of human welfare and well-being is endangered by exhausting resources, and that polluting the water, air, and soil means we have to make increasing efforts to keep the environment livable. As humans we depend on the environment. Environmental problems also have an important social dimension, since some problems hit those lower on the social ladder harder or first. For example, housing near to sources of pollution is cheaper, such as housing near to highways, municipal dumps, major industrial installations, etc. In an international context, economically

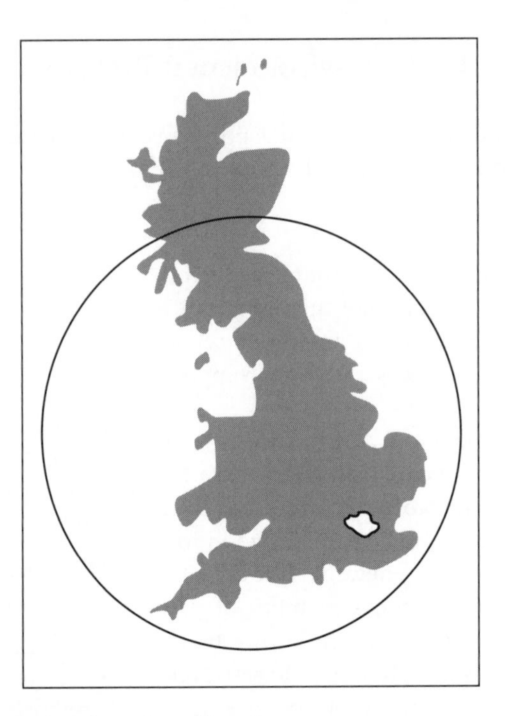

Figure 10.2 The ecological footprint of London.

weaker countries are willing to allow investments or encourage industry at the expense of strict environmental legislation. One example is the demolition of large ships including oil tankers. At high tide these tankers go right onto the beaches (in Bangladesh among other places) at full speed. Following that they are demolished in a primitive way without any regard for the environment. One could say that environmental problems hit those lower on the social ladder harder, but also that the existence of inequalities (and their persistence) supports environmental problems. In this context, the term sustainable development has been mentioned frequently in recent years.

The heart of sustainable development lies in the notion that the development of welfare as we experience it through time (not only in the West but in all countries) must be sustainable. The fact that this is not the case at the moment can be clarified using the term **ecological footprint** (see Figure 10.2). In essence you can calculate the total environmental load of a person's lifestyle and express it as an amount of space required to support this lifestyle.[2] The circle indicates the ecological footprint of London.[3]

Ecological footprint A measure for the total environmental impact of a person's lifestyle expressed in an amount of space required to support this lifestyle.

One important finding is that the size of the ecological footprint greatly differs per continent. The *Living Planet Report* of the World Wildlife Fund indicates that the average ecological footprint in the United States is 9.6 ha per person, 5.0 ha in Western Europe, and 1.4 ha in Asia and Africa. For the present world population, about 1.9 ha is available per person in the shape of productive surface (that is, land surface that offers real resources) (World Wildlife Fund, 2002). According to that same report, the total footprint of the world population is approximately 20 percent too large. In other words, the regenerative ability of the various resources cannot keep up with demand. The obvious conclusion is that we cannot continue along these lines.

10.4 Sustainable Development

10.4.1 The Brundtland definition

The best-known definition of **sustainable development** originates from the Brundtland report (see box).

Sustainable development
Development that meets the needs of the present without compromising the ability of future generations to meet their own needs (Brundtland definition).

Sustainable Development

Sustainable development is development that meets the needs of the present without compromising the ability of future generations to meet their own needs. It contains within it two key concepts:

- the concept of "needs," in particular the essential needs of the world's poor, to which over-riding priority should be given; and
- the idea of limitations imposed by the state of technology and social organization on the environment's ability to meet present and future needs.

(World Commission on Environment and Development, 1987)

Sustainability is not only related to the natural living environment. And in this context, three factors are mentioned that are of importance for sustainable development: next to ecological factors there are social and economic ones.[4] For sustainability this means that certain considerations have to be taken into account: ecological values cannot unrestrictedly be given precedence over social justice or economic achievability. For that matter, we must monitor whether these three aspects do not always clash. It is possible that a measure or design choice can be positive for more than one field at the same time. Take for example a design in which you manage to reduce the amount of material required. This is favorable both from an economic and environmental point of view.

10.4.2 Moral justification

The heart of sustainable development lies in two kinds of justice. The first kind of justice relates to the division of resources between our own generation and future generations: **intergenerational justice**. The question, "can we continue to use fossil fuels simply until we run out and let the next generation find alternatives?" falls under this heading. Next to that, sustainable development requires a just division of resources within our own generation (compare the First and the Third World): **intragenerational justice**. The question, "can we use agricultural resources for producing biofuels even if that makes food more expensive?" falls under this heading. These two types of justice can be justified and described in various ways. We shall discuss three of the theoretical backgrounds. It will make little difference whether we refer to intergenerational or intragenerational justice; the only difference is that the first group is separated by time while the latter is separated by place. Morally speaking these distinctions are negligible. The foundation selected will however have an impact on how sustainability is practiced.

Intergenerational justice Justice that relates to the just distribution of resources between different generations.

Intragenerational justice Justice that relates to the just distribution of resources within a generation.

Property rights

The first possible foundation for sustainable development stems from the historical principle attached to the justification of **property rights**.[5] The traditional question asked about property is whether we can justify that some matters belong to individuals. If we apply this historical principle to the environment and sustainable development, we get the following statement: the environment belongs to all of humanity, but what we mix with labor belongs to us. However, this only works as long as there is enough of the same quality left over for others. This is exactly the point that links up the formulation "without compromising the ability of future generations to meet their own needs" from the Brundtland definition. Only to the extent that we leave enough for future generations can we consider the environment and its resources as the property of this generation, and may we use it as we see fit. Essentially, this is an extension of the **polluter pays principle**[6] or the notion that "the one who breaks something is also expected to mend it." The point of departure is that damage to the environment must be repaired by the party responsible for the damage.

Property right The right to ownership of a specific matter or resource like money, land, or an environmental resource (like clean air).

Polluter pays principle The principle that damage to the environment must be repaired by the party responsible for the damage.

Utilitarianism

The utilitarian approach defends intergenerational justice in the following way. Say that we develop in a way that does not allow us to continue our lifestyle. It would mean that some time in the future the total utility for all would diminish. In turn, the aim of utilitarianism – the greatest happiness for the greatest number – would not be achieved. Development along non-sustainable lines therefore is undesirable. Two points can be mentioned in connection with the argument. First, the total utility must be maximized over an extremely extended period: maximization of utility in the short term will not lead to sustainability. Second, one could dispute this utilitarian reasoning by arguing that the expected aggregated total utility across the entire period would be greatest if we use a lot now and leave the future generations to their own resourcefulness. The veracity of this factual statement is extremely hard to assess. Apart from this, this criticism shows that utilitarianism is insensitive to questions of distribution, both between different groups of people or across time. Morally, this is highly unsatisfactory.

Duty ethics

The duty ethical approach defends intergenerational justice on the basis of the Golden Rule: "Treat other generations as you would have them treat you." (The Golden Rule is somewhat comparable to the universality principle of Kant, see Section 3.8.1). Obviously, we do not need to reason for long to come to the conclusion that we would not wish other generations to have lived a life of abandon leaving us with no resources. Now it might be thought that the Golden Rule cannot be used to defend intergenerational justice, because while we can do plenty of things to make future generations less well off, they cannot make us less well off – so why make an effort? In other words, there is a causal asymmetry which raises the question whether the Golden Rule can be applied to intergenerational justice. However, this is only an apparent problem. The Golden Rule is not about giving others reasons to treat you well, but rather about putting yourself in the other's shoes in order to reflect on how we ought to treat them. The causal asymmetry does not pose any problem to that exercise of the imagination. Therefore, given that we want other generations to treat us well (regardless of whether we think about our actual selves or imagine ourselves to be part of a future generation), we have to treat future generations well.

One can also reason along the lines of the second formulation of the categorical imperative, the reciprocity principle. Future generations should be treated the same as all groups of people; they are not only a means but also an end in themselves. If we were to live a life of abandon so that future generations cannot provide in their own needs, then we have used them as a means to achieve our own ends. Without their permission we deny them the opportunity to strive for their own ends in the way that we did. However, if we strive to maintain the ability to fulfill needs in the future too then we do justice to the fact that future generations will have their own ends that they wish to strive for in a rational manner.

Often two arguments are mentioned against sustainable development. These can be set aside using deontological ethics theories. The first counter argument is whether previous generations made an effort for us too. Apart from a high standard of living, we have also inherited substantial environmental problems. Given that we must use our ingenuity to cope with this, why would it be unjust to desire the same thing from future generations? The Golden Rule however states not that it is relevant how others factually treat us, but rather how we would prefer them to treat us. The fact someone steals my bicycle does not justify my decision to steal somebody else's bicycle.

The second counter argument is that the largest environmental problems are caused by population and consumptive growth in developing countries like China. This can lead to the question why we should make efforts to solve a problem that has its origins elsewhere. However, we need to realize that as far as consumptive growth is concerned these countries still have some catching up to do. In other words, it would be odd to say that the problem does not lie with us. If our own welfare could only be maintained by denying that welfare to others, then we can hardly call the maintenance of such welfare just. We would be using the others as a means and no longer view them as an end.

For that matter, we need to realize that the impact on the environment is largely determined by the combination of economic growth and population growth. Since population growth is influenced by material welfare and schooling, greater intragenerational justice may result in a reduced environmental impact in those countries in the long run, even if in the short run a more equitable distribution of wealth would lead to higher environmental pressure. However, the tension between intergenerational and intragenerational justice, which also surfaced in the discussion about biofuels, remains a serious and difficult issue that cannot be completely solved by means of the above justifications.

10.4.3 Operationalization

Many people feel the Brundtland definition is vague. There are all sorts of concrete ways to fill it in. There has been an explosive growth of such ways and as a result many believe that sustainability has turned into a rather hollow phrase. However, we should not give in to this pessimism. Even though the level of abstraction of the Brundtland definition is high, it does not mean that it cannot be made more concrete in a sensible fashion. We should take the definition for what it is: a foundation for further discussion about how we should take responsibility in time and space. In short, the Brundtland definition requires operationalization: it should be detailed into a number of concrete policy measures if we are talking about environmental policy or into concrete design guidelines when we are talking about sustainable design of technology.

It is important to see that operationalization of sustainability requires normative choices; choices with respect to what the relevant aspects are, and how they should be weighed in relation to each other. Working out what sustainable development should entail means making ethical decisions. And we should not be too hasty to conclude that all further definitions are equally good. There should be ethical discussion about which definition is best. If you denied the necessity of such a discussion, it would lead to normative relativism and all its inherent problems (see Section 3.5.1).

Points for discussion

What are the most important questions that the Brundtland definition raises? Or which points of discussion have been placed on the agenda?

- *Needs*: which needs are we talking about here? Are none of the present needs disputable? Is any development that places limitations on people by definition non-sustainable? We are all becoming increasingly mobile, for example, we have at least one car, we go on holidays several times per year (using a plane), we desire spacious, comfortable, and properly heated living space, we want the latest model of mobile phone, and want to eat vegetables that come from abroad the whole year round. Should all this simply be allowed? If not, which needs are legitimate ones? Are they only the basic ones like food and shelter? And if we feel they go further than basic needs, how do we justify them?

- *Present needs*: whose are they? The less quoted second part of the definition clearly shows that the Brundtland committee wished to have the basic needs of the Third World addressed first. As a result, we can deduce that the needs in the more prosperous parts of the world can only be fulfilled if the environment allows space for them.
- *Needs of future generations*: how do we know what these are? Superficially this point could lead to a statement like, "Of course we don't know what the needs will be in 50 or a 100 years, because 50 to a 100 years ago people did not know what we feel to be obvious now!" This interpretation does not suffice due to the above point of the legitimacy of needs: "obvious" needs are not necessarily legitimate. Nevertheless, this is an important point. Economic and technological developments may turn needs that are legitimate now into illegitimate ones and vice versa.
- "*Without compromising the ability of future generations to meet their own needs*": Should we interpret this in such a way that we may not reduce the options of future generations at all, or should we interpret as giving us the room to leave future generations with some ways of fulfilling their needs? If we take the second interpretation, based on the notion that technology will continue to develop, then to what extent should we assume that future technology will be able to meet needs? As the future level of technology is uncertain, we are running a risk that may be felt to be irresponsible. Moreover, there is another catch: technology does not develop on its own – it is something the present generation has to work on. That is why any argument that states we do not need to do much about sustainability because future technology will be capable of providing in future needs more efficiently is suspicious to say the least.

The first interpretation – in which we may not reduce the options for future generations in any way – is also known as the **stand still principle**. Essentially it states we must not pass on a poorer environment to the next generation than the one we received from the previous generation. The idea is also expressed in the notion that we must not rely on environmental loans, that is, do not create problems that we trust future generations will solve.[7]

Stand still principle The principle that we must not pass on a poorer environment to the next generation than the one we received from the previous generation.

There is at least one main disadvantage to the above definition: what do we mean by poorer? Does it mean that everything has to stay the same, or could the worsening of one environmental aspect be compensated by the improvement of another? The problem here is that the comparison soon becomes very hard. Can the storage of radioactive waste with a very long half-life be compensated by lower CO_2 production as a result of using electricity produced by nuclear power plants? Again it would be wrong to think that all answers to this question are equally good. Some answers are more ethical than others, that is, they lead to a more defendable justification of sustainable development.

Environmental space

One of the terms that is implicitly or explicitly used quite often in an attempt to achieve operationalization is the term **environmental space**. The sixth environmental action program of the European Union states that one of the objectives is to reach a situation in which the use of renewable and non-renewable resources does not exceed the boundaries of what the environment can take. The suggestion of this term is that we have a scientific and objective way to determine the size of the environmental space we can use. Once we have determined this, it is clear what we must do: we must ensure that our environmental impact does not exceed our environmental space. This idea is narrowly connected to the notion of the ecological footprint. Expressing someone's environmental impact using a surface is only possible if we know exactly what the environmental impact is that a given surface can take. Thus, we must know what the **carrying capacity** of the environment is: what damage can we do without the damage being irreversible?

Environmental space The (maximum) amount of use of renewable and non-renewable resources that does not exceed the boundaries of what the environment can take.

Carrying capacity The amount of damage that can be done to the environment without that damage being irreversible.

The assumption that we are able to calculate this carrying capacity and, therefore, make use of the concept of environmental space in the operationalization of sustainability is very problematic. First, there are theoretical problems related to knowledge. Not only is there much for us to learn, but there are things that we will never be able to know. The fact that irreversible damage has been done often becomes clear when it is too late. In smaller, replaceable systems the environmental effect can be tested, but with something as large as the earth this is simply not possible. One can make predictions, but these have to be based on assumptions about how stable a given system is. Is the system quite stable and does it correct itself to a large extent, or is it more the case that above a certain level of disruption increasingly uncontrollable damage occurs? Thus we are forced to make assumptions and the size of the risk of irreversible environmental damage depends on those assumptions. The degree to which these risks are acceptable in turn depends on the desirability or acceptability of the social consequences that ensue from avoiding those risks. How much more are we willing to invest in the environmentally friendly gathering of fossil fuels, and how much in flue gas purification? Such measures could prevent our energy consumption from outstripping the power the environment has to withstand its use. As the costs of generating energy will rise (for the measures do of course involve costs) it is possible that the economically challenged part of the populace will have to spend a lot on their energy bills.

If we base measures aimed at sustainable development on the concept of the environmental space, this could lead to us implicitly leaving the answering of these normative questions to scientists, even though they have no special expertise in that field. The concept of the environmental space may be usable, but we must make sure that normative issues get the attention that they deserve.

The precautionary principle

The precautionary principle, which we introduced in Section 8.7.1, is often mentioned when we consider which risks of irreversible environmental damage are acceptable. As we have seen there in the Rio Declaration the precautionary principle was defined as follows: "Where there are threats of serious or irreversible damage, lack of full scientific certainty shall not be used as a reason for postponing cost-effective measures to prevent environmental degradation." The important connection between sustainable development and the precautionary principle lies in the notion that we should not leave environmental loans to the coming generations. In other words, we should not give future generations problems that we allow to continue because we cannot agree about the question whether there are serious environmental effects. In this way, we are not guilty of falling into the trap of a wait-and-see policy. The main problem of the precautionary principle is that it seems to forbid too much. Opponents point out that a number of important technical innovations that we now consider to be desirable would not have been implemented if we adhered to the precautionary principle. The high speed of train traffic was thought to have all sorts of negative effects, and if the suspicions had been taken seriously at the time then train traffic could never have become so important (see Schivelbusch, 1986). According to them, the precautionary principle places absurdly high demands – in fact they are nonsensical demands since you can never prove that something will not cause damage. It seems more justifiable, though, to demand that we have reasonable grounds rather than incontrovertible proof that there will be no damage. But what is reasonable? Another problem lies in the cost effectiveness. If we are ignorant of which damage will occur and especially what the chances are of such damage occurring, it is particularly difficult to say something about when a measure will be cost effective.

10.5 Can a Sustainable Society be Realized?

It is often said that a sustainable society simply cannot be realized. "Sustainable development is not yet achievable, because it is not cost effective economically speaking," or "consumers just don't want to pay more for sustainable products." We should be careful, however, with the idea that sustainable development is unachievable or impossible. At the very least we should be precise about what we mean here.

If we say that something is possible or impossible, we always have certain boundary conditions in mind. Without boundary conditions it is meaningless to speak about possibilities. And these boundary conditions can be debated too; "I couldn't finish the assignment because I had promised my friends to meet them at the pub" is a less acceptable excuse compared to, "I couldn't finish the assignment because I had to take my mother to hospital." The discussion does not stop at the expected lack of achievability or impossibility, because we must first consider how acceptable the excuse is.

When we try to motivate why a sustainable society is not achievable, we often are confronted with economic boundary conditions. "It is impossible to fulfill our transportation needs by means of third-generation biofuels, since they are too expensive." However, first, these are not set in concrete either. The costs of environmental damage have not yet been sufficiently internalized in the prices of products. In other words, the price of a product does not take all the environmental costs on board. An example on an inconsistent calculation of environmental impact in product prices is the environment tax we have to pay on petrol. The environmental tax does not apply to kerosene, used as fuel for planes. This results in the fact that flying to many places in Europe is cheaper than taking the train, although the train seems to be a more sustainable type of transport than the plane. Second, economic boundary conditions are not laws of nature. You cannot escape laws of nature, but you certainly can choose to do things that are uneconomical. What yield is to be expected from care for the elderly? Even so, we still do it because we feel that we must.

Some sustainable developments are socially unachievable. Not many people would be willing to reduce their present mobility to reduce harm to the environment – which would be the morally desirable thing to do. We want to keep going on holiday to other parts of the world and we want to go to work with our own car. Our mobility has become so important that we do not want to sacrifice it as a contribution to a more sustainable society. However, this too can change, although it probably will not happen by simply coming up with some technology that offers an alternative. Sometimes there are boundary conditions with strong arguments in their favor that make the sustainable option impossible. But it is a good thing to make this explicit, rather than terminating all discussions by saying "this is impossible!" without supplying reasons.

What is related to economic and social achievability is the question whether we will ever reach a sustainable society, given the relatively short time horizon in a democratic system. Politicians are unlikely to take the drastic measures that are needed for sustainability. On the one hand, politicians wish to be re-elected and, on the other, the government's responsibility is not solely focused on realizing a sustainable society to benefit the environment. Politicians are also partly responsible for creating work, which requires an economically attractive climate. Is there an alternative for the democratic system that is more likely to be able to realize sustainable development? We could try to allow policy-making to become more expert-based. However, as we saw earlier in our discussion of the term environmental space, there are no experts – either of technology or environmental care – that can give us the complete story about how to achieve sustainability. Even if their factual knowledge is great, their expertise and authority is no greater than that of an ordinary person concerning normative considerations (see also Section 1.5.2 on technocracy). Would a green dictatorship be an option then? Even if a dictator had the best intentions regarding the realization of sustainability, we have often seen that dictatorships fail to achieve many of their original aims. Thus there is little to be expected from that solution. Moreover, one may well wonder whether enforced behavioral change may not lead to reduced participation by civilians in sustainability, which would have a negative impact in the long term. Finally, the limitation of civil rights and freedoms that this would necessitate is ethically dubious.

Some believe that sustainable technologies will arise because of the demand. So the market will bring a solution, as it has in many other instances. This means that if it is not economically feasible we should not strive for it. An example can be taken from the use of fossil fuels: the reasoning is that when they become scarce and thus more expensive, other sources of energy will automatically acquire a more competitive position.

Apart from what was said about the solidness of economic laws, what we have left untouched is that we are often dealing with situations in which it would be to everybody's advantage to act more sustainably, but that the first person to act that way is placed in a serious competitive disadvantage.[8] By waiting for the market to solve the problem, it is possible that a worse situation will arise for everyone. As a solution to this kind of situation you can imagine regulations for particular branches, such as standards issued by government or by branch organizations. In opposition to this, setting standards may lead to conservatism, which will not result in much innovation. However, it is questionable whether that is the case. The demands made on technology from the viewpoint of the environment will continue to be upheld. So from a costs perspective it is wise to stay one step ahead with the technology you are designing. Making small adjustments each time to meet legal environmental requirements can prove more expensive than an initially major investment in a sustainable design.[9]

10.6 Engineers and Sustainability

Although there is only limited agreement on what a sustainable society is, and on how or whether it will be achieved, engineers can certainly make a contribution. Technical knowledge can be used in different ways to directly or indirectly solve or prevent environmental problems. This means engineers are responsible for making use of those opportunities.

Various technical measures can be taken to satisfy this responsibility for the environment. These measures can be categorized according to level and type.

The levels are as follows:

a. Product level;
b. Process level;
c. Business level.

The types are as follows:

1 Cleaning up pollution;
2 Processing of waste flows (end of pipe);
3 Preventing waste flows.

An example of a b2 measure (processing waste flows at the process level) is the purification of water before it leaves a factory. A catalyst in a car is an example of an a2 measure. Changing to low-energy/cleaner production processes is an example of a b3 (or c3) measure. Now the interest of these two lists is not so much in making fine-grained and exhaustive distinctions, but rather in aiding reflection about increasing the sustainability of technology. It is generally agreed that the list of types is one of increasing effectiveness. In other words, a type 3 measure is often the most effective, followed by a type 2 measure and then a type 1 measure. So, if a measure already taken or conceived is type 1 or 2, it is worth thinking whether a type 3 measure could be conceived of instead. And if we already have an a3 measure, we might try coming up with b3 and/or c3 measures.

In addition, it is interesting to note that a3 measures can be of a varied nature: examples are making cars more energy efficient, making complex devices more modular so that full replacement is not needed if a part becomes defective, the efforts of designers that result in consumers no longer throwing away machines before they become unusable. Much can be achieved by this type of effort, and it is in stark contrast with, for example, the present practice in the field of cell phones. The continuing design of new models in order to convince consumers that they need to dump their old model (even if it is still perfectly functional) is widespread in all sorts of sectors where producers wish to maintain their production despite a level of market saturation.

10.6.1 Points of attention during the design process

As we saw in Chapter 6, a number of moral questions can be asked during the design process and this also applies to sustainability. These questions can concern the environmental impact of a product during its **life phases** (i.e., the production phase, use phase, and removal phase). Each phase raises its own environmental questions. The choices for the environmental impact of one phase can influence the environmental impact of another phase. As an example of how considerations of sustainability work in a practical design context, we shall discuss the design of a heat pump boiler.

Life phases The phases through which a product goes during its "life": production phase, use phase, and removal phase.

Case Heat Pump Boiler

A heat pump boiler is a piece of equipment that provides the household requirements for hot water by means of a thermodynamic cycle process in which heat is extracted from a source to heat the water. It is a kind of reverse fridge.

One important characteristic of such a device is that heat can be pumped in a direction where it normally would not go. The heat pump boiler can extract heat from ground water or from ventilation air and because this normally unused low-quality energy is utilized, the yield (defined as [high-quality energy out/ high quality energy in] * 100%) even rises above 100%. So this would seem to be a truly sustainable device.

During design a number of choices have to be made that have an impact on the sustainability of the device:

- Life span: what life span does the device require?
- Recyclability: this is a term that can be defined in many ways. In principle nearly everything is recyclable. The question is how much energy is

involved in the recycling. Reuse is another option: a heat pump boiler could be designed in a modular fashion, so that components with shorter life spans can be replaced allowing the parts that still work to be used much longer.

- Energy source: should the heat pump boiler work on natural gas or electricity?
- Just like a refrigerator, a heat pump boiler contains a primary refrigerant – this is a substance that has thermodynamic properties suitable for a cycle process. Primary refrigerants often have an impact on the greenhouse effect and the ozone hole, but they differ greatly in the degree to which they display these effects (see Section 6.3).
- The source medium: this is where the heat has to come from.

Note that placing these design choices up front does not even begin to touch on the question whether the heat pump boiler technology as such is sustainable. It is quite possible that some other technology offers a better combination of functions or that applying heat pump boilers leads to more environmental problems than the older technology. Sometimes the greatest sustainability benefits can be gained by scrapping an entire concept instead of holding onto all sorts of small design choices within a given concept. This plays a role in the discussion about the hydrogen cell car. This kind of car may be more sustainable than the present generation of petrol, diesel, and gas cars, but the question of whether sustainable transport should consist of car-like solutions is not answered this way. Sustainable technology development requires that the system boundaries that the design choices are framed in are not ignored unless there is reason to do so. Sustainability is especially suited for radical design choices (see Section 6.4).

How responsible choices can be made in the development of new technology and design choices within tried and tested concepts is explained below. Some examples will be given with the instruments used for making the above choices.

10.6.2 Life cycle analysis

The point of departure for **life cycle analysis** is that you must be able to compare the environmental impact you cause with your design with alternative designs in order to achieve a sustainable design. This can be done by mapping the environmental impact across the entire cycle of *extraction*, *refining*, *production*, *use*, and *disposal*.[10] What is important is the integral nature of these comparisons: fair comparisons of products, or better yet, of ways in which a series of functions can be carried out – this can only be achieved by looking at the whole life cycle. The viewpoint is reflected in a number of instruments and design approaches – some are more quantitative and other more qualitative.

Life cycle analysis An analysis that maps the environmental impact of a product across the entire cycle of production, use, and disposal.

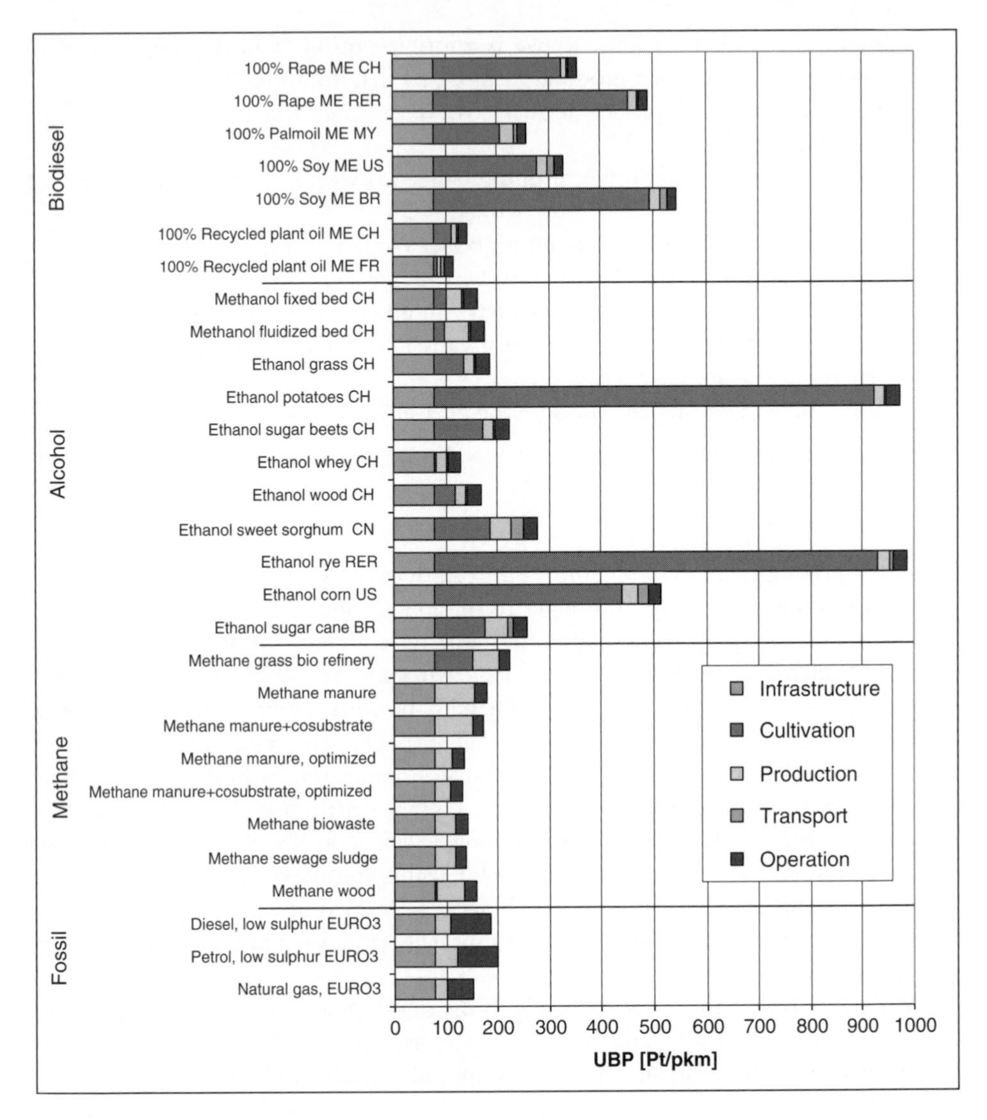

Figure 10.3 Comparison of aggregated environmental impact (method of ecological scarcity, UBP 06) of bio fuels in comparison with fossil fuels (petrol, diesel and natural gas). Reproduced by permission of EMPA.

Quantitative

There are various software packages, like SimaPro and EcoScan, suitable for a quantitative life cycle analysis. These software packages make use of a database in which various environmental consequences of production (obtaining raw materials, processing methods, and transport), use, and scrapping (reuse, recycling, and waste) are included. As a final result, all of the above is summarized in a score that expresses the total environmental impact of the product or design in question. An example of the result of a life cycle analysis is shown in Figure 10.3. To obtain better designs, we are given insight into which phases there is an impact and what is having an impact in the first place. Often conclusions can be drawn from this about the phase in which most can be done to improve environmental impact.

This type of software can be a major benefit to sustainable design, but it also has a number of limitations. One practical limitation is that making a quantitative analysis is time consuming. Thus, the environmental impact of only a restricted number of alternatives can be measured in practice, making the choice of alternatives extremely important. This is especially true when various functions can be grouped in one device and the alternatives differ in how the grouping takes place, which makes further deliberation very labor intensive. For example, in households there is a need for central heating and warm tap water. Both functions can be fulfilled by a high-efficiency boiler. The heat pump boiler cannot perform both jobs efficiently unless a low-temperature central heating unit is installed. So, the seemingly more sustainable heat pump boiler requires a division of functions that are often combined in most households, making it a less favorable option. However, this comparison would require us to make a life cycle analysis on the entire system of heating and supplying hot water and not just the heat pump boiler. In such cases carrying out a more qualitative analysis is the more obvious choice.

A second important limitation is related to the use of one-dimensional eco-indicators used to simplify the output generated by these kinds of packages, such as the Eco-indicator 99,[11] or the newer ReCiPe.[12] These indicators express the environmental impact on a one-dimensional scale (see Figure 10.4). Exhaustion of materials, pollution and degradation are all placed on one heap by means of weighing factors. These weighing factors cannot be altered by the user of the software package. This gives a false sense of factuality and objectivity, while this is not justified. The weighing factors should be based on normative choices related to what should weigh most. The user of the package, the designer, is not given access to those choices, while they are of great importance for the final result.

Let's say that you want to determine when to replace a car on the basis of the environmental impact. If we look at the amount of energy used during the life cycle, we see that it is relatively small in the production phase compared to the energy use in the use phase. You could draw the conclusion that it would be more sustainable to buy a new car more often. The lower energy use of the new car easily compensates for the costs of its production. However, if we look at the output of damaging substances, this is relatively high during production compared to use. This would result in the opposite conclusion: do not replace your car too quickly. So the question is: should raw material exhaustion (use of fossil fuels) weigh heavier than air pollution? This is an important discussion with small variations for many products. However, the use of a quantitative lifecycle analysis with a one-dimensional output will not stimulate this discussion. The results, therefore, seem to be somewhat random, for they are based on the normative points of departure of the eco-indicator used. While there may be merits in concentrating this normative debate in the stage of constructing the eco-indicator, leading to clear-cut lifecycle comparisons between technological choices, we must nonetheless wonder whether the normative debate about weighing factors is best conducted at that level of abstraction, and whether it is a good idea to shield those making concrete lifecycle assessments from this debate,

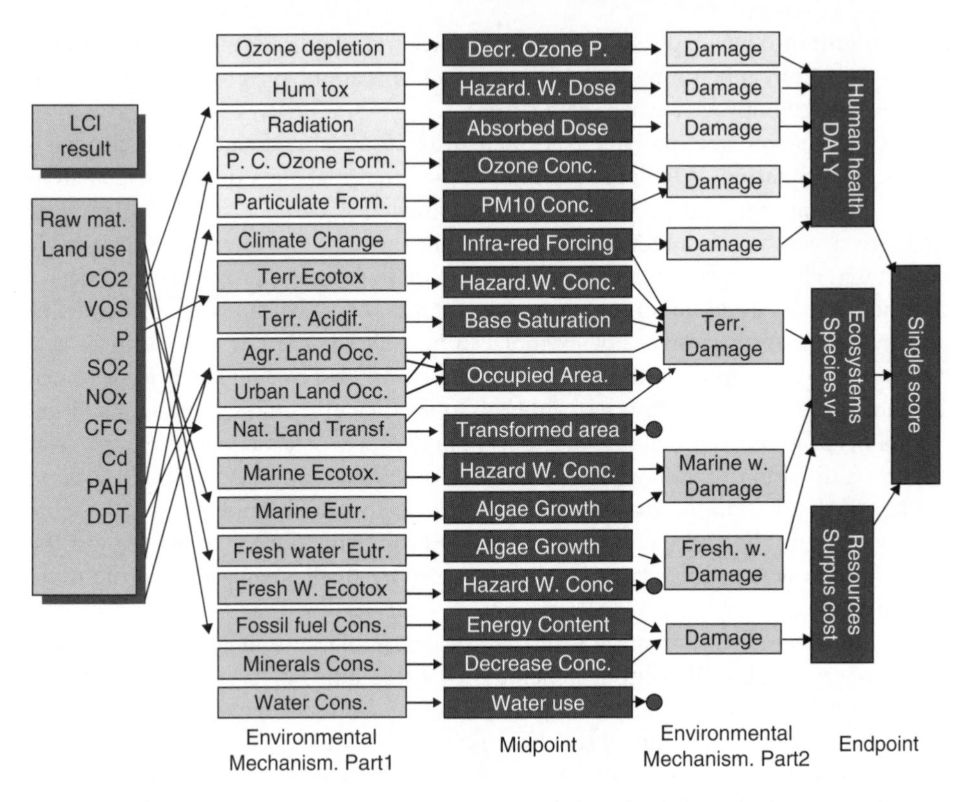

Figure 10.4 The overall structure of ReCiPe methodology for Life Cycle Assessment Impact Assessment. Reproduced by permission of PRé Consultants B.V.

often leaving them puzzled in the face of apparently randomly different results from different eco-indicators.

> **LiDS wheel approach** An abbreviation for Lifecycle Design Strategies. A qualitative oriented life cycle approach.

Qualitative

More qualitatively oriented lifecycle approaches often come to a number of rules of thumb for design. The **LiDS wheel approach** (LiDS is an abbreviation for Lifecycle Design Strategies) focuses on the following eight rules of thumb (Brezet and Van Hemel, 1997):

1 New concept development;
2 Selection of low-impact materials;
3 Reduction of material;
4 Optimization of production techniques;
5 Efficient distribution system;
6 Reduce of the environmental impact in the user stage;
7 Optimization of initial life-time; and
8 Optimization of end-of-life system.

Let us take a brief look what a few of these rules of thumb would mean in practice. Here, we use the example of the heat pump boiler (see box). The first point, new concept development, is in fact the most important point. The clever integration of functions can offer advantages. You could for example envisage air conditioning by means of a heat pump boiler. With one piece of equipment two functions are provided. Consider that the heat pump boiler already divides the functions of providing hot water and central heating, which is highly questionable from the perspective of sustainability. Next to that you can consider using one piece of equipment for more than one household.

Selection of low-impact materials means that non-damaging and less scarce materials are chosen; if possible they have been recycled and can at least be recycled; and they have low energy content. The choice of a certain type of coolant for the heat pump boiler is of importance for issues like the greenhouse effect and the ozone hole.

The environmental impact in the use phase (6) is very important for this kind of machine; it is after all an energy transformer. The yield of the machine depends on a number of things, including the energy source. The energy source can be either gas or electricity. The advantage of gas is that for most households the energy source for warm water remains unchanged. The yields of a gas heat pump are however somewhat lower and the compressor cannot be hermetically closed, so that there will be a loss of coolant having an impact on the environment. Moreover, there is no necessary dependence of fossil energy sources if electricity is used as an energy source; electricity can be generated in various ways. Besides the yield and the loss of coolant, there is another possible source of environmental impact in the use phase: the large scale use of ground water as a heat source may result in it cooling down. People usually think of heating up when the thermal load is mentioned, but cooling could also fall under this heading. The possible consequences have not been mapped yet and for that reason the use of another source of heat may be wiser.

While optimizing the initial lifespan, we can consider a modular build, so that parts can be replaced. But high reliability and other matters that stimulate user friendliness contribute to a piece of equipment not being disposed of too early. For that matter, "optimal" is not the same thing as "as long as possible"; if something has to work for a long time this places additional requirements on the production and materials. This can have consequences for recycling. Moreover, technology may advance so quickly that the difference between the environmental impact of an older machine during use can be many times higher than that of a new machine with the same function, which easily compensates the additional environmental impact of the other two phases. In other words, the machine is best replaced from an environmental point of view.

A qualitative lifecycle approach of a design is a good way to make an inventory of where the biggest improvements can be made regarding sustainability and also what the most important design questions are. A number of sustainability issues occur outside the design framework and therefore are not dealt with in this approach.

It should be noted that a lifecycle analysis focuses exclusively on the environmental impact. The question is whether the other aspects of sustainability – such as intragenerational and intergenerational justice – are sufficiently dealt with. Intergenerational justice is indirectly dealt with as the quality of the environment in the longer run is considered. The intragenerational justice also is indirectly addressed to the extent that environmental damage often is for the account of weaker parties in society, so a lower environmental impact helps that group in particular. However, this is quite something else from "giving priority to the essential needs" of the poorest people on the planet. That is not so strange considering if we remember that we are discussing the introduction of sustainability considerations in the design process. Outside that framework all sorts of steps must be taken to achieve true sustainability. However, lifecycle approaches have hardly any bearing on this.

10.7 Chapter Summary

Two arguments can be given why we should care for the environment. One is anthropocentrism, which states that the environment is only valuable as a means to human well-being. The second argument is that the environment is intrinsically valuable. On both positions, the environment needs to be protected, albeit for different reasons and probably to different extents.

A core notion nowadays in the discussion about the environment is "sustainable development." In 1987, the Brundtland committee gave the following definition: "Sustainable development is development that meets the needs of the present without compromising the ability of future generations to meet their own needs." This definition raises questions about how sustainable development is to be justified and how it is to be operationalized (specified) especially in engineering contexts.

With respect to justification, two basic values lay at the basis of sustainability: intragenerational justice and intergenerational justice. Intragenerational justice refers to the just distribution of resources within a generation, for example between the developed countries like the United States and Europe and developing countries in, for example, Africa. Intergenerational justice refers to the just distribution of resources between generations, for example, between us and our grandchildren. As we saw in the biofuel example at the beginning of the chapter both types of justice may conflict. Biofuels may contribute to intergenerational justice, by decreasing problems like oil shortages and the greenhouse effect for future generations, but at the same time they may very well decrease intragenerational justice, because they will likely result in higher food prices that especially damage the already poor and hungry in the Third World.

With respect to operationalization, the following more specific principles to attain sustainable development have been proposed:

1 The stand still principle, which states that we must not pass on a poorer environment to the next generation than the one we received from the previous generation. A crucial question here is what we mean by "poorer" and whether we can compensate a degradation in one respect by an improvement in another.
2 The notion of environmental space, which is based on a scientific determination of the carrying capacity of the environment that should not be transgressed. We have seen that determining the carrying capacity involves normative choices and cannot be left to scientists and engineers. Otherwise it will result in technocracy (see Section 1.5.2).
3 The precautionary principle, which states that scientific disagreement about environmental effects should not be a reason to postpone measures against

possible irreducible adverse effects, especially if those measures are cost-effective. A possible objection to this principle is that it may well forbid too many developments.

We have seen that technical choices have far-reaching consequences for nature and the environment. They can contribute to the realization of sustainable development, but also to "unsustainability." On this basis, engineers have a special responsibility. This responsibility can be expressed by the development of special environmentally friendly and sustainable techniques. However, engineers can take considerations of the environment and sustainability in standard design processes too. This can be realized by sustainable design, which can be given shape by a lifecycle approach. True sustainable design means discussing the boundaries of the system. It may mean that you have to consider a radical design – to borrow the phrase from Section 6.4.

The responsibility of engineers, however, reaches beyond the design of sustainable products. For a technology to be accepted it is sometimes necessary for changes to be made at a societal level. This responsibility lies not only with the engineers, but in part it does. A more sustainable society is not necessarily sustainable. In fact, there is no guarantee that the steps that are taken to make technology more environmentally friendly will indeed result in a truly more sustainable society: a society that meets a defensible notion of the Brundtland definition. In many cases even more radical steps are necessary, such as the complete cessation of a certain activity or technology. These more radical steps do not always lie within the sphere of influence of engineers. However, the fact that the responsibility for the realization of sustainable development is largely in hands of other actors in society is no reason to be aloof as an engineer. A meaningful societal discussion about what is wise from the perspective of sustainability and what is technically possible can only be held if engineers are willing to contribute their specialized knowledge.

Study Questions

1. What is meant by operationalizing "sustainable development?" Mention three ways in which the Brundtland definition needs to be operationalized?
2. Explain the notion "environmental space." Mention two reasons why determining the environmental space is not just a factual question that can be answered by science.
3. Mention two disadvantages of quantitative life cycle analyses.
4. Why can social and environmental problems not be separated?
5. In Section 8.7.1, it was indicated that the precautionary principle has four dimensions. Indicate what these four dimensions are in the application of the principle to environmental problems that was discussed in this chapter.
6. Explain what the stand still principle implies.
7. Which three kinds of environmental problems can be distinguished? Give an example of each kind. For each problem also indicate how it may be impacted by technology.
8. What is meant by internalizing the costs of environmental damage in the prices of products? Explain how this may contribute to the development of more sustainable technologies.
9. Which justification of sustainable development do you consider most convincing? Try to indicate what this justification implies for the operationalization of the notion?

10 Should engineers be concerned about considerations of intragenerational and intergenerational justice in the case of biofuels discussed in the beginning of the chapter? Which kind of considerations should they give more weight, and why?

Discussion Questions

1 Do you believe that animals or the environment can be bearers of rights? If not: does this mean that humans have no moral obligations to animals and the environment or should and can these obligations be justified otherwise? How then?
2 Is compensation ever due to those harmed by the effect of pollution? Who would have to pay, and how should the amount to be paid be calculated? What problems do you envisage?
3 Have engineers a duty to design sustainable technologies? If so, what is the extent of this duty and on what is it based?

Notes

1 The 2008 spike in food prices was, rightly or wrongly, in part attributed to the drive for biofuels.
2 The Internet provides various simplified calculators for this.
3 www.voetafdruk.be/html/voetafdruk/wat/wat.htm (accessed November 13, 2010).
4 This is also expressed as *People, Planet, Prosperity*. At the World Summit on Sustainable development in Johannnesburg (2002), the strapline "people, planet, prosperity" was adopted to reflect the requirement that sustainable development implies the balancing of economic and social development with environmental protection. In the context of the business sector, this strapline has sometimes been paraphrased as "people, planet, profits," the Three Pillars, or the "triple bottom line."
5 See Singer (2002, ch. 2). He refers to Locke (1986 [1690]).
6 The first mention of the Principle at the international level is to be found in the 1972 Recommendation by the Organization for Economic Co-operation and Development (OECD) Council on Guiding Principles concerning International Economic Aspects of Environmental Policies, where it stated that: "The principle to be used for allocating costs of pollution prevention and control measures to encourage rational use of scarce environmental resources and to avoid distortions in international trade and investment is the Polluter-Pays Principle." It is regarded as a regional custom because of the strong support it has received in most OECD and European Community countries. In international environment law it is mentioned in the Rio Declaration on Environment and Development (Principle 16).
7 In the context of the 2002 World Summit on Sustainable Development in Johannesburg, the stand still principle has been the subject of renewed interest and debate.
8 This kind of situation is studied in game theory under the name of "the prisoner's dilemma."
9 Though, this also entails risks, because one takes a loan how consensus will be developed; one can of course sometimes be wrong at the prediction of the exact content of future norms.
10 Often more phases are distinguished, but here this seems sufficient.
11 See www.pre.nl/eco-indicator99/default.htm (accessed November 6, 2009).
12 See www.lcia-recipe.net/ (accessed November 6, 2009).

Appendix I: Engineering Qualifications and Organizations in a Number of Countries

Table AI.1 Overview

Country	*USA*	*UK*	*Australia*	*Netherlands*	*Europe*
Titles/qualifications	PE	CEng IEng	CPEng RPEQ	Ir. Ing.	EUR ING
Regulatory or licensing body	State licensing boards	Engineering Council	NERB Engineers Australia BPEQ	–	FEANI
Other important organizations	NSPE NAE ABET IEEE ASCE ASME AIChE NIEE OEC	RAEng IET ICE IMechE IChemE IED	ACEA APESMA	KIVI-NIRIA	

United States

Titles and qualifications

Engineers in the United States can get licensure as Professional Engineer (PE). "To become licensed, engineers must complete a four-year college degree, work under a Professional Engineer for at least four years, pass two intensive competency exams and

Ethics, Technology, and Engineering: An Introduction, First Edition.
Ibo van de Poel and Lambèr Royakkers.

earn a license from their state's licensure board. Then, to retain their licenses, PEs must continually maintain and improve their skills throughout their careers." "Only a licensed engineer may prepare, sign and seal, and submit engineering plans and drawings to a public authority for approval, or seal engineering work for public and private clients" (www.nspe.org/Licensure/WhatisaPE/index.html).

For many engineering jobs, especially in industry, licensure is not required.

For more information, see www.nspe.org/Licensure/WhatisaPE/index.html

For an overview of state licensing boards, see www.nspe.org/Licensure/Licensing Boards/index.html

Important organizations

NSPE (National Society of Professional Engineers): www.nspe.org
NAE (National Academy of Engineering): www.nae.edu
ABET (Accreditation Board for Engineering and Technology): www.abet.org
IEEE: www.ieee.org
ASCE (American Society of Civil Engineers): www.asce.org
ASME (American Society of Mechanical Engineers): www.asme.org
AIChE (American Institute for Chemical Engineers): https://www.aiche.org
NIEE: (National Institute for Engineering Ethics): www.niee.org
OEC (Online Ethics Center): www.onlineethics.org

Codes of conduct

ASCE: https://www.asce.org/inside/codeofethics.cfm
ASME: https://files.asme.org/ASMEORG/Governance/3675.pdf
IEEE: www.ieee.org/portal/pages/iportals/aboutus/ethics/code.html
AIChE: www.aiche.org/About/Code.aspx
NSPE: see Appendix II

UK

Titles and qualifications

> "The Engineering Council is the regulatory body for the engineering profession in the UK. We hold the national registers of 235,000 Chartered Engineers (CEng), Incorporated Engineers (IEng), Engineering Technicians (EngTech) and Information and Communications Technology Technicians (ICT*Tech*). In addition, the Engineering Council sets and maintains the internationally recognised standards of professional competence and ethics that govern the award and retention of these titles. This ensures that employers, government and wider society – both in the UK and overseas – can have confidence in the knowledge, experience and commitment of registrants." (www.engc.org.uk)

CEng (Chartered Engineer)

> "The CEng professional qualification is open to anyone who can demonstrate the required professional competences and commitment. These are set out in our professional standard, UK-SPEC, and are developed through education and working experience.

The process will be more straightforward if you have particular academic qualifications, which will also allow you to obtain interim registration. For CEng these are:

- an accredited Bachelors degree with honours in engineering or technology, plus either an appropriate Masters degree accredited by a professional engineering institution, or appropriate further learning to Masters level
- or an accredited integrated MEng degree.

However, you can still become a Chartered Engineer if you do not have these academic qualifications. Further information about the assessment process can be found in UK-SPEC." (www.engc.org.uk/professional-qualifications/chartered-engineer/about-chartered-engineer.aspx)

IEng (Incorporated Engineer)

"The IEng professional qualification is open to anyone who can demonstrate the required professional competences and commitment. These are set out in our professional standard, UK-SPEC, and are developed through education and working experience.

The process will be more straightforward if you have particular academic qualifications, which will also allow you to obtain interim registration. For IEng these are:

- an accredited Bachelors or honours degree in engineering or technology;
- or a Higher National Certificate or Diploma or a Foundation Degree in engineering or technology, plus appropriate further learning to degree level;
- or an NVQ4 or SVQ4 which has been approved for the purpose by a licensed engineering institution.

However, you can still become an Incorporated Engineer if you do not have these academic qualifications. Further information about the assessment process can be found in UK-SPEC." (www.engc.org.uk/professional-qualifications/incorporated-engineer/about-incorporated-engineer.aspx)

Important organizations

ECUK	(Engineering Council UK): www.engc.org.uk
RAEng	(Royal Academy of Engineering): http://www.raeng.org.uk
ICE	(Institution of Civil Engineers): www.ice.org.uk
IET	(Institution of Engineering and Technology): http://www.theiet.org
IMechE	(Institution of Mechanical Engineers): www.imeche.org
IChemE	(Institution of Chemical Engineers): www.icheme.org
IED	(The Institution of Engineering Designers): www.ied.org.uk

For other relevant institutions see: www.engc.org.uk/about-us/our-partners/professional-engineering-institutions.aspx

Codes of conduct

Royal Academy of Engineering: www.raeng.org.uk/societygov/engineeringethics/pdf/ Statement_of_Ethical_Principles.pdf

IET: www.theiet.org/about/ethics/rules

IMechE: www.imeche.org/NR/rdonlyres/4AFBFBCD-69CC-4FB5-8A0C-0DC7101B71F6/0/CodeofConductAugust2009.pdf
IchemE: http://cms.icheme.org/mainwebsite/Resources/File/ByLaws2004.pdf

Australia

Titles and qualifications

> "There is no formal system of regulation for engineers throughout Australia. Engineering services are regulated under a variety of Acts in ad hoc areas, many of which relate to engineers in the building and construction industry. There are also many pieces of subordinate legislation, such as regulations, by-laws and orders-in-council that impose various prescriptive standards and incur unnecessary costs to the engineering industry in complying. Queensland currently is the only state where engineers are required by legislation to be registered (if offering or providing engineering services). In Queensland, persons who are not registered are prohibited from offering or providing professional engineering services. The only exception is for individuals who practise under the direct supervision of registered professional engineers. In other states and territories engineers operate under the self-regulatory system operated by the National Engineering Registration Board (NERB)." (www.engineersaustralia.org.au/nerb/regulatory-schemes/introduction.cfm)

The relevant register for professional engineers is National Professional Engineers Register (NPER). "The requisite qualification for NPER is a four-year engineering qualification accredited by Engineers Australia or equivalent. Applicants must also demonstrate that they are practising competently in the area of practice where they are applying for registration." (www.engineersaustralia.org.au/nerb/applying/professional-engineers/professional.cfm)

The following two titles have requirements that are similar to NPER registration:

- *CPEng (Chartered Professional Engineer).* "Chartered Status is exclusive to Engineers Australia. Professional engineers with Chartered Status enjoy recognition by government, business and the general public worldwide." (www.engineersaustralia.org.au/ieaust/index.cfm?0F9D7A85-BD73–7DEA-B482-F0A8012BC0EA)
- *RPEQ (Registered Professional Engineer in Queensland).* See Board of Professional Engineers of Queensland: www.bpeq.qld.gov.au

Important Organizations

Engineers Australia: www.engineersaustralia.org.au
NERB (National Engineering Registration Board): www.nerb.org.au
BPEQ (Board of Professional Engineers of Queensland): www.bpeq.qld.gov.au
ACEA (Association of Consulting Engineers Australia): www.acea.com.au/

APESMA (Association of Professional Engineers, Scientists and Managers, Australia): www.apesma.asn.au/

Code of conduct

Engineers Australia (code is also adopted by ACEA and APESMA): www.engineers-australia.org.au/shadomx/apps/fms/fmsdownload.cfm?file_uuid=F0647595-C7FE-7720-EA17–70AC27062E0B&siteName=ieaust

The Netherlands

Titles and qualifications

Ir. is the title for engineers holding a Master's degree from a university and *Ing.* for engineers holding a Bachelor's degree from a professional school. There is no system for licensing or registration of engineers in the Netherlands.

Organization

KIVI NIRIA is the Dutch association for engineers and engineering students. With 25 000 members KIVI NIRIA is the largest engineering association in the Netherlands. All engineering disciplines are organized within KIVI NIRIA. See www.kiviniria.nl

Code of Conduct

KIVI NIRIA: www.kiviniria.net/CM/PAG000002804/Gedragscode-2006.html

Europe

Titles and qualifications

European engineers can qualify for the title EUR ING of FEANI (European Federation of National Engineering Associations) with the following requirements:

> "After a secondary education at a high level validated by one or more official certificates, normally awarded at the age of about 18 years, a minimum total period of seven years' formation – education, training and experience – is required by FEANI for the EUR ING title. This formation consists of:
>
> - Minimum three years of engineering education successfully completed by an official degree, in a discipline/programme and given by a university (U) or other recognized body at university level, recognised by FEANI (see FEANI Index).
> - Minimum two years of valid professional experience (E).
> - In case the education and experience together is less than the minimum seven years' formation required, the balance to seven years should be covered by education (U),

experience (E), or training (T) monitored by the approved engineering institutions, or by preliminary engineering professional experience.

In addition to these formation requirements, EUR INGs are required to comply with a Code of Conduct respecting the provisions of the FEANI Position Paper on Code of Conduct: Ethics and Conduct of Professional Engineers." (www.feani.org)

Organization

FEANI (European Federation of National Engineering Associations): www.feani.org

Code of conduct

FEANI: see Appendix III

General Links

Titles and qualifications

en.wikipedia.org/wiki/Professional_Engineer

Codes of conduct

www.onlineethics.org/Resources/ethcodes.aspx
ethics.iit.edu/indexOfCodes.php?cat_id=9
courses.cs.vt.edu/~cs3604/lib/WorldCodes/WorldCodes.html

Appendix II: NSPE Code of Ethics for Engineers

Preamble

Engineering is an important and learned profession. As members of this profession, engineers are expected to exhibit the highest standards of honesty and integrity. Engineering has a direct and vital impact on the quality of life for all people. Accordingly, the services provided by engineers require honesty, impartiality, fairness, and equity, and must be dedicated to the protection of the public health, safety, and welfare. Engineers must perform under a standard of professional behavior that requires adherence to the highest principles of ethical conduct.

I. Fundamental Canons

Engineers, in the fulfillment of their professional duties, shall:

1 Hold paramount the safety, health, and welfare of the public.
2 Perform services only in areas of their competence.
3 Issue public statements only in an objective and truthful manner.
4 Act for each employer or client as faithful agents or trustees.
5 Avoid deceptive acts.
6 Conduct themselves honorably, responsibly, ethically, and lawfully so as to enhance the honor, reputation, and usefulness of the profession.

II. Rules of Practice

1 Engineers shall hold paramount the safety, health, and welfare of the public.
   a. If engineers' judgment is overruled under circumstances that endanger life or property, they shall notify their employer or client and such other authority as may be appropriate.

Ethics, Technology, and Engineering: An Introduction, First Edition.
Ibo van de Poel and Lambèr Royakkers.

b. Engineers shall approve only those engineering documents that are in conformity with applicable standards.
c. Engineers shall not reveal facts, data, or information without the prior consent of the client or employer except as authorized or required by law or this Code.
d. Engineers shall not permit the use of their name or associate in business ventures with any person or firm that they believe is engaged in a fraudulent or dishonest enterprise.
e. Engineers shall not aid or abet the unlawful practice of engineering by a person or firm.
f. Engineers having knowledge of any alleged violation of this Code shall report thereon to appropriate professional bodies and, when relevant, also to public authorities, and cooperate with the proper authorities in furnishing such information or assistance as may be required.

2 Engineers shall perform services only in the areas of their competence.
a. Engineers shall undertake assignments only when qualified by education or experience in the specific technical fields involved.
b. Engineers shall not affix their signatures to any plans or documents dealing with subject matter in which they lack competence, nor to any plan or document not prepared under their direction and control.
c. Engineers may accept assignments and assume responsibility for coordination of an entire project and sign and seal the engineering documents for the entire project, provided that each technical segment is signed and sealed only by the qualified engineers who prepared the segment.

3 Engineers shall issue public statements only in an objective and truthful manner.
a. Engineers shall be objective and truthful in professional reports, statements, or testimony. They shall include all relevant and pertinent information in such reports, statements, or testimony, which should bear the date indicating when it was current.
b. Engineers may express publicly technical opinions that are founded upon knowledge of the facts and competence in the subject matter.
c. Engineers shall issue no statements, criticisms, or arguments on technical matters that are inspired or paid for by interested parties, unless they have prefaced their comments by explicitly identifying the interested parties on whose behalf they are speaking, and by revealing the existence of any interest the engineers may have in the matters.

4 Engineers shall act for each employer or client as faithful agents or trustees.
a. Engineers shall disclose all known or potential conflicts of interest that could influence or appear to influence their judgment or the quality of their services.
b. Engineers shall not accept compensation, financial or otherwise, from more than one party for services on the same project, or for services pertaining to the same project, unless the circumstances are fully disclosed and agreed to by all interested parties.
c. Engineers shall not solicit or accept financial or other valuable consideration, directly or indirectly, from outside agents in connection with the work for which they are responsible.

d. Engineers in public service as members, advisors, or employees of a governmental or quasi-governmental body or department shall not participate in decisions with respect to services solicited or provided by them or their organizations in private or public engineering practice.
e. Engineers shall not solicit or accept a contract from a governmental body on which a principal or officer of their organization serves as a member.

5 Engineers shall avoid deceptive acts.
   a. Engineers shall not falsify their qualifications or permit misrepresentation of their or their associates' qualifications. They shall not misrepresent or exaggerate their responsibility in or for the subject matter of prior assignments. Brochures or other presentations incident to the solicitation of employment shall not misrepresent pertinent facts concerning employers, employees, associates, joint venturers, or past accomplishments.
   b. Engineers shall not offer, give, solicit, or receive, either directly or indirectly, any contribution to influence the award of a contract by public authority, or which may be reasonably construed by the public as having the effect or intent of influencing the awarding of a contract. They shall not offer any gift or other valuable consideration in order to secure work. They shall not pay a commission, percentage, or brokerage fee in order to secure work, except to a bona fide employee or bona fide established commercial or marketing agencies retained by them.

III. Professional Obligations

1 Engineers shall be guided in all their relations by the highest standards of honesty and integrity.
   a. Engineers shall acknowledge their errors and shall not distort or alter the facts.
   b. Engineers shall advise their clients or employers when they believe a project will not be successful.
   c. Engineers shall not accept outside employment to the detriment of their regular work or interest. Before accepting any outside engineering employment, they will notify their employers.
   d. Engineers shall not attempt to attract an engineer from another employer by false or misleading pretenses.
   e. Engineers shall not promote their own interest at the expense of the dignity and integrity of the profession.

2 Engineers shall at all times strive to serve the public interest.
   a. Engineers are encouraged to participate in civic affairs; career guidance for youths; and work for the advancement of the safety, health, and well-being of their community.
   b. Engineers shall not complete, sign, or seal plans and/or specifications that are not in conformity with applicable engineering standards. If the client or employer insists on such unprofessional conduct, they shall notify the proper authorities and withdraw from further service on the project.
   c. Engineers are encouraged to extend public knowledge and appreciation of engineering and its achievements.

d. Engineers are encouraged to adhere to the principles of sustainable development[1] in order to protect the environment for future generations.

3 Engineers shall avoid all conduct or practice that deceives the public.
  a. Engineers shall avoid the use of statements containing a material misrepresentation of fact or omitting a material fact.
  b. Consistent with the foregoing, engineers may advertise for recruitment of personnel.
  c. Consistent with the foregoing, engineers may prepare articles for the lay or technical press, but such articles shall not imply credit to the author for work performed by others.

4 Engineers shall not disclose, without consent, confidential information concerning the business affairs or technical processes of any present or former client or employer, or public body on which they serve.
  a. Engineers shall not, without the consent of all interested parties, promote or arrange for new employment or practice in connection with a specific project for which the engineer has gained particular and specialized knowledge.
  b. Engineers shall not, without the consent of all interested parties, participate in or represent an adversary interest in connection with a specific project or proceeding in which the engineer has gained particular specialized knowledge on behalf of a former client or employer.

5 Engineers shall not be influenced in their professional duties by conflicting interests.
  a. Engineers shall not accept financial or other considerations, including free engineering designs, from material or equipment suppliers for specifying their product.
  b. Engineers shall not accept commissions or allowances, directly or indirectly, from contractors or other parties dealing with clients or employers of the engineer in connection with work for which the engineer is responsible.

6 Engineers shall not attempt to obtain employment or advancement or professional engagements by untruthfully criticizing other engineers, or by other improper or questionable methods.
  a. Engineers shall not request, propose, or accept a commission on a contingent basis under circumstances in which their judgment may be compromised.
  b. Engineers in salaried positions shall accept part-time engineering work only to the extent consistent with policies of the employer and in accordance with ethical considerations.
  c. Engineers shall not, without consent, use equipment, supplies, laboratory, or office facilities of an employer to carry on outside private practice.

7 Engineers shall not attempt to injure, maliciously or falsely, directly or indirectly, the professional reputation, prospects, practice, or employment of other engineers. Engineers who believe others are guilty of unethical or illegal practice shall present such information to the proper authority for action.
  a. Engineers in private practice shall not review the work of another engineer for the same client, except with the knowledge of such engineer, or unless the connection of such engineer with the work has been terminated.
  b. Engineers in governmental, industrial, or educational employ are entitled to review and evaluate the work of other engineers when so required by their employment duties.

c. Engineers in sales or industrial employ are entitled to make engineering comparisons of represented products with products of other suppliers.

8 Engineers shall accept personal responsibility for their professional activities, provided, however, that engineers may seek indemnification for services arising out of their practice for other than gross negligence, where the engineers' interests cannot otherwise be protected.

a. Engineers shall conform with state registration laws in the practice of engineering.

b. Engineers shall not use association with a nonengineer, a corporation, or partnership as a "cloak" for unethical acts.

9 Engineers shall give credit for engineering work to those to whom credit is due, and will recognize the proprietary interests of others.

a. Engineers shall, whenever possible, name the person or persons who may be individually responsible for designs, inventions, writings, or other accomplishments.

b. Engineers using designs supplied by a client recognize that the designs remain the property of the client and may not be duplicated by the engineer for others without express permission.

c. Engineers, before undertaking work for others in connection with which the engineer may make improvements, plans, designs, inventions, or other records that may justify copyrights or patents, should enter into a positive agreement regarding ownership.

d. Engineers' designs, data, records, and notes referring exclusively to an employer's work are the employer's property. The employer should indemnify the engineer for use of the information for any purpose other than the original purpose.

e. Engineers shall continue their professional development throughout their careers and should keep current in their specialty fields by engaging in professional practice, participating in continuing education courses, reading in the technical literature, and attending professional meetings and seminars.

Note 1: "Sustainable development" is the challenge of meeting human needs for natural resources, industrial products, energy, food, transportation, shelter, and effective waste management while conserving and protecting environmental quality and the natural resource base essential for future development.

As Revised July 2007

By order of the United States District Court for the District of Columbia, former Section 11(c) of the NSPE Code of Ethics prohibiting competitive bidding, and all policy statements, opinions, rulings or other guidelines interpreting its scope, have been rescinded as unlawfully interfering with the legal right of engineers, protected under the antitrust laws, to provide price information to prospective clients; accordingly, nothing contained in the NSPE Code of Ethics, policy statements, opinions, rulings or other guidelines prohibits the submission of price quotations or competitive bids for engineering services at any time or in any amount.

Statement by NSPE Executive Committee

In order to correct misunderstandings which have been indicated in some instances since the issuance of the Supreme Court decision and the entry of the Final Judgment, it is noted that in its decision of April 25, 1978, the Supreme Court of the United States declared: "The Sherman Act does not require competitive bidding."

It is further noted that as made clear in the Supreme Court decision:

1 Engineers and firms may individually refuse to bid for engineering services.
2 Clients are not required to seek bids for engineering services.
3 Federal, state, and local laws governing procedures to procure engineering services are not affected, and remain in full force and effect.
4 State societies and local chapters are free to actively and aggressively seek legislation for professional selection and negotiation procedures by public agencies.
5 State registration board rules of professional conduct, including rules prohibiting competitive bidding for engineering services, are not affected and remain in full force and effect. State registration boards with authority to adopt rules of professional conduct may adopt rules governing procedures to obtain engineering services.

As noted by the Supreme Court, "nothing in the judgment prevents NSPE and its members from attempting to influence governmental action …"

Note: In regard to the question of application of the Code to corporations vis-à-vis real persons, business form or type should not negate nor influence conformance of individuals to the Code. The Code deals with professional services, which services must be performed by real persons. Real persons in turn establish and implement policies within business structures. The Code is clearly written to apply to the Engineer, and it is incumbent on members of NSPE to endeavor to live up to its provisions. This applies to all pertinent sections of the Code.

Appendix III: FEANI Position Paper on Code of Conduct: Ethics and Conduct of Professional Engineers

Approved by the FEANI General Assembly on 29 September 2006.

Ethical Principle

The decisions and actions of engineers have a large impact on the environment and on society. The engineering profession thus has an obligation to ensure that it works in the public interest and with regard for health, safety and sustainability.

Framework Statement

National associations of engineers, and FEANI with regard to EURING registrants, have codes of conduct which have much in common and which have the intent of implementing the above ethical principle. As a result of this convergence the European engineering profession as a whole can make a universal statement regarding the conduct of professional engineers.

Individual engineers have a personal obligation to act with integrity, in the public interest, and to exercise all reasonable skill and care in carrying out their work.

In so doing engineers:

- Shall maintain their relevant competences at the necessary level and only undertake tasks for which they are competent
- Shall not misrepresent their educational qualifications or professional titles
- Shall provide impartial analysis and judgement to employer or clients, avoid conflicts of interest, and observe proper duties of confidentiality
- Shall carry out their tasks so as to prevent avoidable danger to health and safety, and prevent avoidable adverse impact on the environment

Ethics, Technology, and Engineering: An Introduction, First Edition.
Ibo van de Poel and Lambèr Royakkers.

- Shall accept appropriate responsibility for their work and that carried out under their supervision
- Shall respect the personal rights of people with whom they work and the legal and cultural values of the societies in which they carry out assignments
- Shall be prepared to contribute to public debate on matters of technical understanding in fields in which they are competent to comment

Codes of Conduct

The pan-European statement on engineering ethics and conduct presented above is best implemented through the codes issued by national engineering associations. These codes can, and in general already do, incorporate the listed objectives in a form which reflects national circumstances and allow additional objectives to be added as required by national practice.

Source: Downloaded from www.feani.org/webfeani/ (accessed October 19, 2009). Reprinted by Permission of FEANI.

Appendix IV: Shell Code of Conduct

Shell General Business Principles

> The Shell general business principles govern how each of the shell companies which make up the Shell group conducts its affairs.

The objectives of the Shell Group are to engage efficiently, responsibly and profitably in oil, gas, chemicals and other selected businesses and to participate in the search for and development of other sources of energy to meet evolving customer needs and the world's growing demand for energy.

We believe that oil and gas will be integral to the global energy needs for economic development for many decades to come. Our role is to ensure that we extract and deliver them profitably and in environmentally and socially responsible ways.

We seek a high standard of performance, maintaining a strong long-term and growing position in the competitive environments in which we choose to operate.

We aim to work closely with our customers, partners and policy makers to advance more efficient and sustainable use of energy and natural resources.

Our Values

Shell employees share a set of core values – honesty, integrity and respect for people. We also firmly believe in the fundamental importance of trust, openness, teamwork and professionalism, and pride in what we do.

Sustainable Development

As part of the Business Principles, we commit to contribute to sustainable development. This requires balancing short- and long-term interests, integrating economic, environmental and social considerations into business decision making.

Ethics, Technology, and Engineering: An Introduction, First Edition.
Ibo van de Poel and Lambèr Royakkers.

Responsibilities

Shell companies recognize five areas of responsibility. It is the duty of management continuously to assess the priorities and discharge these inseparable responsibilities on the basis of that assessment.

To shareholders

To protect shareholders' investment, and provide a long-term return competitive with those of other leading companies in the industry.

To customers

To win and maintain customers by developing and providing products and services which offer value in terms of price, quality, safety and environmental impact, which are supported by the requisite technological, environmental and commercial expertise.

To employees

To respect the human rights of our employees and to provide them with good and safe working conditions, and competitive terms and conditions of employment.

To promote the development and best use of the talents of our employees; to create an inclusive work environment where every employee has an equal opportunity to develop his or her skills and talents.

To encourage the involvement of employees in the planning and direction of their work; to provide them with channels to report concerns.

We recognize that commercial success depends on the full commitment of all employees.

To those with whom we do business

To seek mutually beneficial relationships with contractors, suppliers and in joint ventures and to promote the application of these Shell General Business Principles or equivalent principles in such relationships. The ability to promote these principles effectively will be an important factor in the decision to enter into or remain in such relationships.

To society

To conduct business as reasonable corporate members of society, to comply with applicable laws and regulations, to support fundamental human rights in line with the legitimate role of business, and to give proper regard to health, safety, security and the environment.

Principle 1

Economic

Long-term profitability is essential to achieving our business goals and to our continued growth. It is a measure both of efficiency and of the value that customers place on Shell products and services. It supplies the necessary corporate resources for the continuing investment that is required to develop and produce future energy supplies to meet customer needs. Without profits and a strong financial foundation, it would not be possible to fulfil our responsibilities.

Criteria for investment and divestment decisions include sustainable development considerations (economic, social and environmental) and an appraisal of the risks of the investment.

Principle 2

Competition

Shell companies support free enterprise. We seek to compete fairly and ethically and within the framework of applicable competition laws; we will not prevent others from competing freely with us.

Principle 3

Business integrity

Shell companies insist on honesty, integrity and fairness in all aspects of our business and expect the same in our relationships with all those with whom we do business. The direct or indirect offer, payment, soliciting or acceptance of bribes in any form is unacceptable. Facilitation payments are also bribes and should not be made. Employees must avoid conflicts of interest between their private activities and their part in the conduct of company business. Employees must also declare to their employing company potential conflicts of interest. All business transactions on behalf of a Shell company must be reflected fairly and accurately in the accounts of the company in accordance with established procedures and are subject to audit and disclosure.

Principle 4

Political activities

Of companies
Shell companies act in a socially responsible manner within the laws of the countries in which we operate in pursuit of our legitimate commercial objectives.

Shell companies do not make payments to political parties, organizations or their representatives. Shell companies do not take part in party politics. However, when dealing with governments, Shell companies have the right and responsibility to make our position known on any matters which affect us, our employees, our customers, our shareholders or local communities in a manner which is in accordance with our values and the Business Principles.

Of employees
Where individuals wish to engage in activities in the community, including standing for election to public office, they will be given the opportunity to do so where this is appropriate in the light of local circumstances.

Principle 5

Health, safety, security and the environment

Shell companies have a systematic approach to health, safety, security and environmental management in order to achieve continuous performance management.

To this end, Shell companies manage these matters as critical business activities, set standards and targets for improvement, and measure, appraise and report performance externally.

We continually look for ways to reduce the environmental impact of our operations, products and services.

Principle 6

Local communities

Shell companies aim to be good neighbours by continuously improving the ways in which we contribute directly or indirectly to the general well-being of the communities within which we work.

We manage the social impacts of our business activities carefully and work with others to enhance the benefits to local communities and to mitigate any negative impacts from our activities.

In addition, Shell companies take a constructive interest in societal matters, directly or indirectly related to our business.

Principle 7

Communication and engagement

Shell companies recognize that regular dialogue and engagement with our stakeholders is essential. We are committed to reporting of our performance by providing full relevant information to legitimately interested parties, subject to any overriding considerations of business confidentiality.

In our interactions with employees, business partners and local communities, we seek to listen and respond to them honestly and responsibly.

Principle 8

Compliance

We comply with all applicable laws and regulations of the countries in which we operate.

Living by our Principles

Our shared core values of honesty, integrity and respect for people, underpin all the work we do and are the foundation of our Business Principles.

The Business Principles apply to all transactions, large or small, and drive the behaviour expected of every employee in every Shell company in the conduct of its business at all times.

We are judged by how we act. Our reputation will be upheld if we act in accordance with the law and the Business Principles. We encourage our business partners to live by them or by equivalent principles.

We encourage our employees to demonstrate leadership, accountability and teamwork and through these behaviours, to contribute to the overall success of Shell.

It is the responsibility of management to lead by example, to ensure that all employees are aware of these principles, and behave in accordance with the spirit as well as with the letter of this statement.

The application of these principles is underpinned by a comprehensive set of assurance procedures which are designed to make sure that our employees understand the principles and confirm that they act in accordance with them.

As part of the assurance system, it is also the responsibility of management to provide employees with safe and confidential channels to raise concerns and report instances of non-compliance. In turn, it is the responsibility of Shell employees to report suspected breaches of the Business Principles to Shell.

The Business Principles have for many years been fundamental to how we conduct our business and living by them is crucial to our continued success.

Source: The general business principles are pages 8–11 of the *Shell Code of Conduct. How to Live by the Shell General Business Principles*, Shell 2006, available from the Shell website: www.shell.com/home/content/aboutshell/who_we_are/our_values/code_of_conduct/code_of_conduct_30032008.html,

Appendix V: DSM Values and Whistle Blowing Policy

DSM Values[1]

Introduction

With operations at more than 200 sites worldwide, DSM is globally active in the fields of advanced chemical and biotechnological products and performance materials. As such, DSM is very much part of society. We are aware that companies are increasingly under public scrutiny and that DSM, as an integral part of the community, has major responsibilities. We understand that public acceptance of our activities is a necessary condition for our success.

Because of this, we are guided by the DSM Values. We are stating these values clearly so that everyone – both in and outside the company – knows what DSM is and what we stand for. What you will find here is not a new set of rules but an updated version of the values underlying our code of conduct, organized with three key audiences in mind: our customers, our employees and the communities where we do business.

The DSM Values guide our choices and decisions and influence the way we conduct our business. They are also the standard against which the company's conduct and that of its employees is judged.

The DSM Values apply to all DSM employees, regardless of where they are based. They also apply to companies or businesses acquired by DSM – which are required to achieve compliance with the DSM Values within a set period of time. In forging structural relationships with other companies, we try to ensure that these partners respect the DSM Values in all joint endeavours.

Peter Elverding Chairman of the Board

Ethics, Technology, and Engineering: An Introduction, First Edition.
Ibo van de Poel and Lambèr Royakkers.

Core values

- Our activities are aimed at creating value: value for our customers and shareholders, as well as for our employees and the communities in which we operate. We achieve this goal by combining entrepreneurial drive with an awareness of the need for continuity and a strong sense of responsibility.
- We serve the interests of our customers, employees, shareholders and business partners. To a large extent, our success depends on their success. Our relationships with our customers and other business partners are VALUABLE PARTNERSHIPS. We take into account the interests of the communities in which we operate and the demands and requirements of local, regional, national and international authorities and relevant interest groups.
- People are the key to the success of any business, and this is no different in a science and technology based company like DSM. For this reason, RESPECT FOR PEOPLE forms a cornerstone of the DSM Values. Moreover, we know that we cannot succeed without a "licence to operate" which we can only secure through GOOD CORPORATE CITIZENSHIP.
- We pursue a policy of transparency and openness, providing clear information about our activities, strategy, financial policy, organizational structure and the impact thereof on society and the environment. We periodically publish reports in which we account for our performance with respect to financial results, social policy and safety, health and sustainable development. These reports also contain an evaluation of our compliance with our own DSM Values. We strive for an active dialogue with the public at large and with the communities in which we operate. Our communication with the world around us takes the shape of direct contact with interest groups and indirect communication through the media.

Valuable partnerships

Within our goal to create value, DSM's challenge lies in contributing to the success of our customers and the end users of our products.

- We reject any restrictions to free trade other than duly enacted national and international laws.
- In accordance with the principles of product stewardship, we identify, manage and minimize the risks attached to our products during their entire lifecycle. In this connection, we share relevant knowledge, expertise and experience with our suppliers, customers and other parties.
- When considering a business partnership, we base our choice of partner not only on economic considerations, but also on the prospective partner's track record in the field of safety, health and environmental management and sustainable development.
- In making decisions, we take into account the views of our shareholders, customers, suppliers and employees.

- Our employees will not give or accept gifts that could compromise or raise doubts about the neutrality of the decisions made by either of the parties involved.
- Our employees are required to contact management if there are any indications that a business partner is conducting illegal practices or is consistently infringing the DSM Values.

Respect for people

Openness, fairness and trust form the foundation on which employer-employee relations at DSM are based. We encourage our employees to be capable, reliable, empowered and responsive. It goes without saying that we acknowledge fundamental human rights as defined by the United Nations. Respect for employees and employee integrity are the cornerstones of our human resources policy.

- We invest in the knowledge and skills of our employees on an ongoing basis to ensure their long-term employability.
- We create an atmosphere of candour and stimulate openness and accountability by involving our employees in the development and execution of our business objectives.
- We provide our employees with coaching and mentoring for growth and personal development.
- We pursue a fair and competitive remuneration policy with due recognition for performance.
- We recognize our employees' right to organize themselves in order to protect their own interests.
- We seek to create an incident- and injury-free work environment. At all levels, our employees play an active role in identifying and rectifying unsafe situations.
- We do our utmost to prevent the occurrence of occupational illness and health problems associated with the company's activities.
- We do not use child labour or forced labour.
- We do not discriminate in any manner on the basis of race, ethnic background, age, religion, gender, sexual orientation or disability.

Good corporate citizenship

To ensure our future and secure our "licence to operate", we want our operations to be not only profitable but also socially acceptable. As part of this social commitment, we endorse the obligations formulated in the chemical industry's international Responsible Care Programme.

- We are keenly aware of our responsibility for the environment and we endorse the importance of sustainable entrepreneurship. To us, in our corporate role, this means conducting our activities in a way that meets today's needs without compromising the ability of future generations to meet their needs.
- Our choice of production processes and products is guided by our commitment to promoting sustainability and safety. We exercise great prudence in developing new

technologies, taking public opinion seriously into account. Moreover, in line with our policy of transparency and openness, we provide our customers and the general public with clear information about our products and production processes.
- We make an ongoing effort to minimize the use of raw materials and energy in our production processes.
- We continually evaluate and improve our working methods, production processes, products and services so as to ensure that they are safe and acceptable from the point of view of our employees, our customers, the public at large and the environment.
- We abide by the laws and regulations in force. If these leave room for practices that clash with the DSM Values, employees are required to report this to company management.
- Our employees are aware of and show respect for local traditions and customs.
- Our employees are prohibited from seeking to influence the political decision making process by granting favours or giving gifts.
- In emergency situations such as natural disasters and public disturbances, we give top priority to the safety of our employees and residents living near our production sites.
- We encourage our employees to adopt a civic-minded and socially responsible attitude.
- Our employees are to avoid even the suggestion of a conflict of interest between their official functions on behalf of the company and their conduct as private citizens that might compromise their integrity in their official capacity or compromise the integrity of the company.
- DSM employees who possess "inside information" are prohibited from dealing in or recommending that third parties deal in DSM securities. Employees who through DSM have non-public information about other companies are likewise prohibited from dealing in shares of those companies.

DSM Whistle Blowing Policy[2]

This document sets out the policy and procedure adopted by DSM to support individuals to express their concerns about suspected serious misconduct at DSM (also referred to as whistleblowing).

1.0 Policy

1.1 DSM is committed to high standards of openness, decency and integrity in its work. To maintain these standards, DSM encourages its employees who have concerns about suspected serious misconduct to come forward and express these concerns without fear of punishment or unfair treatment.

1.2 Suspected serious misconduct includes any activity by DSM or a DSM employee that violates:

- Laws or regulations
- DSM Values
- DSM Corporate Requirements

1.3 Adopted by the Managing Board of DSM on June 14, 2004.

2.0 Background

Governments all over the world recognize that employees, from time to time, have concerns about what is happening at work but are afraid to report those concerns. In order to provide protection to those individuals who do report concerns, laws, regulations and codes have been prepared, such as certain paragraphs in the Dutch Corporate Governance Code.

2.1 Paragraphs within the Dutch Corporate Governance Code

- The management board (of the company) shall ensure that employees can report alleged acts of misconduct of a general, operational and financial nature within the company to the chairman of the management board or to an official designated by him, without jeopardizing their legal position.
- Alleged acts of misconduct concerning the functioning of management board members shall be reported to the chairman of the supervisory board.
- The arrangements for whistleblowers shall in any event be posted on the company's website.

2.2 Pointers

- Reports about possible misconduct are not limited to fraud, theft or corruption, but cover a much wider range of bad practices, including behavior that is not in line with the DSM Values.
- Such bad practice can have happened, be happening, or be likely to happen.
- These procedures are designed to encourage employees to voice concerns internally and promptly so as to prevent or remedy acts of misconduct.

DSM encourages employees to engage in a discussion with colleagues who display behavior that is or could be violating any law or DSM value, if at all possible. If a discussion is not a realistic option, then the employee should report internally and voice his /her concerns outside the organization only if he/she is unable to report internally.

3.0 DSM's position

3.1 DSM has formulated a code of conduct – the DSM Values. These values form the basis upon which choices are made, determining the framework for the way DSM does business. The DSM Values are used to evaluate the manner in which DSM conducts its business operations. The DSM Values apply to every DSM employee. In addition, DSM complies with the relevant laws and regulations that apply to the company and its employees. DSM is dedicated to the prevention, avoidance, detection and investigation of all forms of non-compliance, fraud, theft and corruption.

3.2 DSM realizes that employees are often the first to notice that there may be something wrong within the company. However, the employees may not express their concerns due to feelings of disloyalty to colleagues, or fear of punishment or unfair treatment.

3.3 DSM acknowledges that most concerns are of relatively minor nature and can be resolved through the normal channels relatively easily. However, where the concerns are more serious, and especially where they involve serious misconduct such as criminal acts, or financial misconduct, or in situations where employees, the public, or the environment may be subject to danger, it can be difficult for the employee to know what to do and to whom to report such concerns.

3.4 DSM wishes to make it clear that individuals can raise such serious concerns without fear of punishment or unfair treatment. This is to encourage individuals to report concerns so that management can take appropriate action to prevent or stop intolerable behavior that harms employees, the public, the environment or DSM.

3.5 This procedure has been introduced to give guidance on how to raise concerns at an early stage.

4.0 Aims and scope of the procedure

4.1 The aim of this procedure

This procedure is intended to:

- Provide avenues for employees to raise concerns and define a way to handle these concerns.
- Enable management to be informed at an early stage about acts of misconduct.
- Reassure employees that they will be protected from punishment or unfair treatment for disclosing concerns in good faith in accordance with this procedure.
- Help develop a culture of openness, accountability and integrity.

4.2 The scope of this procedure

This procedure is separate from and in addition to DSM's other existing or future, more specific grievance/complaint procedures. Employees who wish to voice a grievance relating to their employment, or any other complaint covered by a more specific procedure should use that specific procedure

5.0 Safeguards

5.1 Prevention of punishment or unfair treatment

DSM does not tolerate punishment or unfair treatment when concerns are raised in good faith and will take action to protect staff.

5.2 Confidentiality

DSM recognizes that some individuals will wish to raise a concern in confidence under this policy. Confidentiality will be maintained to the largest extent possible.

Therefore, DSM will protect the identity of an employee who discloses concerns according to this procedure.

DSM does however acknowledge that in some circumstances it may be obvious who has raised the concern and filed the report, or the investigation process may lead to the point where a statement is required or the individual is called to provide evidence. In such circumstances, where finding the truth is hindered by maintaining complete confidentiality, DSM can not guarantee complete confidentiality to the reporting employee.

DSM wants to avoid anonymous reports, as it can make the investigation of the allegations much more difficult. However, if an employee feels there is no other way than making an anonymous allegation, then that allegation will be acted upon appropriately.

5.3 Untrue allegations

DSM encourages people to raise concerns in good faith. However, if upon investigation some of these concerns cannot be confirmed or may not have substance, no action will be taken against employees raising concerns in good faith.

Investigations, however, are a costly, time consuming and potentially damaging process. If reported allegations are to be judged malicious and without any factual foundation, DSM may take appropriate action against employees making such malicious allegations.

6.0 Raising a concern

6.1 Who to report to

6.1.1 As a rule concerns should be raised with the employee's line manager (or the supervisor of the line manager), who will look into the matter and provide a solution. If for some reason the employee does not feel able to report through the line manager, then he/she can raise their concerns directly with the designated officer, the officer (specifically assigned by the Chairman of the Managing Board) for whistleblowing.

6.1.2 The DSM Alert Officer is:
Title: Director, Corporate Operational Audit
Name: Hans van Suijdam
Department: COA

6.1.3 If the allegation is about the designated officer, then the reporting employee should report directly to the Chairman of the Managing Board.

6.1.4 If the allegation is about any member of the Managing Board, then the reporting employee should report directly to the Chairman of the Supervisory Board.

6.2 How to report

6.2.1 Concerns may be voiced in a face-to-face meeting. If that is not possible, they may be reported through the web site, by telephone or fax, by email, or in a face-to-face meeting, providing the background, history and reason for the concern, together with names, dates, places and as much information as possible. DSM will always arrange for ways to report in the native language, if so desired.

6.2.2 Individuals will not be expected to prove truth of an allegation but they should be able to demonstrate that there are sufficient grounds to have a reasonable belief that something is wrong.

6.2.3 Individuals are encouraged to express their concerns at the earliest possible stage so that timely action can be taken.

7.0 DSM's response

7.1 The DSM Alert Officer will:

7.1.1 Perform / Arrange an initial confidential interview with the reporting employee to:

- Reassure them they will be protected from possible punishment or unfair treatment;
- Determine if there is a wish for confidentiality and explain the level of confidentiality that can be aimed for;
- Determine if they wish to make an oral or written statement; and
- Write a brief summary of the interview that should be agreed by both parties

7.1.2 Maintain a record that a report has been filed

7.1.3 Inform the DSM Alert Committee (DAC) that a disclosure has been made and highlights of the allegation. The name of the reporting employee is kept confidential if that is desired.

If the allegation is about anybody in the DSM Alert Committee then the designated officer will inform the Chairman of the Managing Board.

7.2 Some concerns may be resolved by agreed action without the need for an investigation. Concerns about allegations, which fall within the scope of specific procedures (for example grievance procedures), will normally be referred for consideration under those procedures.

7.3 The DSM Alert Committee consists of representatives from Corporate Operational Audit, Corporate Secretariat, Corporate Legal Affairs and Corporate Human Resources. The DSM Alert Officer acts as chairperson.

7.4 The DSM Alert Committee will:

- Confirm the initial assessment from the DSM Alert Officer; and
- Based upon the results of the initial assessment decide if and what further action, such as a full investigation, is required. Investigation may be done by the DSM Alert Officer, a representative of the DSM Alert Committee or another person appointed by the DSM Alert Committee.

7.5 The person(s) performing the initial investigation may need to speak to the reporting employee to clarify the information provided or to seek additional information.

7.6 All concerns received will be acknowledged to the employee in writing within 10 working days. Wherever possible the acknowledgement will:

- Indicate the proposed way forward with regard to the matter
- Advise whether initial investigations have been made
- Advise whether further investigations are to take place, and if not why not and
- Give an estimate of how long it may take to provide a final response.

7.7 If the DSM Alert Committee decides that a full investigation is required, the type of investigation will depend upon the nature of the concern. The matters raised may be:

- Investigated internally
- Referred to an external investigator
- Referred to the police

DSM recognizes that the individual raising the concern needs to be assured that the matter has been properly addressed. Thus, subject to legal constraints, individuals making a complaint will be kept informed during the investigation and be informed about the outcome of any investigation.

7.8 Once a complaint has been adequately handled in the opinion of the DSM Alert Committee, the DSM Alert Officer will prepare a brief report and the case will be closed.

Notes

1 Document *DSM Values*, DSM March 2002. Available from the DSM website: www.dsm.com/en_US/downloads/about/Values_English.pdf. Reprinted by Permission of DSM.

2 *DSM Alert: Whistleblowing Policy & Procedure for expressing concerns about suspected serious misconduct at DSM,* July 2009. Available at www.dsm.com/en_US/downloads/governance/whistleblower_policy_en.pdf. Reprinted by Permission of DSM.

Glossary

Absolutism A rigid form of universalism in which no exceptions to rules are possible.

Acceptable risk A risk that is morally acceptable. The following considerations are relevant for deciding whether a risk is morally acceptable: (1) the degree of informed consent with the risk; (2) the degree to which the benefits of a risky activity weigh up against the disadvantages and risks; (3) the availability of alternatives with a lower risk; and (4) the degree to which risks and advantages are justly distributed.

Accountability Backward-looking responsibility in the sense of being held to account for, or justify one's actions towards others.

Act utilitarianism The traditional approach to utilitarianism in which the rightness of actions is judged by the (expected) consequences of those actions.

Active responsibility Responsibility before something has happened referring to a duty or task to care for certain state-of-affairs or persons.

Actor Any person or group that can make a decision how to act and that can act on that decision.

Advisory codes A code of conduct that has the objective to help individual professionals or employees to exercise moral judgments in concrete situations.

Ambiguity The property that different interpretations or meanings can be given to a term.

Animal tests Tests for determining dose-response relationships by exposing animals to various dosages and assessing their response.

Anthropocentrism The philosophical view that the environment has only instrumental value, that is, only value for humans and not in itself.

Anticipating mediation by imagination Trying to imagine the ways technology-in-design could be used. This insight is then used to deliberately shape user operations and interpretations.

Argument A set of statements, of which one (the conclusion) is claimed to follow from the others (the premises).

Argumentation by analogy A type of non-deductive argumentation. An argumentation based on comparison with another situation in which the judgment is clear. The judgment is supposed also to apply to the analogous situation.

Argumentation theory An interdisciplinary study of analyzing and evaluating arguments.

Ethics, Technology, and Engineering: An Introduction, First Edition.
Ibo van de Poel and Lambèr Royakkers.

Aspirational code A code that expresses the moral values of a profession or company.

Best available technology As an approach to acceptable risk (or acceptable environmental emissions), best available technology refers to an approach that does not prescribe a specific technology but uses the best available technological alternative as yardstick for what is acceptable.

Biocentrism The viewpoint that the environment has intrinsic value (value of its own).

Black-and-white-strategy A strategy for action in which only two options for actions are considered: doing the action or not.

Blameworthiness Backward-looking responsibility in the sense of being a proper target of blame for one's actions or the consequences of one's actions. In order for someone to be blameworthy, usually the following conditions need to apply: wrong-doing, causal contribution, foreseeability, and freedom.

Care ethics An ethical theory that emphasizes the importance of relationships, and which holds that the development of morals does not come about by learning general moral principles.

Carrying capacity The amount of damage that can be done to the environment without that damage being irreversible.

Categorical imperative A universal principle of the form "Do A" which is the foundation of all moral judgments in Kant's view.

Causality argumentation A type of non-deductive argumentation. An argumentation in which an expected consequence is derived from certain actions.

Certification The process in which it is judged whether a certain technology meets the applicable technical codes and standards.

Characteristic-judgment argumentation A type of non-deductive argumentation. An argument based on the assumption that a certain judgment about a thing or person can be derived from certain characteristics of that thing or person.

Code of conduct A code in which organizations (like companies or professional associations) lay down guidelines for responsible behavior of their members.

Collective responsibility The responsibility of a collective of people.

Collective responsibility model The model in which every member of a collective body is held responsible for the actions of the other members of that same collective body (and for the responsibility of the collective).

Collective risks Risks that affect a collective of people and not just individuals, like the risks of flooding.

Collingridge dilemma This dilemma refers to a double-bind problem to control the direction of technological development. On the one hand, it is often not possible to predict the consequences of new technologies already in the early phases of technological development. On the other hand, once the (negative) consequences materialize it often has become very difficult to change the direction of technological development.

Common sense method The method that weighs the available options for actions in the light of the relevant values.

Conceptual design stage The stage in which the designer or the design team generates concept designs. The focus is on an integral approach to the design problem.

Conclusion of an argument The statement that is affirmed on the basis of the premises of the argument.

Confidentiality duties Duties on employees to keep silent certain information.

Conflict of interest The situation in which one has an interest (personal or professional) that, when pursued, can conflict with meeting one's professional obligations to an employer or to (other) clients.

Consequentialism The class of ethical theories which hold that the consequences of actions are central to the moral judgment of those actions.

Constructive Technology Assessment (CTA) Approach to Technology Assessment (TA) in which TA-like efforts are carried out parallel to the process of technological development and are fed back to the development and design process.

Contingent validation An approach to express values like safety or sustainability in monetary units by asking people how much they are willing to pay for a certain level of safety or sustainability (for example, the preservation of a piece of beautiful nature).

Corporate code Code of conduct that is formulated by a company.

Corporate liability Liability of a company (corporation) when it is treated as a legal person.

Corporate Social Responsibility The responsibility of companies towards stakeholders and to society at large that extends beyond meeting the law and serving shareholders' interests.

Cost-benefit analysis A method for comparing alternatives in which all the relevant advantages (benefits) and disadvantages (costs) of the options are expressed in monetary units and the overall monetary cost or benefit of each alternative is calculated.

Creativity The virtue of being able to think out or invent new, often unexpected, options or ideas. Creativity is an important professional virtue for designers.

Critical loyalty Giving due regard to the interest of the employer, insofar as this is possible within the constraints of the employee's personal and professional ethics.

Critical questions Questions belonging to a certain type of non-deductive argumentation to check the degree of plausibility of a conclusion.

Decision stage The stage of the design process in which various concept designs are compared with each other and a choice is made for a design that has to be detailed.

Deductive argument An argument which has a conclusion that is enclosed in (implied by) the premises.

Degradation Structural damage to the environment. An example is soil erosion.

Descriptive ethics The branch of ethics that describes existing morality, including customs and habits, opinions about good and evil, responsible and irresponsible behavior, and acceptable and unacceptable action.

Descriptive judgment A judgment that describes what is actually the case (the present), what was the case (the past), or what will be the case (the future).

Design criteria A kind of design requirements which are formulated in such a way that products meet them to a greater or lesser extent. Design criteria are often used to compare and choose between different concept designs.

Design process An iterative process in which certain functions are translated into a blueprint for an artifact, system, or service. Often the following six stages are distinguished: problem analysis and formulation; conceptual design; simulation; decision; detail design; and prototype development and testing.

Design requirements Requirements that a good or acceptable design has to meet.

Detail design stage The stage in which a chosen design is elaborated on and detailed.

Development risks In the context of product liability: Risks that could not have been foreseen given the state of scientific and technical knowledge at the time the product was put into circulation.

Disciplinary code A code that has the objective to achieve that the behavior of all professionals or employees meets certain values and norms.

Discount rate The rate that is used in cost-benefit analysis to discount future benefits (or costs). This is done because 1 dollar now is worth more than 1 dollar in 10 years time.

Distribution of responsibility The ascription or apportioning of (individual) responsibilities to various actors.

Distributive justice The value of having a just distribution of certain important goods, like income, happiness, and career.

Duty ethics Also known as deontological ethics. The class of approaches in ethics in which an action is considered morally right if it is in agreement with a certain moral rule (law, norm, or principle).

Duty of care The legal obligation to adhere to a reasonable standard of care when performing any acts that could foreseeably harm others.

Ecological footprint A measure for the total environmental impact of a person's lifestyle expressed in an amount of space required to support this lifestyle

Effectiveness The extent to which an established goal is achieved.

Effectiveness requirement The moral requirement that states that responsibility should be so distributed that the distribution has the best consequences, that is, is effective in preventing harm (and in achieving positive consequences).

Efficiency The ratio between the goal achieved and the effort required.

Engineering design The activity in which certain functions are translated into a blueprint for an artifact, system, or service that can fulfill these functions with the help of engineering knowledge.

Environmental space The (maximum) amount of use of renewable and non-renewable resources that does not exceed the boundaries of what the environment can take.

Epidemiological research Research in which population data is used to find out what the relationship is between the occurrence of certain diseases or certain mental deviations and certain factors that may cause these deviations.

Equality postulate The prescription to treat persons as equals, that is, with equal concern and respect.

Ethical cycle A tool in structuring and improving moral decisions by making a systematic and thorough analysis of the moral problem, which helps to come to a moral judgment and to justify the final decision in moral terms.

Ethics The systematic reflection on morality.

Event tree Tree of events in which one starts with a certain event and considers what events will follow.

Exhaustion A type of environmental problem in which something valuable is removed from the environment that cannot, or at least not easily, be renewed.

External auditing Assessing of a company in terms of its code of conduct by an external organization.

Failure mode Series of events that may lead to the failure of an installation.

Fallacy An error or deficiency in an argument.

Fault tree Tree of events in which we move backwards from an unwanted event (a fault) to the events that could lead to the undesirable event.

Freedom principle The moral principle that everyone is free to strive for his/her own pleasure, as long as they do not deny or hinder the pleasure of others.

Global code of conduct A code of conduct that is believed to apply worldwide.

Good will A central notion in Kantian ethics. According to Kant, we can speak of good will if our actions are led by the categorical imperative. Kant believes that the good will is the only thing that is unconditionally good.

Hazard Possible damage or otherwise undesirable effect.

Hedonism The idea that pleasure is the only thing that is good in itself and to which all other things are instrumental.

Hierarchical responsibility model The model in which only the organization's top level of personnel is held responsible for the actions of (people in) the organization.

"Hired gun" Someone who is willing to carry out any task or assignment from his employer without moral scruples.

Honesty Telling what one has good reasons to believe to be true and disclosing all relevant information.

Hypothetical consent Hypothetical consent refers to a form of informed consent in which people do not actually consent to something but are hypothetically supposed to consent if certain conditions are met, for example that it would be rational for them to consent or in their own interest.

Hypothetical norm A condition norm, that is, a norm which only applies under certain circumstances, usually of the form "If you want X do Y."

Ideals Ideas or strivings which are particularly motivating and inspiring for the person having them, and which aim at achieving an optimum or maximum.

Ignorance Lack of knowledge. Refers to the situation in which we do not know what we do not know.

Ill-structured problem A problem that has no definitive formulation of the problem, may embody inconsistent problem formulations, and can only be defined during the process of solving the problem.

Incommensurability Two (or more) values are incommensurable if they cannot be expressed or measured on a common scale or in terms of a common value measure.

Individual responsibility model The model in which each individual is held responsible insofar as he or she meets the conditions for individual responsibility.

Inductive argumentation A type of non-deductive argumentation. Argumentation from the particular to the general.

Informed consent Principle that states that activities (experiments, risks) are acceptable if people have freely consented to them after being fully informed about the (potential) risks and benefits of these activities (experiments, risks).

Inherently safe design An approach to safe design that avoids hazards instead of coping with them, for example by replacing substances, mechanisms and reactions that are hazardous by less hazardous ones.

Instrumental value Something that is valuable in as far as it is a means to, or contributes to something else that is intrinsically good or valuable.

Integrity Living by one's own (moral) values, norms and commitments.

Interests Things actors strive for because they are beneficial or advantageous for them.

Intergenerational justice Justice that relates to the just distribution of resources between different generations.

Interval scale A measurement scale in which in addition to the order of items also the distance between the items has meaning.

Intragenerational justice Justice that relates to the just distribution of resources within a generation.

Intrinsic value Value in and of itself.

Intuitivist framework The ethical framework in which options for action are evaluated on basis of one's view about what is intuitively most acceptable and that formulates arguments for this statement.

Invitation-inhibition structure The fact that mediating technology invited specific actions, while other actions are inhibited.

Liability Legal responsibility: backward-looking responsibility according to the law. Usually related to the obligation to pay a fine or repair or repay damages.

LiDS wheel approach An abbreviation for Lifecycle Design Strategies. A qualitative oriented life cycle approach.

Life cycle analysis An analysis that maps the environmental impact of a product across the entire cycle of production, use, and disposal.

Life phases The phases through which a product goes during its "life": production phase, use phase, and removal phase.

Limited liability The principle that the liability of shareholders for the corporation's debts and obligations is limited to the value of their shares.

Marginal utility The additional utility that is generated by an increase in a good or service (income for example).

Means-end argumentation A type of non-deductive argumentation. An argumentation in which from a given end the means are derived to realize that end.

Mediation of action The influence of artifacts on human action.

Mediation of perception The influence of artifacts on human perception, that is, the sensory relationship with reality.

Models for dose-response relationships Models that presuppose or predict a certain relationship between dose and response.

Modus ponens Form of a valid argument in which the conclusion "q" follows from the premises "p" and "if p then q."

Modus tollens Form of a valid argument in which the conclusion "not-p" follows from the premises "if p then q" and "not-q."

Moral autonomy The view that a person himself or herself should (be able to) determine what is morally right through reasoning.

Moral balance sheet A balance sheet in which the costs and benefits (pleasures and pains) for each possible action are weighed against each other. Bentham proposed the drawing up of such balance sheets to determine the utility of actions. Cost-benefit analysis is a more modern variety of such balance sheets.

Moral deliberation An extensive and careful consideration or discussion of moral arguments and reasons for and against certain actions.

Moral dilemmas A moral problem with the crucial feature that the agent has only two (or a limited number of) options for action and that whatever he chooses he will commit a moral wrong.

Moral fairness requirement The requirement that a distribution of responsibility should be fair (just). In case of passive responsibility, this can be interpreted as that a person should only be held responsible if that person can be reasonably held responsible according the following conditions wrong-doing; causal contribution; foreseeability; and freedom of action. In terms of active responsibility it can be interpreted as implying that persons should only be allocated responsibilities that they can live by.

Moral problem Problem in which two or more positive moral values or norms cannot be fully realized at the same time.

Moral responsibility Responsibility that is based on moral obligations, moral norms or moral duties.

Morality The totality of opinions, decisions, and actions with which people express, individually or collectively, what they think is good or right.

Moralization of technology The deliberate development of technologies in order to shape moral action and decision-making.

Multiple criteria analysis A method for comparing alternatives in which various decision criteria are distinguished on basis of which the alternatives are scored. On basis of the score of each of the alternatives on the individual criteria, usually a total score is calculated for each alternative.

Multiple independent safety barriers A chain of safety barriers that operate independently of each other so that if one fails the others do not necessarily also fail.

Multistability The phenomenon that a technology can have several "stabilities," depending on the way it is embedded in a use context.

Negative feedback mechanism A mechanism that if a device fails or an operator loses control assures that the (dangerous) device shuts down.

Negligence Not living by certain duties. Negligence is often a main condition for legal liability. In order to show negligence for the law, usually proof must be given of a duty owed, a breach of that duty, an injury or damage, and a causal connection between the breach and the injury or damage.

No harm principle The principle that one is free to do what one wishes, as long as no harm is done to others. Also known as the freedom principle.

Non-renewable resources Natural resources that cannot be renewed or reproduced. An example is fossil fuel.

Normal design Design in which the normal configuration and working principle of the product remain the same.

Normative ethics The branch of ethics that judges morality and tries to formulate normative recommendations about how to act or live.

Normative judgment Judgment about whether something is good or bad, desirable or undesirable, right or wrong.

Normative relativism An ethical theory that argues that all moral points of view – all values, norms, and virtues – are equally valid.

Norms Rules that prescribe what actions are required, permitted, or forbidden.

Ordinal scale A measurement scale in which only the order of the items of the scale has meaning.

Organizational deviance Norms that are seen as deviant or unethical outside the organization are seen within the organization as normal and legitimate.

Overlapping consensus An agreement on the level of moral judgments, while there may be disagreement on the level of moral principles and background theories. Each of the participants should be able to justify the overlapping consensus in terms of his or her own wide reflective equilibrium.

Passive responsibility Backward-looking responsibility, relevant after something undesirable occurred; specific forms are accountability, blameworthiness, and liability.

Paternalism The making of (moral) decisions for others on the assumption that one knows better what is good for them than those others themselves.

Personal risks Risks that only affect an individual and not a collective. For example, the risk of smoking. The relevant distinction with collective risk is whether the individual can stop or avert the risks for him or her individually. We can individually decide not to smoke but cannot individually prevent flooding for ourselves.

Plausibility principle The principle that enumeration and supplementary argumentation in a non-deductive argumentation can make the conclusion plausible (acceptable).

Polluter pays principle The principle that damage to the environment must be repaired by the party responsible for the damage.

Pollution Environmental problems in which something undesirable or damaging is added to the environment.

Practical wisdom The intellectual virtue that enables one to make the right choice for action. It consists in the ability to choose the right mean between two vices.

Precautionary principle Principle that prescribes how to deal with threats that are uncertain and/or cannot be scientifically established. In it most general form the precautionary principle has the following general format: If there is (1) a threat, which is (2) uncertain, then (3) some kind of action (4) is mandatory. This definition has four dimensions: (1) the threat dimension; (2) the uncertainty dimension; (3) the action dimension; and (4) the prescription dimension.

Premises The statements, which are affirmed (or assumed) as providing support or reasons for accepting the conclusion.

Prima facie norms Prima facie norms are the applicable norms, unless they are overruled by other more important norms that become evident when we take everything into consideration.

Problem analysis stage The stage of the design process in which the designer or the design team analyses and formulates the design problem, including the design requirements.

Problem of many hands The occurrence of the situation in which the collective can reasonably be held morally responsible for an outcome, while none of the individuals can be reasonably held responsible for that outcome.

Product liability Liability of manufacturers for defects in a product, without the need to proof that those manufacturers acted negligently.

Profession Often mentioned characteristics of a profession include: 1) use of specialized knowledge and skills; 2) a monopoly on the carrying out of the occupation; 3) assessment only possible by peers. In addition the following two requirements are also sometimes mentioned: 4) service orientation to society; and 5) ethical standards.

Professional autonomy The ideal that individual professionals achieve themselves moral conclusions by reasoning clearly and carefully.

Professional code Code of conduct that is formulated by a professional association.

Professional ideals Ideals that are closely allied to a profession or can only be aspired to by carrying out the profession.

Professional responsibility The responsibility that is based on one's role as professional in as far it stays within the limits of what is morally allowed.

Proof from the absurd A deductive argumentation in which a certain proposition is proved by showing that the negation of the proposition leads to a contradiction.

Property right The right to ownership of a specific matter or resource like money, land, or an environmental resource (like clean air).

Radical design The opposite of normal design. Design in which either the normal configuration or the working principle (or both) of an existing product is changed.

Ratio scale A measurement scale in which the ratio between items on a scale has meaning.

Reciprocity principle Second formulation of the categorical imperative: Act as to treat humanity, whether in your own person or in that of any other, in every case as an end, never as means only.

Regulation A legal tool that can forbid the development, production, or use of certain technological products, but more often it formulates a set of the boundary conditions for the design, production, and use of technologies.

Regulators Organizations who formulate rules or regulations that engineering products have to meet such as rulings concerning health and safety, but also rulings linked to relations between competitors.

Regulatory framework The totality of (product-specific) rules that apply to the design and development of a technology.

Renewable resources Natural resources that can be renewed or reproduced.

Risk A risk is a specification of a hazard. The most often used definition of risk is the product of the probability of an undesirable event and the effect of that event.

Risk assessment A systematic investigation in which the risks of a technology of an activity are mapped and expressed quantitatively in a certain risk measure.

Risk communicators Specialists that inform, or advise how to inform, the public about risks and hazards.

Risk-cost-benefit analysis This is a variant of regular cost-benefit analysis. The social costs for risk reduction are weighed against the social benefits offered by risk reduction, so achieving an optimal level of risk in which the social benefits are highest.

Role responsibility The responsibility that is based on the role one has or plays in a certain situation.

Rule utilitarianism A variant of utilitarianism that judges actions by judging the consequences of the rules on which these actions are based. These rules, rather than the actions themselves, should maximize utility.

Safety The condition that refers to a situation in which the risks have been reduced as far as reasonably feasible and desirable.

Safety factor A factor or ratio by which an installation is made safer than is needed to withstand either the expected or the maximum (expected) load.

Script A prescription how to act that is built (designed) into an artifact.

Separatism The notion that scientists and engineers should apply the technical inputs, but appropriate management and political organs should make the value decisions.

Simulation stage The stage of the design process in which the designer or the design team checks through calculations, tests, and simulations whether the concept designs meet the design requirements.

Social ethics of engineering An approach to the ethics of engineering that focuses on the social arrangements in engineering rather than on individual decisions. If these social arrangements meet certain procedural norms the resulting decisions are considered acceptable.

Societal experiments We speak of the introduction of new technology in society as a societal experiment if the (final) testing of possible hazards and risks of a technology and its functioning take place by the actual implementation of a technology in society.

Sound argumentation An argumentation for which the corresponding critical questions can be answered positively and which therefore makes the conclusion plausible if the premises are true.

Stakeholder principles Principles that guide the relationship between a company and its stakeholders.

Stakeholders Actors that have an interest ("a stake") in the development of a technology.

Stand still principle The principle that we must not pass on a poorer environment to the next generation than the one we received from the previous generation.

Strategy of cooperation The action strategy that is directed at finding alternatives that can help to solve a moral problem by consulting other stakeholders.

Strict liability A form of liability that does not require the defendant to be negligent.

Structure of amplification and reduction The fact that mediating technologies amplify specific aspects of (the perception of) reality while reducing other aspects.

Sustainable development Development that meets the needs of the present without compromising the ability of future generations to meet their own needs (Brundtland definition).

Technical codes and standards Technical codes are legal requirements that are enforced by a governmental body to protect safety, health, and other relevant values. Technical standards are usually recommendations rather than legal requirements that are written by engineering experts in standardization committees.

Technocracy Government by experts.

Technological enthusiasm The ideal of wanting to develop new technological possibilities and taking up technological challenges.

Technological mediation The phenomenon that when technologies fulfill their functions, they also help to shape the actions and perceptions of their users.

Technology Assessment (TA) Systematic method for exploring future technology developments and assessing their potential societal consequences.

Test The execution of a technology in circumstances set and controlled by the experimenter, and in which data are gathered systematically about how the technology functions in practice.

The good life The highest good or *eudaimonia*: a state of being in which one realizes one's uniquely human potential. According to Aristotle, the good life is the final goal of human action.

Threshold The minimal level of a (design) criterion or value that an alternative has to meet in order to be acceptable with respect to that criterion or value.

Trade-off Compromise between design criteria. For example, you trade off a certain level of safety for a certain level of sustainability.

Tripartite model A model that maintains that engineers can only be held responsible for the design of products and not for wider social consequences or concerns. In the tripartite model three separate segments are distinguished: the segment of politicians; the segment of engineers; and the segment of users.

Trumping (of values) If one value trumps another any (small) amount of the first value is worth more than any (large) amount of the second value.

Type I error The mistake of assuming that a scientific statement is true while it actually is false. Applied to risk assessment: The mistake that one assumes a risk when there is actually no risk.

Type II error The mistake of assuming that a scientific statement is false while it actually is true. Applied to risk assessment: The mistake that one assumes that there is no risk while there actually is a risk.

Uncertainty A lack of knowledge. Refers to situations in which we know the type of consequences, but cannot meaningfully attribute probabilities to the occurrence of such consequences

Uncritical loyalty Placing the interests of the employer, as the employer defines those interests, above any other considerations.

Universalism An ethical theory that states that there is a system of norms and values that is universally applicable to everyone, independent of time, place, or culture.

Universality principle First formulation of the categorical imperative: Act only on that maxim which you can at the same time will that it should become a universal law.

Users People who use a technology and who may formulate certain wishes or requirements for the functioning of a technology.

Utilitarianism A type of consequentialism based on the utility principle. In utilitarianism, actions are judged by the amount of pleasure and pain they bring about. The action that brings the greatest happiness for the greatest number should be chosen.

Utility principle The principle that one should choose those actions that result in the greatest happiness for the greatest number.

Valid argument An argument whose conclusion follows with necessity from its premises: if the premises are true, the conclusion must be true.

Value conflict A value conflict arises if (1) a choice has to be made between at least two options for which at least two values are relevant as choice criteria, (2) at least two different values select at least two different options as best, and (3) the values do not trump each other.

Value Sensitive Design An approach that aims at integrating values of ethical importance in a systematic way in engineering design.

Values Lasting convictions or matters that people feel should be strived for in general and not just for themselves to be able to lead a good life or to realize a just society.

Virtue ethics An ethical theory that focuses on the nature of the acting person. This theory indicates which good or desirable characteristics people should have or develop to be moral.

Virtues A certain type of human characteristics or qualities.

Whistle-blowing The disclosure of certain abuses in a company by an employee in which he or she is employed, without the consent of his/her superiors, and in order to remedy these abuses and/or to warn the public about these abuses.

Wide reflective equilibrium Approach that aims at making coherent three types of moral beliefs 1) considered moral judgments; 2) moral principles; and 3) background theories. Also the resulting coherent set of moral beliefs is often called a wide reflective equilibrium.

Window-dressing Presenting a favorable impression that is not based on the actual facts.

Working principle The (scientific) principle on which the working of a product is based.

References

Achterberg, W. (1994). *Samenleving, natuur en duurzaamheid. Een inleiding in de milieufilosofie* [Society, nature and sustainability. An introduction to environmental philosophy], Van Gorcum, Assen.

Achterhuis, H.J. (1995). De moralisering van de apparaten [The moralization of apparatuses]. *Socialisme en Democratie*, 52 (1), 3–12.

Achterhuis, H.J. (1998). *De erfenis van de utopie* [The legacy of utopia], Ambo, Amsterdam.

Adorno, R. (2004). The precautionary principle: A new legal standard for a technological age. *Journal of International Biotechnology Law*, 1 (1), 11–19.

Akrich, M. (1992). The de-scription of technical objects, in *Shaping Technology/Building Society* (eds W.E. Bijker and J. Law), MIT Press, Cambridge, pp. 205–224.

Anderson, R. M., Otten, J. and Schendel, D.E. (1983). The Bay Area Rapid Transit (BART) Incident, in *Engineering Professionalism and Ethics* (eds J.H. Schaub, K. Pavlovic and M.D. Morris), John Wiley & Sons, New York, pp. 373–380.

Anderson, R. M., Perucci, R., Schendel, D.E. and Trachtman, L.E. (1980). *Divided Loyalities – Whistle-Blowing at BART*. Purdue Research Foundation, West Lafayette, IN.

Arendt, H. (1965). *Eichmann in Jerusalem: A Report on the Banality of Evil*, Viking Compass, New York.

Aristotle (1980) [350 BC]. *Nicomachean Ethics* (translated by W.D. Ross), Clarendon Press, Oxford.

Arkin, R.C. (2007). *Governing Lethal Behavior: Embedding Ethics in a Hybrid Deliberative/ Reactive Robot Architecture* (Technical report GIT-GVU-07–11), Georgia Institute of Technology, Atlanta.

Baeyens, E.F. (1992). *Herald of Free Enterprise. Hét verslag van de ramp.* [*Herald of Free Enterprise*. The Story of the Disaster], Van Geyt Productions, Hulst.

Baron, J. and Spranca, M. (1997). Protected values. *Organizational Behavior and Human Decision Processes*, 70 (1), 1–16.

Baudet, H. (1986). *Een vertrouwde wereld: 100 jaar innovatie in Nederland* [A well-known world: 100 years of innovation in the Netherlands], Bert Bakker, Amsterdam.

Baxter, W.F. (1974). *People or Penguins. The Case for Optimal Pollution*. Columbia University Press, New York.

Ethics, Technology, and Engineering: An Introduction, First Edition.
Ibo van de Poel and Lambèr Royakkers.

Beauchamp, T.L. (1984). On eliminating the distinction between applied ethics and ethical theory. *Monist*, 67, 514–531.

Beder, S. (1993). Engineers, ethics and etiquette. *New Scientist*, 25 (September), 36–41.

Begley, T.H., White, K., Honigfort, P., Twaroski, M.L., Neches, R. and Walker, R.A. (2005). Perfluorochemicals: Potential sources of and migration from food packaging. *Food Additives and Contaminants*, 22 (10), 1023–1031.

Bentham, J. (1948) [1789]. *An Introduction to the Principles of Morals and Legislation*, Hafner Press, Oxford.

Birch, D. and Fielder, J.H. (eds.) (1994). *The Ford Pinto Case: A Study in Applied Ethics, Business and Technology*, State University of New York Press, Albany.

Boele, R., Fabig, H. and Wheeler, D. (2001). Shell, Nigeria and the Ogoni. A study in unsustainable development: I. The story of Shell, Nigeria and the Ogoni people – environment, economy, relationships: conflict and prospects for resolution. *Sustainable Development*, 9, 74–86.

Boers, C. (1981). *Wetenschap, techniek en samenleving. Bouwstenen voor een kritische wetenschapstheorie* [Science, technology and society. Building stones for a critical theory of science], Boom, Meppel.

Bovens, M. (1998). *The Quest for Responsibility. Accountability and Citizenship in Complex Organisations*, Cambridge University Press, Cambridge.

Brady, F.N. (1990). *Ethical Managing. Rules and Results*, Macmillan Publishing Company, New York.

Brezet, H. and van Hemel, C. (1997). *Ecodesign: A Promising Approach to Sustainable Production and Consumption*, UNEP, Paris.

Brumsen, M. (2006). Het ongeluk met de ICE-hogesnelheidstrein bij Eschede [The accident with the ICE high speed train at Eschede], in *Bedrijfsgevallen. Morele beslissingen van ondernemingen* (eds W. Dubbink and H. Van Luijk), Van Gorcum, Assen, pp. 45–55.

Chang, R. (ed.) (1997). *Incommensurability, Incomparability, and Practical Reasoning*, Harvard University Press, Cambridge, MA.

Collingridge, D. (1980). *The Social Control of Technology*, Frances Pinter, London.

Congressional Testimony of Schrage: The internet in China (2006). http://googleblog.blogspot.com/2006/02/testimony-internet-in-china.html (accessed November 13, 2010).

Copi, I.M. and Burgess-Jackson, K. (1990). *Informal Logic*, Prentice-Hall, New Jersey.

Cottril, K. (2000). Global codes of conduct. *Journal of Business Strategy*, 17 (3), 55–59.

Covello, V.T. and Merkhofer, M.W. (1993). *Risk Assessment Methods. Approaches for Assessing Health and Environmental Risks*, Plenum, New York.

Cranor, C.F. (1990). Some moral issues in risk assessment. *Ethics*, 101, 123–143.

Cross, N. (1989). *Engineering Design Methods*, John Wiley & Sons, Chichester.

Cumming, M.L. (2006). Automation and accountability in decision support system interface design. *Journal of Technology Studies*, 32 (1), 23–31.

Dancy, J. (1993). *Moral Reasons*, Blackwell Publishers, Oxford.

Daniels, N. (1979). Wide reflective equilibrium and theory acceptance in ethics. *Journal of Philosophy*, 76, 256–282.

Daniels, N. (1996). *Justice and Justification. Reflective Equilibrium in Theory and Practice.* Cambridge University Press, Cambridge.

Dann G.E. and Haddow, N. (2008). Just doing business or doing just business: Google, Microsoft, Yahoo! and the business of censoring China's internet. *Journal of Business Ethics*, 79, 219–234.

Darley, J.M. (1996). How organizations socialize individuals into evildoing, in *Codes of Conduct. Behavioral Research into Business Ethics* (eds M. Messick and A.E. Tenbrunsel), Russell Sage Foundation, New York, pp. 13–43.

Davis, M. (1998). *Thinking Like an Engineer. Studies in the Ethics of a Profession*, Oxford University Press, New York and Oxford.

Davis, M. (2006). Engineering ethics, individuals, and organizations. *Science and Engineering Ethics*, 12, 223–231.

De Boer, E. (1994). Zestig jaar Deltawerken: Dordrecht als opening en sluitpost [Sixty years Delta Works. Dordrecht as start and finish], in *Wonderen der techniek: Nederlandse ingenieurs en hun kunstwerken 200 jaar civiele techniek* (eds M.L. ten Horn-van Nispen, H. Lintsen and A.J. Veenendaal), Zutphen, Walburg, pp. 197–210.

De George, R.T. (1990) *Business Ethics*, Macmillan, New York.

De George, R. (1993). *Competing with Integrity in International Business*, Oxford University Press, New York.

Derby, S.L. and Keeney, R.L. (1990). Risk analysis. Understanding "how safe is safe enough?", in *Readings in Risk* (eds T.S. Glickman and M. Gough), Resources for the Future, Washington, pp. 43–49.

Devon, R. (1999). Towards a social ethics of engineering: The norms of engagement. *Journal of Engineering Education*, January, 87–92.

Devon, R. (2004). Towards a social ethics of technology: A research prospect. *Techne*, 8 (1), http://scholar.lib.vt.edu/ejournals/SPT/v8n1/devon.html (accessed November 13, 2010).

Devon, R. and Van de Poel, I. (2004). Design ethics: The social ethics paradigm. *International Journal of Engineering Education*, 20 (3), 461–469.

Disco, C. (1990). *Made in Delft. Professional Engineering in the Netherlands*, UvA, Amsterdam.

Dorst, C.H.M. and Royakkers, L.M.M. (2006). The design analogy: a model for moral problem solving. *Design Studies*, 27, 633–656.

Downey, G.L., Lucena, J.C. and Mitcham, C. (2007). Engineering ethics and identity: Emerging initiatives in comparative perspective. *Science and Engineering Ethics*, 13, 463–487.

Dworkin, R. (1977). *Taking Rights Seriously*, Harvard University Press, Cambridge.

Eco, U. (1983). *The Name of the Rose*, Helen & Kurt Wolff, New York.

Eddy, P., Potter, E. and Page, B. (1976). *Destination Disaster: From the Tri-Motor to the DC-10, the Risk of Flying*, Quadrangle, New York.

Eilperin, J. (2006). Harmful PTFE chemical to be eliminated by 2015. *Washington Post* 2006–01–26, http://www.washingtonpost.com/wp-dyn/content/article/2006/01/25/AR2006012502041.html (accessed November 13, 2010).

Elkington, J. (1994). Towards the sustainable corporation: Win-win-win business strategies for sustainable development. *California Management Review*, 36 (2), 90–100.

Fahlquist, J.N. (2006a). Responsibility ascriptions and public health problems. Who is responsible for obesity and lung cancer. *Journal of Public Health*, 14 (1), 15–19.

Fahlquist, J.N. (2006b). Responsibility ascriptions and vision zero. *Accident Analysis and Prevention*, 38, 1113–1118.

FCC (2009). *V-chip: Viewing Television Responsibly* [Internet]. Federal Communications Commission, http://www.fcc.gov/cgb/consumerfacts/vchip.html (accessed March 25, 2009).

Felt, U., Wynne,B., Callon, M., Gonçalves, M.E., Jasanoff, S., Jepsen, M., Joly, P.-B., Konopasek, Z., May, S., Neubauer, C., Rip, A., Siune, K., Stirling, A. and Tallacchini, M. (2007). *Taking European Knowledge Society Seriously*. Report of the expert group on science and governance to the science, economy and society directorate, Directorate-General for Research, European Commission, Directorate-General for Research, Science, Economy and Society. Brussels.

Fischhoff, B., Lichtenstein, S. and Slovic, P. (1981). *Acceptable Risk*, Cambridge University Press, Cambridge.

Florman, S.C. (1976). *The Existential Pleasures of Engineering*, St. Martin's Press, New York.

Frankel, M.S. (1989). Professional codes: Why, how, and with what impact? *Journal of Business Ethics*, 8 (2), 109–115.

Frankena, W.K. (1973). *Ethics*. Prentice-Hall, Englewood Cliffs.

Friedman, B., Kahn, P.H., Jr. and Borning, A. (2006). Value sensitive design and information systems, in *Human-Computer Interaction in Management Information Systems: Foundations* (eds P. Zhang and D. Galletta), M.E. Sharpe, Armonk, pp. 348–372.

Friedman, M. (1962). *Capitalism and Freedom*, University of Chicago Press, Chicago.

Gambatese, J.A., Behm, M. and Rajendran, S. (2008). Design's role in construction accident causality and prevention: Perspectives from an expert panel. *Safety Science*, 46 (4), 675–691.

Gert, B. (1984). Moral theory and applied ethics. *Monist*, 67, 532–548.

Gilligan, C. (1982). *In a Different Voice: Psychological Theory and Women's Development*, Harvard University Press, Cambridge, MA.

Goldberg, S. (1987). The space shuttle tragedy and the ethics of engineering. *Jurimetrics Journal*, 27 (Winter), 155–159.

Grauls, M. (1993). *Uitvinders van het dagelijks leven 2* [Inventors of daily life 2], CODA, Antwerp.

Grunwald, A. (2001). The application of ethics to engineering and the engineer's moral responsibility: Perspectives for a research agenda. *Science and Engineering Ethics*, 7 (3), 415–428.

Habermas, J. (1981). *Theorie des kommunikativen Handelns* [Theory of communicative action], Suhrkamp Verlag, Frankfurt am Main.

Hansson, S.O. (2003). Ethical criteria of risk acceptance. *Erkenntnis*, 59, 291–309.

Hansson, S.O. (2004a). Fallacies of risk. *Journal of Risk Research*, 7 (3), 353–360.

Hansson, S.O. (2004b). Weighing risks and benefits. *Topoi*, 23, 145–152.

Hansson, S.O. (2007a). Safe design. *Techne*, 10 (1), 43–49.

Hansson, S.O. (2007b). Philosophical problems in cost-benefit analysis. *Economics and Philosophy*, 23, 163–183.

Hansson, S.O. (2009). Risk and safety in technology, in *Handbook of the Philosophy of Science. Vol. 9: Philosophy of Technology and Engineering Sciences* (ed. A. Meijers), Elsevier, Amsterdam, pp. 1069–1102.

Hare, R.M. (1982). Ethical theory and utilitarianism, in *Utilitarianism and Beyond* (eds. A. Sen and B. Williams), Cambridge University Press, Cambridge.

Hare, R.M. (1988). Why do applied ethics?, in *Applied Ethics and Ethical Theory* (eds D.M. Rosenthal and F. Shehadi), University of Utah Press, Salt Lake City, pp. 71–83.

Harris, C.E. (1998). Engineering responsibilities in lesser-developed nations: The welfare requirement. *Science and Engineering Ethics*, 4, 321–331.

Harris, C.E., Pritchard, M.S. and Rabins, M.J. (2005). *Engineering Ethics. Concepts and Cases* (3rd edn), Wadsworth, Belmont.

Hassink, H., de Vries, M.and Bollen, L. (2007). A content analysis of whistleblowing policies of leading European companies. *Journal of Business Ethics*, 75 (1), 25–44.

Health Council of the Netherlands (2006). *Health Significance of Nanotechnologies*, Health Council of the Netherlands, The Hague.

Herkert, J.R. (1999). ABET's engineering criteria 2000 and engineering ethics: Where do we go from here? Presented at the *OEC International Conference on Ethics in Engineering and Computer Science*, March 1999, http://onlineethics.org/CMS/edu/instructessays/herkert2.aspx#fl (accessed October, 19, 2009).

Herkert, J.R. (2001). Future directions in engineering ethics research: Microethics, macroethics and the role of professional societies. *Science and Engineering Ethics*, 7 (3), 403–414.

Hummels H. and Karssing, E. (2007). Organising ethics, in *Ethics and Business* (ed R.J.M. Jeurissen), Van Gorcum, Assen, pp. 249–275.

Hunter, T.A. (1997). Designing to codes and standards, in *ASM Handbook. Vol. 20. Materials Selection and Design* (eds G.E. Dieter, S.D. Henry and S.R. Lampman), ASM, Materials Park, pp. 66–71.

Hursthouse, R. (1991). Virtue theory and abortion. *Philosophy and Public Affairs*, 20, 223–246.

Ihde, D. (1990). *Technology and the Lifeworld*, Indiana University Press, Bloomington.

Ihde, D. (1991). *Insrumental Realsim: The Interface between Philosophy of Science and Philosophy of Technology*, Indiana University Press, Bloomingon.

Inderwildi, O.R. and King, D.A. (2009). Quo vadis biofuels? *Energy & Environmental Science*, 2 (4), 343–346.

International Crisis Group (2008). *Nigeria: Ogoni Land after Shell.* Africa Briefing no. 54 (September 18, 2008), http://www.unhcr.org/refworld/docid/48d359552.html (accessed September 30, 2008).

Jeurissen, R.J.M. and van de Ven, B. (2007). Values and norms in organisations, in *Ethics and Business* (ed. R.J.M. Jeurissen), Van Gorcum, Assen, pp. 54–92.

Johnson, B.B. (1999). Ethical issues in risk communication. Continuing the discussion. *Risk Analysis*, 19 (3), 335–348.

Johnson, D.G. (2001). *Computer Ethics*, Prentice Hall, Upper Saddle River.

Jungermann, H. (1996). Ethical dilemmas in risk communication, in *Codes of Conduct. Behavioral Research into Business Ethics* (eds M. Messick and A.E. Tenbrunsel), Russell Sage Foundation, New York, pp. 300–317.

Jungk, R. (1958). *Brighter than a Thousand Suns*, Penquin Books, Middlesex.

Kant, I. (1990) [1784]. Beantwortung der Frage: Was ist Aufklärung?, translated as "An Answer to the Question: What is Enlightenment?" in *Foundations of the Metaphysics of Morals and What is Enlightenment*, Macmillan, New York.

Kant, I. (2002) [1785]. *Grundlegung zur Metaphysik der Sitten*, translated as *Groundwork for the Metaphisics of Morals*, Oxford University Press, Oxford.

Kaptein, M. (2004). Business codes of multinational firms: What do they say? *Journal of Business Ethics*, 50, 13–31.

Kienpointer, M. (1992). *Alltagslogik* [Everyday logic], Frommann-Holzboog, Stuttgart.

Kneese, A.V., Ben-David, S. and Schulze, W.D. (1983). The ethical foundations of benefit-cost analysis, in *Energy and the Future* (eds D. MacLean and P.G. Brown), Rowman and Littefield, Totowa, pp. 59–74.

Knippenberg (1999). De antikinderarbeidlobby is failliet [The anti child lobby is morally bankrupt], *NRC*, 19–10–1999.

Korsgaard, C.M. (1996). *Creating the Kingdom of Ends*, Cambridge University Press, Cambridge.

Kraakman, R., Davies, P., Hansmann, H., Hertig, G., Hopt, K., Kanda, H. and Rock, E. (eds) (2004). *The anatomy of Corporate Law: A Comparative and Functional Approach*, New York, Oxford University Press.

Kremer, E. (2002). (Re)examining the Citicorp case: Ethical paragon or chimera. *Practice*, 6 (3), 269–276.

Krohn, W. and Weyer, J. (1994). Society as a laboratory: The social risks of experimental research. *Science and Public Policy*, 21 (3), 173–183.

Ladd, J. (1991). The quest for a code of professional ethics. An intellectual and moral confusion, in *Ethical Issues in Engineering* (ed. D.G. Johnson), Prentice Hall, Englewood Cliffs, pp. 130–136.

Latour, B. (1992). Where are the missing masses? The sociology of a few mundane artifacts, in *Shaping Technology / Building Society* (eds W.E. Bijker and J. Law), MIT Press, Cambridge, pp. 225–258.

Latour, B. (1994). On technical mediation – Philosophy, sociology, genealogy. *Common Knowledge*, 3, 29–64.

Latour, B. (2002). Morality and technology: The end of the means. *Theory, Culture & Society*, 19 (5–6), 247–260

Lave, L.B. (1984). Eight frameworks for regulation, in *Technological Risk Assessment* (eds P.F. Ricci, L.A. Sagan and C.G. Whiplle), Martinus Nijhoff, The Hague, pp. 169–190.

Layton, E.T. (1971). *The Revolt of the Engineers. Social Responsibility and the American Engineering Profession*, The Press of the Case Western Reserve University, Cleveland and London.

Lenk, H. and Ropohl, G. (eds.) (1987). *Technik und Ethik*, Reclam, Stuttgart.

Lloyd, P.A. and Busby, J.A. (2003). "Things that went well – no serious injuries or deaths" Ethical reasoning in a normal engineering design process. *Science and Engineering Ethics*, 9 (4), 503–516.

Locke, J. (1986) [1690]. *Second Treatise of Government*, Prometheus Books, New York.

Luegenbiehl, H.C. (2004). Ethical autonomy and engineering in a cross-cultural context. *Techné*, 8 (1), 57–78.

Luegenbiehl, H.C. (2010). Ethical principles for engineering in a global environment, in *Philosophy and Engineering* (eds I. van de Poel and D. Goldberg), Springer, Dordrecht, pp. 147–160.

MacCollum, D.V. (1995). *Construction Safety Planning*, John Wiley & Sons, New York.

MacIntyre, A. (1984a). *After Virtue*, Notre Dame University Press, Notre Dame.

MacIntyre, A. (1984b). Does applied ethics rest on a mistake? *The Monist*, 67, 498–513.

Malin, M. (1983). Legal protection for whistleblowers, in *Beyond Whistleblowing. Defining Engineers' Responsibilities* (ed. V. Weil), Center for the Study of Ethics in the Professions Illinois Institute of Technology, Chicago, pp. 11–32.

Martin, K.E. (2008). *Google, Inc., in China* (Case BRI-1004), Business Roundtable, Institute for Corporate Ethics, Washington.

Martin, M.W. and Schinzinger, R. (1996). *Ethics in Engineering*, 3rd edn, McGraw-Hill New York.

McLinden, M.O. and Didion, D.A. (1987). Quest for alternatives. *ASHRAE Journal*, 29, 32–34.

Mill, J.S. (1859). *On Liberty*, John W. Parker and Son, London.

Mill, J.S. (1979) [1863]. *Utilitarianism*, Collins, London.

Mishan, E.J. (1975). *Cost-Benefit Analysis. An Informal Introduction*, Allen & Unwin, London.

Moberg, D. (1997). Virtuous peers in work organizations. *Business Ethics Quarterly*, 7 (1), 67–85.

Morgan, M.G. and Lave, L. (1990). Ethical considerations in risk communication practice and research. *Risk Analysis*, 10 (3), 355–358.

Morgenstern, J. (1995). The fifty-nine-story crisis. *The New Yorker*, May 29, 45–53

Mostert, P. (1982). Reactorveiligheid [Reactor safety], in *Kernenergie in beweging. Handboek bij vraagstukken over kernenergie* [Nuclear energy on the move. Handbook for nuclear energy issues], (eds C.D. Andriesse and A. Heertje), Keesing Boeken, Amsterdam, pp. 61–73.

Mulish, H. (1962). *De zaak 40l61* [The case 40/61], De Bezige Bij, Amsterdam.

Naylor, R.L., Liska, A.J., Burke, M.B., Falcon, W.P., Gaskell, J.C., Rozelle, S.D. and Cassman, K.G. (2007). The ripple effect: Biofuels, food security, and the environment. *Environment*, 49 (9), 30–43.

Nelson, D. (1980). *Frederick W. Taylor and the Rise of Scientific Management*, University of Wisconsin Press, London.

Neufeld, M.J. (1995). *The Rocket and the Reich. Peenemunde and the Coming of the Ballistic Missile Era*, Harvard University Press, Cambridge, MA.

Noble, D.F. (1977). *America by Design. Science, Technology and the Rise of Corporate Capitalism*, Alfred A. Knopf, New York.

Oberdöster, G., Oberdöster, E. and Oberdöster, J. (2005). Nanotoxicology: An emerging discipline evolving from studies of ultrafine particles. *Environmental Health Perspectives*, 113 (7), 823–829.

Otway, H.J. and von Winterfeldt, D. (1982). Beyond acceptable risk. On the social acceptability of technologies. *Policy Sciences*, 14, 247–256.

Paine, L.S. (2000). Does ethics pay? *Business Ethics Quarterly*, 10 (1), 319–330.

Petroski, H. (1982). *To Engineer is Human. The Role of Failure in Successful Design*, St. Martin's Press, New York.

Piszkiewicz, D. (1995). *The Nazi Rocketeers. Dreams of Space and Crimes of War*, Praeger, Westport, CT.

Platform Ethiek en Techniek TU Delft (1996). *Werkconferentie Ethische aspecten van de ingenieurswetenschappen* [Working conference ethical aspects of the engineering sciences], April 19, Delft (The Netherlands).

Pöttgens, J.J.E. (1988). Schade door delfstofwinning [Damage by mining]. *Heidemijtijdschrift*, 2, 41–43.

Powley, C.R., Michalczyk, M.J., Kaiser, M.A. and Buxton, L.W. (2005). Determination of perfluorooctanoic acid (PFOA) extractable from the surface of commercial cookware under simulated cooking conditions by LC/MS/MS. *The Analyst*, 130 (9), 1299–1302.

Pritchard, M.S. (2001). Responsible engineering. The importance of character and imagination. *Science and Engineering Ethics*, 7 (3), 391–402.

Pritchard, M.S. (2009). Professional standards in engineering practice, in *Handbook of the Philosophy of Science. Vol. 9: Philosophy of Technology and Engineering Sciences*, (ed. A. Meijers), Elsevier, Amsterdam, pp. 953–971.

Radin, T.J. (2004). The effectiveness of global codes of conduct: Role models that make sense. *Business and Society Review*, 109 (4), 415–447.

Raffensperger, C. and Tickner, J. (eds) (1999). *Protecting Public Health and the Environment: Implementing the Precautionary Principle*, Island Press, Washington, DC.

Ravenzwaaij, A. (1994). *Risico-informatie in het veiligheidsbeleid* [Risk information in safety policy], Universiteit Utrecht, Utrecht.

Rawls, J. (1971). *A Theory of Justice*, Harvard University Press, Cambridge, MA.

Rawls, J. (1993). *Political Liberalism*, Columbia University Press, New York.

Rawls, J. (2001). *Justice as Fairness. A Restatement.* The Belknap Press of Harvard University Press, Cambridge, MA.

Raz, J. (1986). *The Morality of Freedom*, Oxford University Press, Oxford.

Renn, O. (2005). *White Paper on Risk Governance – Towards an Integrative Approach*, International Risk Governance Council, Geneva.

Rittel, H.W.J. and Webber, M.M. (1984). Planning problems are wicked problems, in *Developments in Design Methodology* (ed. N. Cross), John Wiley & Sons, Chichester, pp. 135–144.

Roddis, W.M.K. (1993). Structural failure and engineering ethics. *Journal of Structural Engineering*, 119 (5), 1539–1555.

Ross, W.D. (1930). *The Right and the Good*, Clarendon Press, Oxford.

Royakkers, L.M.M. and van Est, R. (2010). The cubicle warrior: The marionette of digitalized warfare. *Ethics and Information Technology*, 12, 289–296.

Ryan, B.L.V. (1991). Conflicts inherent in corporate codes. *International Journal of Value-Based Management*, 4 (1), 119–136.

Sandin, P. (1999). Dimensions of the precautionary principle. *Human and Ecological Risk Assessment*, 5 (5), 889–907.

Schellens, P.J. (1985). *Redelijke argumenten*, Foris, Dordrecht.

Schinzinger, R. and Martin, M.W. (2000). *Introduction to Engineering Ethics*, McGraw Hill, New York.

Schivelbusch, W. (1986). *The Railway Journey; The Industrialization of Time and Space in the 19th Century*, University of California Press, Berkeley.

Schot, J.W. (1992). Constructive technology assessment and technology dynamics: The case of clean technologies. *Science, Technology & Human Values*, 17 (1), 36–57.

Schot, J.W. and Rip, A. (1997). The past and future of constructive technology assessment. *Technology Forecasting and Social Change*, 54 (2/3), 251–268.

Schuler, D. and Namioka, A. (eds) (1993). *Participatory Design: Principles and Practices*, Lawrence Erlbaum, Hillsdale, NJ.

Sclove, R.E. (1995). *Democracy and Technology*, The Guilford Press, New York.

Sethi, S.P. and Williams, O.F. (2000). Creating and implementing global codes of conduct: An assessment of the Sulivan principles as a role model for developing international codes of conduct – Lessons learned and unlearned. *Business and Society Review*, 105 (2), 169–200.

Shrader-Frechette, K.S. (1985). *Risk Analysis and Scientific Method. Methodological and Ethical Problems with Evaluating Societal Hazards*, Reidel, Dordrecht.

Shrader-Frechette, K.S. (1991). *Risk and Rationality. Philosophical Foundations for Populist Reform*, University of California Press, Berkeley.

Sidgwick, H. (1877). *Methods of Ethics*, Macmillan, London.

Simon, H.A. (1973). The structure of ill-structured problems. *Artificial Intelligence*, 4, 181–201.

Simon, H.A. (1974). Spurious correlation: A causal interpretation, in *Causal Models in the Social Sciences* (ed. H.M. Blalock), Macmillan, London and Basingstoke, pp. 1–17.

Singer, P. (2002). *One World. The Ethics of Globalization*, Yale University Press, London.

Slovic, P., Fischhoff, B. and Lichtenstein, S. (1990). Rating the risks, in *Readings in Risk* (eds T.S. Glickman and M. Gough), Resources for the Future, Washington, pp. 61–74.

Smart, J.J.C. (1973). An outline of a system of utilitarian ethics, in *Utilitarianism for and Against* (eds J.J.C. Smart and B. Williams), Cambridge University Press, Cambridge, pp. 3–74.

Smith, A. (1776). *An Inquiry into the Nature and Causes of the Wealth of Nations*, Clarendon Press, Oxford.

Smith, A. (2003). Do You Believe in Ethics? Latour and Ihde in the Trenches of the Sciences Wars, in *Chasing Technoscience: Matrix for Materiality* (eds D. Ihde and E. Selinger), Indiana University Press, Bloomington and Indianapolis, pp. 182–194.

Starr, C. (1990). Social benefit versus technological risk, in *Readings in Risk* (eds T.S. Glickman and M. Gough), Resources for the Future: Washington, pp. 183–193.

Steg, L. (1999). *Verspilde energie? Wat doen en laten Nederlanders voor het milieu* (SCP Cahier no. 156). [Spilt energy? What the Dutch do and don't for the environment], Sociaal en Cultureel Planbureau, The Hague.

Stern, P.C. and Feinberg, H.V. (1996). *Understanding Risk: Informing Decisions in a Democratic Society*, National Academy Press, Washington.

Stuhlinger, E. and Ordway, F.I., III (1994). *Wernher von Braun, Crusader for Space: A Biographical Memoir*, Malabar, Krieger.

Styron, W. (1979). *Sophie's Choice*, Random House, London.

Sunstein, C.R. (2005). Cost-benefit analysis and the environment. *Ethics*, 115, 351–385.

Swierstra, T. (2000). *Kloneren in de polder. Het maatschappelijk debat over kloneren in Nederland Februari 1997-Oktober 1999* [Cloning in the polder. The societal debate on cloning in the

Netherlands February 1997–October 1999], Rathenau Instituut (Studie no. 39), The Hague.

Tavani, H.T. (2004). *Ethics & Technology. Ethical Issues in an Age of Information and Communication Technology.* John Wiley & Sons, Hoboken.

Taylor, F.W. (1911). *The Principles of Scientific Management*, Harper Bros, New York.

Ten Horn-van Nispen, M.-L. (2002). Johan van Veen (1893–1959) in *Biografisch woordenboek van Nederland* (ed. Gabriëls), Instituut voor Nederlandse Geschiedenis, The Hague.

Tetlock, P.E. (2003). Thinking the unthinkable: Sacred values and taboo cognitions. *Trends in Cognitive Sciences*, 7 (7), 320–324.

The Royal Society & The Royal Academy of Engineering (2004). *Nanoscience and Nanotechnologies: Opportunities and Uncertainties*, The Royal Society & The Royal Academy of Engineering, London.

Thompson, D.F. (1980). Moral responsibility and public officials. *American Political Science Review*, 74, 905–916.

Tronto, J.C. (1993). *Moral Boundaries. A Political Argument for an Ethic of Care*, Routledge, New York.

Tversky A. and Kahneman, D. (1981). The framing of decisions and the psychology of choice. *Science*, 211, 453–458.

Unger, S.H. (1994). *Controlling Technology: Ethics and the Responsible Engineer*, John Wiley & Sons, New York.

Valenti, J. and Wilkins, L. (1995). An ethical risk communication protocol for science and mass communication. *Public Understanding of Science*, 4 (18), 177–194.

Van de Poel, I. (1998). *Changing Technologies. A Comparative Study of Eight Processes of Transformation of Technological Regimes*, Universiteit Twente, Enschede.

Van de Poel, I. (2001). Investigating ethical issues in engineering design. *Science and Engineering Ethics*, 7 (3), 429–446.

Van de Poel, I. (2007). Ethics in engineering practice, in *Philosophy in Engineering* (eds S. Hylgaard Christensen, M. Meganck and B. Delahousse), Academica, Aarhus, Denmark, pp. 245–262.

Van de Poel, I. (2009a). Values in engineering design, in *Handbook of the Philosophy of Science. Vol. 9: Philosophy of Technology and Engineering Sciences* (ed. A. Meijers), Elsevier, Amsterdam, pp. 973–1006.

Van de Poel, I. (2009b). The introduction of nanotechnology as a societal experiment, in *Technoscience in Progress: Managing the Uncertainty of Nanotechnology* (eds S. Arnaldi, A. Lorenzet and F. Russo), IOS Press, Amsterdam, pp. 129–142.

Van de Poel, I. and Royakkers, L. (2007). The ethical cycle. *Journal of Business Ethics*, 71 (1), 1–13.

Van de Poel, I. and van Gorp, A. (2006). Degrees of responsibility in engineering design. Type of design and design hierarchy. *Science, Technology & Human Values*, 31 (3), 333–360.

Van de Poel, I., Zandvoort, H. and Brumsen, M. (2001). Ethics and engineering courses at Delft University of Technology: Contents, educational setup and experiences. *Science and Engineering Ethics*, 7 (2), 267–282.

Van de Poel, I. and Zwart, S.D. (2010). Reflective equilibrium in R&D networks. *Science, Technology & Human Values*, 35, 174–199.

Van den Hoven, J. and Vermaas, P.E. (2007). Nano-technology and privacy: On continuous surveillance outside the panopticon. *Journal of Medicine and Philosophy*, 32 (3), 283–297.

Van der Burg, S. and Van de Poel, I.R. (2005). Teaching ethics and technology with Agora, an electronic tool. *Science and Engineering Ethics*, 11 (2), 277–297.

Van der Ham, W. (2003). *Meester van de zee: Johan Van Veen, waterstaatsingenieur 1893–1959.* [Master of the sea: Johan van Veen, civil engineer], Balans, Amsterdam.

Van Eemeren, F.H. and Grootendorst, R. (1992). *Argumentation, Communication and Fallacies: A Pragma-Dialectical Perspective.* Lawrence Erlbaum, Hillsdale, NJ:

Van Gorp, A. (2005). *Ethical Issues in Engineering Design. Safety and Sustainability*, Simon Stevin Series in the Philosophy of Technology, Delft.

Van Gorp, A., and van de Poel, I.R. (2001). Ethical considerations in engineering design processes. *IEEE Technology and Society Magazine*, 20 (3), 15–22.

Van Gorp, A. and Van de Poel, I. (2008). Deciding on ethical issues in engineering design, in *Philosophy and Design. From Engineering to Architecture* (eds P. E. Vermaas, P. Kroes, A. Light and S. A. Moore), Kluwer, Dordrecht, pp. 77–90.

Van Hinte, E. (ed.) (1997). *Eternally Yours: Visions on Product Endurance.* 010 Publishers, Rotterdam.

Van Poortvliet, A. (1999). *Risks, Disasters and Management.* Eburon, Delft.

Van Veen, J. (1962). *Dredge, Drain, Reclaim. The Art of a Nation*, 5th edn. Martinus Nijhoff, The Hague.

Vaughan, D. (1996). *The Challenger Launch Decision*, The University of Chicago Press, Chicago.

Verbeek, P.P. (2005). *What Things Do – Philosophical Reflections on Technology, Agency, and Design*, Pennsylvania State University Press. University Park.

Verbeek, P.P. (2006a). Materializing morality – Design ethics and technological mediation. *Science, Technology and Human Values*, 31 (3), 361–380.

Verbeek, P.P. (2006b), The morality of things – A postphenomenological inquiry, in *Postphenomenology: A Critical Companion to Ihde* (ed. E. Selinger). State University of New York Press, New York, pp. 117–130.

Verbeek, P.P. (2008). Morality in design: Design ethics and the morality of technological artifacts, in *Philosophy and Design: From Engineering to Architecture* (eds P.E. Vermaas, P. Kroes, A. Light and S.A. Moore), Springer, Dordrecht, pp. 91–103.

Vincenti, W. (1990). *What Engineers Know and How They Know It*, Johns Hopkins University Press. Baltimore.

Weckert, J. and Moor, J. (2007). The precautionary principle in nanotechnology in *Nanoethics: The Social and Ethical Implications of Nanotechnology* (eds F. Allhoff, P. Lin, J. Moor and J. Weckert), John Wiley & Sons, Hoboken, pp. 133–146.

Weegink, R.J. (1996). *Basisonderzoek elektriciteitsverbruik kleinverbruikers BEK'95.* [Basic research electricity consumption of citizens. BEK'95], Arnhem: EnergieNed.

Weil, V. (1998). Professional standards: Can they shape practice in an international context? *Science and Engineering Ethics*, 4, 303–314.

Whitbeck, C. (1998a). *Ethics in Engineering Practice and Research*, Cambridge University Press, Cambridge.

Whitbeck, C. (1998b). *The Philosophical Theory Underlying "Ethics in Engineering Practice and Research"* from http://www.washingtonpost.com/wp-dyn/content/article/2006/01/25/AR2006012502041.html ethics.org/bib/part1.html?text

Whitehead, T. (2008). 13,000 people wrongly branded criminals. *Telegraph*, November 12, http://www.telegraph.co.uk/news/newstopics/politics/lawandorder/3449207/13000-people-wrongly-branded-criminals.html.

Williams, B. (1973). *Problems of the Self. Philosophical Papers 1956–1972*, Cambridge University Press, Cambridge.

Winner, L. (1980). Do artifacts have politics? *Daedalus*, 109, 121–136.

Wirtz, R. (2007). Moral responsibility in organisations, in *Ethics & Business* (ed. R.J.M. Jeurissen), Van Gorcum, Assen, pp. 24–43.

World Commission on Environment and Development (1987). *Our Common Future*, Oxford University Press, Oxford.

World Wildlife Fund (2002). *Living planet report* from http://www.wwf.org.uk/filelibrary/pdf/livingplanet2002.pdf (accessed November 13, 2010).

Wynne, B. (1989). Frameworks of rationality in risk management: Towards the testing of naive sociology, in *Environmental Threats: Perception Analysis, and Management* (ed. J. Brown), London, Bellhaven, pp. 33–47.

Zah, R., Böni, H., Gauch, M., Hischier, R., Lehmann, M. and Wäger, P. (2007). *Life Cycle Assessment of Energy Products: Environmental Assessment of Biofuels.* Empa.

Zandvoort, H. (2000). Codes of conduct, the law, and technological design and development, in *The Empirical Turn in the Philosophy of Technology. Vol. 20 Research in Philosophy and Technology* (eds P. Kroes and A. Meijers), Elsevier/JAI Press, London, pp. 193–205.

Index of Cases

Ethics, Technology, and Engineering: An Introduction, First Edition.
Ibo van de Poel and Lambèr Royakkers.

Index

Note 1: page numbers in italics refer to illustrations, figures or tables
Note 2: page numbers in bold denote glossary entries

Ethics, Technology, and Engineering: An Introduction, First Edition.
Ibo van de Poel and Lambèr Royakkers.